DROPS AND BUBBLES IN CONTACT WITH SOLID SURFACES

PROGRESS IN
COLLOID AND INTERFACE SCIENCE

Series Editors
Reinhardt Miller and Libero Liggieri

Drops and Bubbles in Contact with Solid Surfaces
Michele Ferrari, Libero Liggieri, and Reinhardt Miller

Bubble and Drop Interfaces
Reinhardt Miller and Libero Liggieri

Interfacial Rheology
Reinhardt Miller and Libero Liggieri

DROPS AND BUBBLES IN CONTACT WITH SOLID SURFACES

Edited by

M. Ferrari
L. Liggieri
R. Miller

CRC Press
Taylor & Francis Group
Boca Raton London New York

CRC Press is an imprint of the
Taylor & Francis Group, an **informa** business

CRC Press
Taylor & Francis Group
6000 Broken Sound Parkway NW, Suite 300
Boca Raton, FL 33487-2742

First issued in paperback 2019

© 2013 by Taylor & Francis Group, LLC
CRC Press is an imprint of Taylor & Francis Group, an Informa business

No claim to original U.S. Government works

ISBN-13: 978-1-4665-7545-5 (hbk)
ISBN-13: 978-0-367-38078-6 (pbk)

Visit the Taylor & Francis Web site at
http://www.taylorandfrancis.com

and the CRC Press Web site at
http://www.crcpress.com

Foreword

Progress in Colloid and Interface Science (PCIS)

This is the third volume in the book series dedicated to colloids and inter-faces. The first volume dealt with the mechanical properties of liquid interfaces and discussed the experimental and theoretical state of the art of elastic and viscous properties of liquid surface layers. This book is obviously the first comprehensive description of the 2D rheology of liquid interfaces. Volume two represents the progress made in utilizing drops and bubbles for a quantitative characterization of liquid interfacial properties.

This third volume contributes to the field of characterization of solid surfaces. The chapters cover the theoretical and experimental aspects of wetting and wettability, liquid–solid interfacial properties, spreading dynamics. Interesting peculiarities are discussed, such as the phenomena of super-spreading and super-hydrophobicity.

Some sections deal with specific solid surfaces. For example, reactions and wetting of liquid metals at high temperatures is discussed, and also the interaction between nano-bubbles at solid surface and nano-particles at liquid interfaces, respectively, is investigated. The book also includes a chapter on electro-wetting.

We want to encourage all colleagues to contact the series editors in case they have ideas for further books in the field of colloid and interface sciences.

Michele Ferrari, Libero Liggieri and Reinhard Miller
(Editors)

Content

I.   F. Barberis - Contact Angle Measurements over Macro, Micro and Nano Dimensioned Systems   1

II.   J. Radulovic, K. Sefiane, V. M. Starov, N. Ivanova, M. E. R. Shanahan - Review on Kinetics of Spreading and Wetting by Aqueous Surfactant Solutions   37

III.   V. Dutschk, A. Calvimontes, M. Stamm - Wetting Dynamics of Aqueous Surfactant Solutions on Polymer Surfaces   71

IV.   B. Cichocki, M. L. Ekiel-Jeżewska, G. Nägele, E. Wajnryb - Hydrodynamic Interactions Between Solid Particles at a Fluid–Gas Interface   93

V.   E. Bonaccurso, R. Pericet-Camara, H.-J. Butt - Microdrops Evaporating from Deformable or Soluble Polymer Surfaces   105

VI.   M. Antoni, K. Sefiane - Modeling Approaches and Challenges of Evaporating Sessile Droplets   129

VII. V. S. J. Craig, X. Zhang, J. Hu - Nanobubbles at Hydrophobic Surfaces   159

VIII. A. J. B. Milne, K. Grundke, M. Nitschke, R. Frenzel, A. Amirfazli - Model and Experimental Study of Surfactant Solutions and Pure Liquids Contact Angles on Complex Surfaces   175

IX. A. V. Nguyen, M. Firouzi - Collision and Attachment Interactions of Single Air Bubbles with Flat Surfaces in Aqueous Solutions   211

X.   R. Sedev, C. Priest, J. Ralston - Electrowetting of Ionic Liquids in Solid–Liquid–Liquid Systems   241

XI. V. Bertola, M. Marengo - Single Drop Impacts of Complex Fluids: A Review   267

XII. A. Passerone, F. Valenza, M. L. Muolo - Wetting at High Temperature   299

Contact Angle Measurements over Macro, Micro and Nano Dimensioned Systems

Fabrizio Barberis

Department of Civil, Environmental and Architectural Engineering, Materials Engineering Section, Faculty of Engineering, University of Genoa, Piazzale J.F. Kennedy 1, Pad D, 16129, Genoa, Italy.
E-mail: fabrizio.barberis@unige.it

Contents

A. Science 1
B. History 2
C. Young Equation and True Equilibrium 3
D. The Equation of State Approach 5
E. The Acid-Base Approach 7
F. Experimental Application 9
G. Equilibrium and Adsorbed Films 13
H. Microscopic and Nanoscopic Contact Angles 19
I. Conclusions 25
J. References 25

A. Wetting Science

Wetting science, in its more classical definition, studies the way of interaction of a fluid with a solid surface, which determines the specific shape that liquids assume when they are deposited on a solid substrate. T. Young in 1805 [1] stated that a liquid drop placed on a solid surface assumes a geometrical shape that can be described by his famous formula (1), that became the milestone of wetting studies:

$$\cos\theta = (\gamma_{sv} - \gamma_{sl})/\gamma_{lv}, \tag{1}$$

where θ is the static contact angle, γ_{sv} and γ_{sl} are the solid–vapor and solid–liquid surface energies and γ_{lv} is the liquid surface tension (Fig. 1). The *Young Equation* (YE) therefore describes how, in a true equilibrium condition, the contact angle between a fluid and a solid depends only upon the intrinsic nature of the materials engaged in the wetting experience. No dimension or gravity constraints are apparently present in it. Due to its apparent simplicity YE is still considered as the

Drops and Bubbles in Contact with Solid Surfaces
© Koninklijke Brill NV, Leiden, 2012

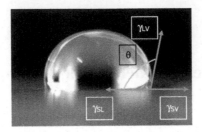

Figure 1. Static Cantact Angle: 1 mm diameter water drop on a polymer solid substrate.

basic formula to manage common industrial wetting problems where, at least from a qualitative and macroscopic point of view, it is quite useful to deduce an idea of the behavior of the evaluated liquid–solid system by taking some pictures of the profile of a static drop. Therefore, by observing the results provided on a macroscopic scale by this visual approach, originated with the Ramé–Hart goniometric technique and arriving at very precise modern tensiometer systems, wetting science seems to be an easy piece. While this simple point of view was rather accepted till some years ago the introduction of micro and nanotechnologies made it now difficult to be supported even for first sight industrial applications.

B. History

The deep complexity features of wetting problems arose at general attention in the mid sixties of the last century after the fundamental work done by Zisman [2] that followed other papers published with Fox [3], Ellison [4], Shafrin [5] and Bernett [6]. Later on in the eighties De Gennes, who established another milestone with his review [7] about the state of the art of wetting science, Brochard-Wyart, Cazabat and co-authors studied [8–13] the influence of long range forces, spreading phenomena and rough substrates, while Neumann working with Good, Kwok and others, moved toward the definition of the *Equation of State Approach* (ESA) [14, 15] following a path that may be appreciated in [16].

During the nineties till to the present days the studies concerning wetting problems became further specialized in their topics, according to a sensible evolution of the available experimental techniques. This fact determined a consequent improvement of the theoretical basis upon which wetting studies may rely, as it may be appreciated in the following authors that, as well as many co-authors and those who treated micro and nano wetting issues, have produced an impressive amount of papers impossible to be mentioned. Their citation here has therefore to be intended only as a mere guide and not as an exhaustive panorama of their activity or of the available literature. The ***spreading phenomena*** on a micro dimensioned solid–liquid system were evaluated for example by Gu *et al.* [17, 18], Erickson *et al.* [19], De Coninck *et al.* [20, 21], Blake *et al.* [22–24], De Ruijter *et al.* [25], Voue' *et al.* [26], Shanahan [27, 28], Starov *et al.* [29, 30], Lee *et al.* [31], Wasan *et* *al.* [32], as well as Cazabat [33] and co-authors [34–42] that also highlighted many features of the film spreading issues.

Many Authors, coming from different thinking schools [43–59], studied the fundamental aspects of the ***triple-line and line-tension effects*** while ***wetting transition phenomena*** were studied by Bonn [60], Humfeld *et al.* [61], Kefiane *et al.* [62], Long *et al.* [63, 64], Chibowsky [65, 66], Extrand *et al.* [67–71], Marmur *et al.* [72–78], Drelich *et al.* [79], Tavana *et al.* [80], Chau [81], pointed their attention mainly to different typologies of ***heterogeneous surfaces and hysteresis*** while Ramos *et al.* [82–84] studied the effects of ***nanoroughed surfaces*** and Daniel *et al.* [85, 86] evaluated ***wettability gradient surfaces***. ***Hydrophobic and super-hydrophobic behaviors*** have been analyzed by He and Patankar [87–90], Marmur [91], Ferrari *et al.* [92], Krasowska *et al.* [93], Vinogradova [94], Christenson *et al.* [95], Yaminsky, Von Bahr and co-authors [96–99], who also studied the influence of ***long range forces and capillarity effects*** on wetting issues. Reviews of this topic may be also found in Shirtcliffe *et al.* [100], Bhushan *et al.* [101], Nosonovsky *et al.* [102]. The stability of ***thin wetting films*** as well as ***dewetting*** problems of polar/nonpolar substrates have been studied, among the others, by Sharma, Jameel, Khanna *et al.* [103–108], Manev *et al.* [109], Saramago [110], Bertrand *et al.* [111]. Lyklema [112, 113], Chen *et al.* [114], Babak [115], Yeh *et al.* [116, 117] focused their attention to ***thermodynamic aspects*** as like Churaev, Starov *et al.* [118–123], Li *et al.* [124], Boinovich *et al.* [125, 126], Toshev [127], Djikaev *et al.* [128] that produced very interesting papers evaluating ***surface force interactions*** in a scenario of energy minimization. Apart of classic static wetting problems, also ***dynamic wetting*** [7] became a fundamental issue of liquid–solid interactions. As a simple reminder Rame' *et al.* [129, 130], De Ruijter *et al.* [131], Thiele *et al.* [132] Della Volpe, Siboni and co-workers [133–135], Lam, Neumann and co-workers [136–138] deeply improved the dynamic wetting statements, either by drop deposition experiments either by the Wilhelmy [139] technique. Periodic very interesting reviews of the state of the art of wetting problems are provided by Neumann and co-authors [140–145].

C. Young Equation and True Equilibrium

The previous mentioned Authors, as like many more in this field, gave fundamental contributions to create the necessary bridge between pure thermodynamic statements and real experiments, in order to reach an effective application of the YE principles. On this ground it could be helpful to emphasize that frequently the different scientific background of those who studied these topics provided quite various solutions, that still nowadays are acknowledged by the wetting community. In the last 10–15 years, on the other hand, the experimental accessibility to nanoscale phenomena revealed new complex aspects of the wetting science.

On a pure scientific ground the *Young Equation* (1), developed by macroscopic observations, has to be considered as a full thermodynamic tool able to provide a deep insight over the energetic connections existing among liquid–solid interfaces, pointing out that any other claimed external influence (gravity, mass, etc.) should not interfere within the core of the wetting problem. This formula in fact explicitly requires that:

- The solid–liquid system has to be in true equilibrium.

- The solid surface has to be rigid, flat and chemically homogeneous.

- Both liquid and solid have to be 'perfect' and their physical and chemical properties must be accurately determined before any wetting experience.

- The liquid and the solid phases do not undergo any transformation during the experiment.

- No chemical interactions between liquid, solid and environment take place.

- No external dynamic effects capable of altering the reached equilibrium position are allowed.

The fulfilling of all these requirements is considered very difficult to be achieved on a practical experimental scenario. At a first glance, by following the classical wetting statements, it appears intuitive that natural, polycrystalline and multiphase materials would be quite difficult to be evaluated in a wetting experience [79, 81] while engineered materials usually are described as much more predictable in their wetting behavior. As indicated in Fig. 1, YE states that the contact angle may be deduced by the configuration of three vectors originating from the same point placed on the border among the three insisting phases of the drop. YE seems therefore to be satisfactorily applicable when a pure mechanical equilibrium is achieved by the fluid–liquid–solid system as depicted by Eq. (2):

$$\sum_{i=1}^{n} F_i = 0, \qquad \sum_{i=1}^{n} M_i = 0. \tag{2}$$

This requirement is yet necessary but indeed not sufficient by itself to fulfill the need of the *true equilibrium* as prescribed by the YE that, as many Authors emphasized [179–181], strictly requires the coexistence of mechanical *and* thermodynamic equilibrium of the experimental set up, i.e., the equivalence of all the chemical potentials of all the substances involved in the experiments as (3):

$$\mu_1^i + kT \log X_1 = \mu_2^i + kT \log X_2 = \cdots = \mu_n^i + kT \log X_n = \mu$$
$$= \text{constant for all states/substances } n = 1, 2, 3 \ldots, \tag{3}$$

where μ_X^i is the energy contribution that accounts for the specific molecular interactions of a region of the system, T is the temperature, and X is the molecular concentration in the region. The term $k \log X_n$ is usually known as the *entropy of mixing*, so that finally μ gives the total free energy per molecule. This definition includes in the concept of *true equilibrium* also the absence of molecular and thermal fluxes and assures that no further internal causes may determine spontaneous evolutions of the evaluated system. The accomplishment of true equilibrium conditions over a real surface, unfortunately, is not an instantaneous set up and usually it occurs, if possible, after a sequence of steps performed on different time scales, strictly related to the specific liquid–solid system features and the adopted experimental technique [155–158]. It is important to stress that a visual 'steady position' of a liquid over a substrate does not have to be automatically considered as the final equilibrium position. This means that there is no *a priori* certainty that a drop placed on a real solid substrate has reached its deepest and therefore more stable potential energy well. In the microscopic size range the presence of heterogcneities or the evaluation of viscous fluids makes highly probable the existence of multiple partial equilibrium positions [44, 72, 87, 117, 135, 191] that, independently of their major or minor degree of stability, may provide misleading results, often difficult to perceive.

Two main theoretical approaches, briefly described in the following, were born after Young and still hold in the wetting community. A third version has to be recognized in the works of the Russian school of Churaev, Starov *et al.* [118–123] who developed, on the basis of the Derjaguin statements, a proper way to deal with wetting problems by adopting a rigorous mathematical approach of the liquid–solid energy minimization issue. Due to the specificity of this issue a direct consultancy of the Authors papers is recommended.

D. The Equation of State Approach

Zisman and his collaborators [2–6] in the mid sixties performed very important experimental and theoretical revisions of the contact angle problem. His analysis provided the wide known correlation among the contact angles of homologous series of fluids (n-alkanes) and their liquid surface tension values. This correspondence was found linearly decreasing. By measuring series of liquids with different properties he obtained band diagrams by which he determined the concept of *critical surface tension of wetting*. He approached the concept that later Neumann *et al.*, by further implementing and expanding his data, developed and proposed in the *Equation of State Approach—ESA* [15]. To correctly apply the YE an extra equation was in fact needed and this was therefore formulated as (4):

$$\gamma^{sl} = \frac{\sqrt{\gamma^{lv}} - \sqrt{\gamma^{sv}}}{1 - 0.015\sqrt{\gamma^{lv}\gamma^{sv}}}, \tag{4}$$

where γ^{sl}, γ^{sv} are respectively the *surface-liquid* and *surface-vapor* energies and γ^{lv} is the liquid surface tension. By combining this definition of the solid–liquid surface energy with the YE Neumann and colleagues derived an expression in which the solid–vapor surface energy was able to be determined by both the experimental measurable parameters, i.e., the liquid surface tension and the contact angle value. This formula may be appreciated in (5):

$$\cos\theta = \frac{(0.015\gamma^{sv} - 2)\sqrt{\gamma^{sv}\gamma^{lv}} + \gamma^{lv}}{\gamma^{lv}(0.015\sqrt{\gamma^{lv}\gamma^{sv}} - 1)}. \tag{5}$$

Later on the same Authors got aware of a discontinuity problem in the form of (4) related to certain surface tension values able to make zero the numerator of the formula. The supporting expressions of the ESA were therefore reformulated in the fashion that became later known as (6) and (7):

$$\gamma_{sl} = \gamma_{lv} + \gamma_{sv} - 2\sqrt{\gamma_{lv}\gamma_{sv}}e^{-\beta(\gamma_{lv}-\gamma_{sv})^2}, \tag{6}$$

$$\cos\theta = -1 + 2\sqrt{\frac{\gamma_{sv}}{\gamma_{lv}}}e^{-\beta(\gamma_{lv}-\gamma_{sv})^2}. \tag{7}$$

The complete review of the birth and genesis of this formula would also imply to mention the studies based upon other Authors, like Antonow [198], Berthelot [199], and may be fully appreciated in the wide scientific production of the Neumann Group and his numerous students and colleagues. The validity of the ESA approach depends upon the fact of being able to correctly evaluate the β factor, that recently has been slightly corrected and expressed as an average value equivalent to 0.000125 $(mJ/m^2)^{-2}$. This number was derived by a huge amount of measurements by which Neumann *et al.* determined the behaviour of numerous surfaces with different liquids. This numerical dependence is also the most common cause of disagreement with the ESA approach. By explicit admission of these Authors in fact solid surfaces have to be distinguished in three families:

- The perfect ones, either chemical either physical, that may be correctly evaluated by the ESA approach and therefore can be interpreted by the YE.

- Those that even being chemically heterogeneous are physically perfect. These substrates may be analysed, even with special care.

- Those who present physical irregularities. By the Authors these surfaces have no chance to be reported in a YE/ESA meaningful frame.

The problem concerning nonperfect surfaces arises from the appearance, over the most of real samples, of the contact angle hysteresis phenomenon, i.e., the simultaneously presence of an *Advancing* and a *Receding* contact angles, both different from the unique, single, real thermodynamic one prescribed by the YE rule on ideal substrates. This terminology comes through the experiences within tilting drops [132] where a drop placed on a tilted plate shows a profile with a relevant difference between the front (*Advancing*) contact angle and the rear (*Receding*) one [7]. A simple practical example of this fact is the shape of a water drop falling down on a window glass. In the sense of YE and ESA approaches any condition able to create hysteresis is a deviation from the real equilibrium and consequently a severe obstacle to correct wetting evaluations. In the case of flat but chemical inhomogeneous surfaces the ESA accepts to consider the advancing contact angle as a good approximation of the effective equilibrium contact angle, while in the case of physical irregularities this assumption is considered not satisfactory. The necessity to deal with perfect, noninteracting, materials is a full requirement for this theoretical approach that recommends not only to exclude polar surfaces from the analyzed systems, due to their intrinsic capability to modify the liquid–solid status, but also needed to develop a special experimental technology, known as ADSA—(Axisymmetric Drop Shape Analysis) [149], able to further reduce any dynamic deposition influence to the evaluation activities. By adopting this measurement technique Gu *et al.* [200] proposed a *Modified Young Equation* (11) in order to correlate the effects of line tension (i.e., the effective contact line among liquid drop, solid surface and vapour) and drop curvature radius to the contact angle value:

$$\cos \theta = \cos \theta_\infty - \frac{\sigma}{\gamma_{lv}} \frac{1}{R}, \tag{8}$$

where θ is the contact angle corresponding to a finite contact radius drop and θ_∞ is the contact angle of a infinitely large drop. The same formula was also implemented to be able to take account of eventual local inclination of the solid supporting material, sounding therefore as (12)

$$\cos \theta = \cos \theta_\infty - \frac{\sigma}{\gamma_{lv}} \frac{\cos \beta}{\rho}, \tag{9}$$

where β is the inclination angle and $1/\rho$ the drop curvature [200, 201].

The Modified YE admits a direct dependence of contact angles by the line tension effects. Even if the features and the sign of line tension are still a point of debate [43–56] among scientists, the direct consequence of it is a *contact angle dependence also by the size of the drop*. This is a very important point in the framework to create a link between the micro–nano observations and macro experiences. The argument of the size effects dependence on contact angle values has been debated by many Authors, approaching at final results sometimes contrasting each other [146–158].

E. The Acid–Base Approach

Starting from a quite different point of view, that frequently determined cross criticisms [167, 180, 181], the *Acid–Base Theory* shares with the ESA approach the same aim to provide an explicit form of the solid–liquid energy γ_{sl} by which it is possible to calculate the *work of adhesion*, reported in (10) with its correlation with the YE:

$$W_{ADH} = \gamma_{lv} + \gamma_{sv} - \gamma_{sl} = \gamma_{lv}(1 + \cos \theta). \tag{10}$$

The Theory is quite complex in all its details and the direct consultancy of original works has to be considered mandatory for those who do not rely on a specific chemical background. Briefly speaking the problem to detect real contact angles is here more focused on the solid surface free energy that is assumed to be considered as a result of two main independent contributes [168–170] (11):

$$\gamma^{\mathrm{TOT}} = \gamma^{\mathrm{LW}} + \gamma^{\mathrm{AB}}, \tag{11}$$

where LW stands for the Lifschitz–van der Waals dispersive components and AB are the Acid–Base ones. The AB components may be further expressed by following two main different approaches. The first is the one supported by Fowkes [170] that makes a link between the W_{ADH}, intended as a free energy term, and the entropic contribution (temperature) to the ΔH^{AB} enthalpy calculated following Drago [171] accordingly to (12):

$$W_{\mathrm{ADH}} = 2\sqrt{\gamma_{\mathrm{lv}}^{\mathrm{LW}} \gamma_{\mathrm{sv}}^{\mathrm{LW}}} + fN(-\Delta H^{\mathrm{AB}}), \tag{12}$$

where f is a parameter expressing the entropic component and N is the number of the acid–base interactions involved on the interface. The second (13) has been provided by Good–van Oss–Chaudury (GvOC) [172] and works directly on the free energy term, due its need of contact angle values determined at a known assigned temperature.

$$W_{\mathrm{ADH}} = 2(\sqrt{\gamma_{\mathrm{lv}}^{\mathrm{LW}} \gamma_{\mathrm{sv}}^{\mathrm{LW}}} + \sqrt{\gamma_{\mathrm{lv}}^{+} \gamma_{\mathrm{sv}}^{-}} + \sqrt{\gamma_{\mathrm{lv}}^{-} \gamma_{\mathrm{sv}}^{+}}). \tag{13}$$

Such approaches therefore depend upon the contact angle data that one may collect by submitting the same solid surface to several liquids, featured by different polar and dispersive characteristics. While the Fowkes version appears difficult to be applied due to the necessity to evaluate contact angles at different temperatures the GvOC seems apparently easier to be used. Equation (13) in principle let one to calculate the LW contribution, the electron-donor γ^{-} (Lewis base) and electron acceptor γ^{+} (Lewis acid) of a solid surface by measuring the contact angle values with three liquids of known surface free energy components. When indeed applied on polymer surfaces some Authors [173, 174], by comparing the results provided by inverse gas chromatography or ζ potentials, found discrepancies with the results provided by GvOC. In particular they noticed a constant overestimation of the basic component of the surface free energy that were labeling these polymers surfaces as basically featured contrarily to experimental data. The Authors of GvOC Theory got aware of this problem and suggested that the observed influence could have been ascribed to the effects of monopolar basics [175] but this hint has not been fully accepted. Della Volpe and Siboni [176–181] highlighted the fact that the source of misunderstandings was probably lying on a wrong choice of referencing fluids, known as *the triplet*. In particular they stressed that the choice made by the original GvOC to compare acidic and basic components to water caused a systematic ill-conditioned situation by which the obtained experimental results appear wholly shifted to basic features. These Authors, recommending that only a direct comparison among same featured components (i.e., acidic with acidic, basic with basic) of different materials let to GvOC to be correctly applied, proposed a matrix

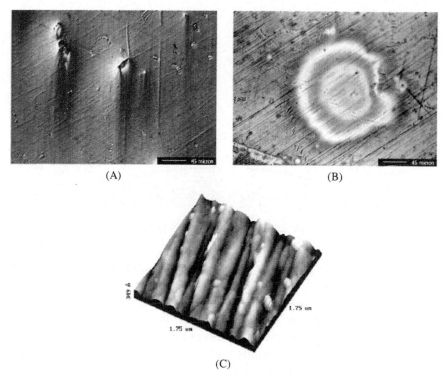

(A)   (B)

(C)

Figure 2. Microscopic *vs.* nanoscopic comparison of a natural and mechanically polished metal surface.

reformulation of this Theory followed by a best fit approach able to determine the acid-base parameters of both liquid and solid without any further information on materials involved in the experimental activity.

F. Experimental Application

Apart of mica [$(KAl_2Si_3AlO_{10}(OH)_2]$ and silicon wafers for electronic application most of the materials cannot be considered as 'flat and homogeneous' as they may eventually look in a macroscopic analysis. When macroscopic (centimetric) wide droplets lie on a surface different from a clearly rough one the problem of the real surface morphology seems far away. When, on the other hand, we deal with micro/nanoscopic drops or, even easier, we focus our attention to the *triple line*, i.e., the visible perimeter contour between solid surface-liquid-environmental fluid, the morphology topic turns up.

Figure 2 illustrates the problem as it may show up for a bulk metal surface (molybdenum in this case). At a first macroscopic insight the substrate sample was originally looking flat but on a microscopic analysis (Fig. 2A) several bumps and defects were found emerging by the surface. After a normal smoothing treatment (Fig. 2B) the surface appeared to be bump free but evident regular scratches

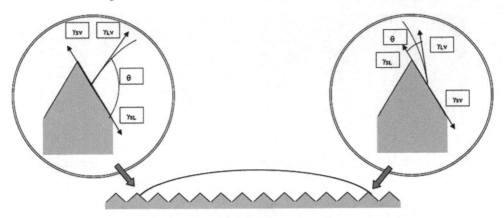

Figure 3. Metastable equilibrium status over a technical solid surface.

originated by the adopted polishing powders were visible. A further cerium oxide powder refining step provided the flattest possible surface but, even if appearing as a mirror on a macroscopic examination, when explored by contact AFM it was found looking like Fig. 2C, where, as usual in AFM practice, $X-Y$ lengths are expressed in microns and Z height is in Angstroms. Regular parallel scratches are carved on the surface while also several polishing particles may be detectable enclosed in the metal surface. By Fig. 2 it may be argued that on such a kind of irregular surfaces the triple line of micro sized drops stands on different places along its path, so appearing indented and irregular (Fig. 3). As a consequence of it the determination of the effective contact angle value on microscopic case may be doubtful due to the presence of various and different liquid–solid profiles.

This condition is described in Fig. 3 that shows how metastable equilibrium has to be thought, on technical substrates, as a default condition if no special attention is reserved to surface preparation or depositing techniques. It is in fact sufficient that liquid drops or layers have dimensions of the same order of the average solid surface defects to determine a situation favorable to multiple equilibrium conditions.

Real materials indeed have bumps, holes and pores so real surfaces, when evaluated from the microscopic point of view that fits with wetting science need, usually are much rougher than a regular succession of channels.

The presence of this typology of defects may obviously deviate the profile of the drop from its natural energetic path but also the opposite situation may happen. Figure 4 shows an alumina sample surface, observed at optical microscope, in which a wetting gap causes a discontinuity in the structure of a thin paraffin oil film placed on it. This effect is generated by chemical heterogeneities, that are usually not as easy to detect as physical discontinuities. Real surfaces present a mixed condition of both these defects. As an extra disturbing issue, one has also to be concerned with the environmental air (endemically liophobic) bubbles or films [160, 161] eventually trapped between the solid surface and the liquid profile. In such a case the liquid surface does not entirely match the solid substrate profile. This lat-

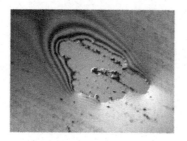

Figure 4. Chemical heterogeneity: wetting hole.

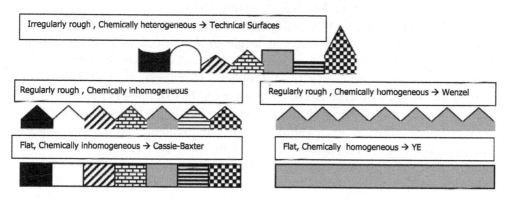

Figure 5. Real surface morphology *vs.* ideal one as required by YE.

ter condition, originally observed in wetting of textiles, can be easily encountered nowadays also with composites or cellular solids. Figure 5 gives a qualitative idea of the normal condition of the average solid surface as it may appear in real experimental conditions. This problem was perceived from the beginning of systematic wetting studies. Wenzel, as well as Cassie and Baxter proposed to modify the YE in order to better meet these requirements and their Equations (14) and (15) are still now reported in the literature [88, 179, 181].

$$\cos \theta_{\mathrm{W}} = \frac{r(\gamma_{\mathrm{sv}} - \gamma_{\mathrm{sl}})}{\gamma_{\mathrm{lv}}}, \tag{14}$$

$$\gamma_{\mathrm{lv}} \cos \theta_{\mathrm{c}} = f_1(\gamma_{\mathrm{s_1v}} - \gamma_{\mathrm{s_1l}}) + f_2(\gamma_{\mathrm{s_2v}} - \gamma_{\mathrm{s_2l}}). \tag{15}$$

Both of these equations work in a statistical fashion based upon the determination of coefficients (r the roughness factor for Wenzel; f_1 and f_2 intended as the surface fractions belonging to different chemical issues for Cassie Baxter).

Indeed apart of regular artificial materials, it is quite difficult to classify a real surface in terms of these coefficients, even only for the mere reason that one never deal with a surface either *physically* or *chemically* heterogeneous.

Leaving aside polymers, generally endowed with very low energy surfaces [162], metals are usually considered to have, among materials, highly reactive surfaces not only for their important Hamaker constant values but also for their short range in-

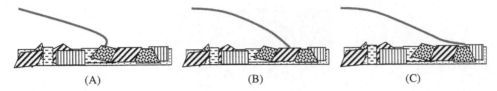

(A)   (B)   (C)

Figure 6. Most diffuse (A and C) deviations observable at microscopic level of the liquid–solid profiles over a technical surface.

teractions [182]. In term of YE, liquids spread the best on materials that show a solid-surface energy higher than their liquid–vapor surface tension. This feature is a common guide that indeed works as a rule of thumb (the chance that a liquid effectively spreads over a solid surface depends also on polar/nonpolar features, chemisorption phenomena, line tension effects, roughness, etc.). It must be remarked that normally metals heavily interact with the external environment. Apart of the obvious case of iron and its rust, more technological metals, such as inox steels, titanium alloys or even aluminum, have the tendency to passivate, i.e., to autonomously create a thin native oxide layer on their surface, that in brief may be thought closer to a ceramic-like phase. Unfortunately, this coating cannot be considered as uniformly distributed due to surface cracking phenomena or differential corrosion events. By the cooperative action of these facts it can be stated that, from the microscopic wetting point of view, the presence of surface heterogeneities, and consequently different surface energy values, gives rise to *different micro-wetting conditions over the same sample*. It can be therefore expected that, by following the YE, liquids assume at microscopic size over technical substrates different shapes depending on the local energy level. This occurrence, even if negligible in a macroscopic observation, assumes a fundamental importance in the microscopic scale. Figure 6 shows a scheme of how may look, on the micro scale, the very end of different liquid–solid interfaces. While the liquid profile of Fig. 6B in its microscopic final end just follows the profile that already presents in its macroscopic behavior, this does not happen for profiles A and C.

In nearly all of the cases the microscopic profile of drops do not ends with the same tendency that showed on a macroscopical level.

Profile A shows a condition in which the liquid–solid system finally appears to be more liophobic while on the contrary profile B is typical of a more liophilic behavior. These kind of situations, specially the B one, are quite diffuse when the wetting analysis is shifted on a lower dimensional case. It has been come in use to talk about *pseudo-partial wetting* conditions [10, 11]. Figure 7 illustrates the situation.

The chemical–physical and morphological conditions (capillarity) of the surface give rise to a sort of *dynamic defects*, in the sense that a slow but evident movement of the liquid profile generates what is known as *precursor film* and lately as *protruding foot*, that looks like a thin crown around the drop. This phenomenon is

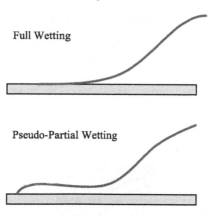

Figure 7. Wetting *vs.* pseudo-partial.

due to spreading events and is quite common when, for example, wetting problems concerning viscous fluids like oils or glycerol are studied.

The *Coefficient of Spreading* sounds like (16)

$$S = \gamma_{sv} - \gamma_{lv} - \gamma_{sl} \tag{16}$$

and takes into account, on a thermodynamic basis, the chance of a fluid to move simply due to energetic difference between itself and the surrounding materials. On a practical ground it can be therefore highlighted that these effects create an energetic metastable condition that, on a microscopic scale, provides at least two problems to be dealt with:

(a) An effective achievement of *True Equilibrium conditions*.

(b) A wise discrimination among *Multiple contact angles* collection.

G. Equilibrium and Adsorbed Films

A problem that gained an exceptional importance in the last years shows up when the chemical equilibrium is affected by *liquid–solid* or even *liquid–atmospheric fluid* interactions.

The lack of chemical equilibrium, originally intended as a 'simple' condition by which bulk phases involved in the experiments were allowed to exchange molecules between their interfaces, was usually intended as an important but not fundamental boundary condition of massive wetting issues. Recently indeed this problem became *The Problem* since researchers focused their attention at dimensions at which interfaces are *the whole system*, i.e., micro and nanowetting issues. The full importance of liquid and solid surface contaminations, beneath evident in its importance to those who normally deal with chemi or physi-sorption tasks [186], is not equally stressed by most of those who, time to time, deal with wetting issues. Figure 8 reports the diagram of surface tension variation of high purity water when exposed to normal laboratory atmosphere in which no known air pollutants as solvents or

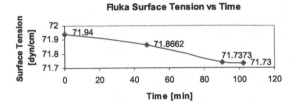

Figure 8. Highly pure water surface tension variation.

Figure 9. Water evaporation in air.

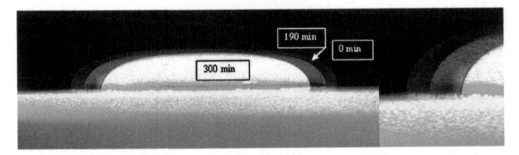

Figure 10. Water evaporation over a Teflon substrate in a fluent Argon environment.

hydrocarbons were dispersed. It can be appreciated that the simple effect of CO_2 derived by human breathing is quite able to reduce in the first exposure minutes the original surface tension value.

A simple example of a macroscopic application aiming to show the combined effects of atmosphere contaminants and evaporation kinetic is reported in Figs 9 and 10. Figure 9 shows the temporary evolution of a high purity macroscopic water drop placed upon a flat Teflon substrate (static contact angle = 109–112 (deg) [184]) by evaluating two superimposed photos of the same drop collected during a 210 minutes long evaporation path in normal air. Profile 1 (back) is the $t = 0$ drop while profile 2 (front) represents how the drop was looking after 210 minutes. The photographic set remained untouched and pictures were obtained by using remote controls. During this time the drop reduced its volume but also the overall shape changed, giving the false impression that the *effective* contact angle value moved down from an original true hydrophobic condition to an apparent more hydrophilic one.

Figure 10 on the other hand shows how does it look the same experience when the whole experimental set was placed in a closed transparent box in a saturated Argon environment at same final pressure and temperature of the normal air experience of Fig. 9. Three different pictures, at $t = 0$, $t = 190$ and $t = 300$ minutes were recorded and superimposed. The mere qualitative analysis that here matters indicates that the drop reduced its volume following a sort of scaling-factor that maintained the original contact angle value. This apparent contact angle variation has been illustrated by McHale, Erbil *et al.* [186, 187].

Apart of eventual solid substrate dissolution phenomena, usually to be considered as a slow phenomenon depending upon many factors like the pH or the polar/nonpolar nature of the engaged substances, other factors turn out to be more effective on wetting experiments. One has to take care not only of *liquid surface tension variation* due to atmospheric pollution of the drop but also to the substrate solid surface over which various pollutants and dust condense and deposit, changing the original support status. In fact in most of the papers concerning specific wetting problems the Authors stresses that a valid and effective *cleaning procedure* has to be adopted to prepare solid surfaces. This fact has an overwhelming importance. Generally it can be stated that, even by accurately controlling the experimental conditions, the more pure the adopted materials the more easy these phases can be contaminated by the external environment. The presence of pollutants over fresh clean materials is therefore endemic and has to be ascribed to the concept of free energy variation [179], by which the higher the energetic difference between different interacting materials the easier a contamination, occurs. A simple example of that may be appreciated in Fig. 9, where the triple line of profile 2 looks pinned due to the presence over the solid surface of deposited air dust, able to interfere with the drop kinetic evolution.

Cleaning procedure indeed have to be effective with the specific sample but they also have to preserve the original surface feature. Mechanically polishing effects have been already indicated in Fig. 2. Chemical washing techniques usually are considered less invasive, even if some best practice indications have to be bear in mind. When dealing with chemical cleaning procedures a general rule of thumb states that *the last substance that leaves the sample surface is the one that it will be later met by the analyzed fluid.* So when one has to clean a solid surface by using chemical solvents it has to figure out at a compatible sequence of cleaning steps, where the first one effectively cleans the solid surface and the next coming ones will progressively remove the previous till arriving to high purity grade water.

In terms of YE this fact is very important, especially if one is interested to evaluate the material surface energy features. Figure 11 illustrates the comparison between the physics of interacting substances as intended by YE point of view and the real conditions with whom one has usually to deal with. The YE recognizes it by admitting that chemical potentials of interacting substances have to be in equilibrium. By applying this fact to the case of a solid surface it can be stated that the existence of a γ_{sv} term, being the solid–air surface energy, has to be thought as the

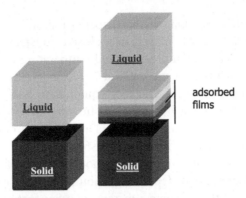

Figure 11. YE hypothesis *vs.* real surfaces.

signal that the substrate has to be in equilibrium with the saturated vapor pressure (P^0) of the external environment, i.e., that a thin adsorbed film of pressure π^0 thermodynamically exists on the surface. Obviously the higher the solid surface energy the more probable the existence of adsorbed films. This fact it was described [2, 155–158] by the *equilibrium spreading pressure* as expressed by (17)

$$\pi = \gamma_s - \gamma_{sv}. \tag{17}$$

Surface adsorption phenomena may happen either by chemical or by physical path. Chemi-adsorbed films are generally thin, stable and not as easy to remove as like physi-adsorbed ones. This issue depends upon the fact that, briefly speaking, chemi-adsorbed films rely on chemical bonds between the 'pollutant' molecule and the solid surface substrate. On the other hand physisorbed films are generally due to molecules temporarily trapped, generally by Van der Waals forces, on the solid support, able therefore to be more easily removed with usual cleaning procedures. The striking difference between these two conditions may be appreciated, for example, with the BET Adsorption Isotherms Technique [185, 188, 189]. By working with powdered samples usually presenting a convenient surface area it is possible, by vacuum and heating treatments, to remove any gas or surface pollutant and replacing it with a known gas like nitrogen or water vapor. Even if this technique is usually adopted to study surface porosity problems [162] the experimental results may provide an interesting profile of the solid surface energy conditions by the evaluation of the adsorbing curves. Apart of the adsorbed molecular quantity it isimportant to stress how a monolayer–Langmuir-like profile, equivalent to a chemisorption, differs from a multilayer system that, indeed, may be approximated to a thick physisorption phenomenon. When dealing with wetting issues the eventual presence of adsorbed mono or multilayer of chemical compounds over the solid surfaces could dramatically change the experimental data even at macroscopic dimensions [163–166].

The work of Gee *et al.* [165] may provide an interesting example of this tendency. By BET and Ellipsometry techniques they measured the contact angle values

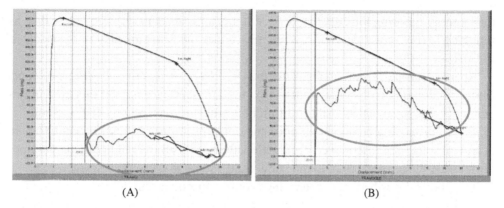

(A)   (B)

Figure 12. (A) DCA on ultrasonic cleaned sample. (B) DCA on a ultrasonic + plasma cleaned.

of water over high energetic quartz surfaces, differently treated in order to reach various hydroxylation grades. It came out that a water condensed film of various thicknesses, dependent upon the surface hydroxylation grades, formed over these substrates and macroscopic contact angles greatly changed as a function of it. Specifically the contact angle was found decreasing as like the condensed ice-like structured film was increasing its thickness.

As a simple example of how these effects matter also at macroscale size Fig. 12 shows how dynamic wetting tests performed with Wilhelmy Apparatus may change as a function of the surface energy status. Figure 12A shows the result for the sample as it behaved after a simple cleaning procedure by ultrasonic bath in pure water (Advancing contact angle close to 88 deg) while Fig. 12B shows the behavior to the same test of the same sample after a 30 sec, 60 W oxygen plasma cleaning procedure at 13.56 MHz (advancing contact angle close to 46 deg). The advancing profile looks much more disturbed in Fig. 12B, a signal that stronger and more specific liquid–solid interactions exists. The expected consequence is a reduced contact angle value, as effectively appears. Advancing contact angle reduction confirms the greater interaction between the fluid and the solid surface, so the liquid spreads better. The plasma treatment effectively removed some adsorbed pollutants revealing the inner surface energy features without causing the damages that some aggressive cleaning fluids like the wide known 'Piranha Solution' (H_2SO_4 + H_2O_2) may cause more on sample surfaces.

On the other hand the same oxygen plasma treatment when inadequately performed [190] may determine misleading results. As an example, when applied over ceramic and glasses it may greatly increase the number of superficial –OH groups (yet removable by high temperature heating), determining a more hydrophilic behavior than expected while over pure metallic surfaces the same treatment may reduce the wetting features due to a lower surface energy uniformly oxidized surface transformation [191, 192].

In these conditions sample preparation problems are always possible but it is highly preferable to deal with a defined, known, even if eventually modified surface than simply ignoring the effective surface status and presenting the collected data as referring to the 'real' material. An example of this mechanisms may be appreciated in [162].

Apart of properly autophobic effects [193, 194] or reactive wetting problems (indeed belonging to specific branches ex. high temperature issues) this kind of dependence has been tested also for different applications by Searcy [195], Beruto *et al.* [196], proposing a way based upon the concept of *partial equilibrium* to put in correlation the macroscopic contact angle values with an overall energetic parameter λ able to take in account all the solid–fluid interactions of the interface zone.

The above described facts determines some consequences:

(1) If cleaning procedures do not work in the proper fashion wetting test do not evaluate the expected liquid–solid interactions but instead the adsorbed film-liquid interactions.

(2) Wetting conditions have to be analyzed by carefully evaluating the dimensional ratio between drops/films and solid surface defects.

(3) True equilibrium is a target to be achieved and not expected as automatically given.

(4) Any kinetic effect due to deposition technique (sessile drop, ADSA, Wilhelmy etc.) may false the real equilibrium.

(5) Macroscopic wetting test may hidden the real solid–liquid interactions that act on smaller scale.

These facts therefore unwraps some problem with the correct application of YE to real laboratory experiments. While thermodynamic, by simply evaluating the energetic possibility that a reaction might take place or not [182] YE does not care about time. Indeed real systems evolve, i.e., undergo to physical and/or chemical interactions that may determine a transformation of the initial experimental conditions.

Moreover YE application faces the well known actual impossibility to get, as experimental data, the numerical values of the solid–liquid (γ_{sl}) and solid–vapor (γ_{sv}) surface energies [7, 183–185]. Actually surface tension (γ_{lv}), is commonly experimentally accessible by different techniques as like Wilhelmy technique, drop weight, drop shape, etc. [7, 139, 184, 197], while the solid surface energies are not directly measurable. This fact makes YE an under-determined condition in which one equation deals with two unknowns. This problem was originally empirically fixed by considering the difference of the terms ($\gamma_{sv} - \gamma_{sl}$), i.e., the numerator of the YE, as a single variable. By using this stratagem and assuming that in the most of the cases the liquid–solid system reaches a quasi-static equilibrium close to a true-equilibrium condition, it has been possible to make YE work but everybody were aware of these fundamental theoretical weakness.

H. Microscopic and Nanoscopic Contact Angles

The brief and necessarily incomplete remind to the main Theories that nowadays hold in the wetting science community has been here reported to highlight the nontrivial issues that affect 'usual' contact angle measurements. When moving down to micro and nano dimensioned liquid–solid systems experimental evaluations become more 'unusual'. This fact comes directly from two important issues:

(1) At smaller dimensions defects, whatever are, change their status in the importance rank gained in the scenario of contact angle measurements, moving from simple disturbing tasks to fundamental problems.

(2) Experimental results appear often difficult to be framed in some classic theoretical schemes. Today it is in fact possible to directly and experimentally evaluate the influences of what were simply known as *boundary conditions*, greatly reducing the room available for generic theoretical speculations.

Microfluidics studies finalized to MEMS, Lab-on-chip and thin films definitely indicated, as first achievement, that contact angles are not size-independent and scale factors are tremendously important in wetting science [146–158].

A further goal has to become aware that a strict application of the YE, that formally states that a liquid may wet ($\theta < 90$ deg) or not wet ($\theta > 90$ deg) a given solid surface, at smaller dimensions has to be wisely applied. At micro and nano scale in fact drops and thin films of the same liquid are often found coexisting at the same time over the solid substrate. Both these facts formally hardly fit with pure YE predictions and therefore they arose some fundamental theoretical issues concerning the effectiveness of portability of YE to micro and nano dimensions. Indeed these problems are not new. They were already known in their importance since the very beginning of the wetting science. For a quite long time they have been left somehow aside, mainly due to the macroscopic efficacy of the visual approach and also because of the presumption that the macroscopic wetting knowledge would also fit micro and nano systems. The contributes of the last decades unequivocally demonstrated that a correct application of the YE is far to be a simple task.

Now it is possible by the experimental point of view to reproduce the ideal solid–liquid systems that YE statements prescribes. In fact at nano and micro dimensions gravity effects can be effectively considered of secondary importance if not completely negligible while artificial solid supports may be considered perfect. Moreover the liquid–solid systems put under observation have a sensible tendency to achieve in a very fast way a stable thermodynamic condition. So for example, eventual Marangoni effects due to surface tension gradients [159] are greatly reduced while on the other hand volatility phenomena due to the specific vapor tension of liquids become very important. Besides the appearance of some specific technical problems it can be affirmed that small-sized experimental activities greatly helped to reduce several indetermination causes since long time affecting wetting issues. A great number of papers concerning the microscopic behavior of solid–liquid sys-

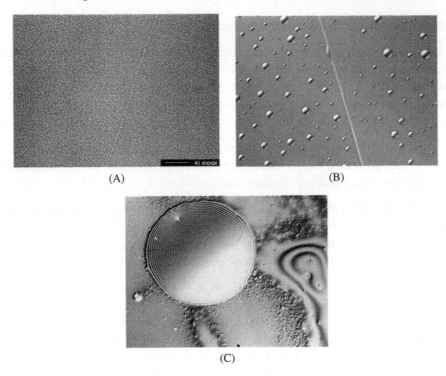

(A)   (B)

(C)

Figure 13. (A) Oil/silanized silicon wafer. (B) Oil/mica. (C) Oil/metal sputtered wafer.

tems as like the microscopic origin of macroscopic wetting phenomena may be collected in the huge production of the same Authors already cited in these pages [211–214] among many others that focused their attention, for example, over the friction problems of spreading phenomena of layered wetting structures. It may be anyway useful to underline some points that affect almost each microscopic wetting experience:

- the sample surface status,

- the deposition methodology,

- the contact angle value determination.

As an example Fig. 13(A, B, C) shows how do look, at metallographic microscope analysis and the same magnifying factor, paraffin oil drops deposited by the author with vaporizing and condensing technique over three different solid surfaces, respectively: silanized silicon wafer (Fig. 13A), muscovite mica (Fig. 13B) and metal sputtered silicon wafer (Fig. 13C). By the simple visual evaluation of these three very unlike solid surfaces it can be immediately perceived how the same fluid, almost inert over all of the analyzed substrates, provides completely different behaviors as a function of the solid surface energy features. Silanes (Fig. 13A), when correctly deposited over a solid surface either by dip-coating either by vaporization techniques, make a geometrical and energetic uniform coating layer able to deeply change the usual wetting features of the bulk material upon which they are placed. This feature may be either directly detected, for example with devices as like contact/noncontact AFM, able to provide the morphological map of the substrate, either indirectly by submitting coated samples to wetting tests.

Figure 13A clearly shows that the obtained surface is energetically and morphologically uniform, causing the liquid to be displaced as a myriad of drops, dimensionally equal, all over the solid substrate. By the point of view of the YE it can be therefore stated that this surface has effectively no evident chemical or physical defects and it is featured by a unique determined surface energy value, being therefore very close to the specifications of an ideal support as needed by YE itself or Neumann's ESA. The sensibility of this visual evaluation procedure may be yet appreciated by considering the straight line of drops evident in the picture. This effect is a consequence of a slight planar misalignment of the atomic planes. This kind of defects are practically completely hidden over usual technical surfaces by the presence of much more important disturbs. Over planar, near perfect, surfaces they may yet be detected by AFM contact analysis that unfortunately has a very local operational area, greatly reducing its efficacy when dealing with huge surfaces evaluations.

Figure 13B shows the same paraffin oil over a freshly cleaved muscovite mica substrate. This material is known to be the flattest available substrate and therefore it is often adopted to perform high sensibility AFM measurements [207]. Its sensibility to the environmental humidity threshold causes a sharp change of the own electrical conductivity behavior due to moisture-activated ion mobility. This feature may create some extra effects when the substrate is wetted by a polar fluid but does not interfere with oil wetting tests as here reported.

It can be clearly seen that even if perfect rounded drops are lying on the surface their dimensions range in a wider interval if compared to a silanized surface. This behavior can be assumed as an example of a solid surface that even if being totally flat does not completely fit, by the energetic uniformity, with the requirements of the pure YE as illustrated in Fig. 5. The sharp fracture line visible over the sample substrate is an endemic feature of this material that suffers of delamination problems, causing fluid infiltration inside the material and providing the appearance of 'internal stains' clearly appreciable in their multicolor fashion at optical microscope.

Apart of the different drop dimension existing over the observed samples it has to be here highlighted that no apparent liquid film appears to be lying over the solid substrates. This fact may provide further support to the concept that both of these surfaces are very close to the YE statements, showing a behavior that fits with the threshold of contact angle existence, i.e., showing a wetting behavior fully deducible by YE statements.

This is not the case of Fig. 13C, where the same oil is displaced on a silicon wafer substrate (i.e., extremely flat) over which a 300 (nm) thick molybdenum layer was previously sputtered. This solid surface, initially explored as clean and dry by contact AFM analysis, showed a uniform roughness with an average peak value equivalent to 30 (nm) along the Z-axis. Sputtering technique permit to get a metal surface able to be studied by nanoscopic wetting point of view. As it can be seen, the liquid morphology is now completely different by the previous illustrated cases. The visible big rounded drop coexist with some liquid unshaped structures on its right and all of the visible objects lie over a substrate that looks like a granular roof. This visible diffused substrate is indeed formed by the same oil that, initially, did not make any drop but created a film all over the surface. The presence of this film was proofed by performing multiple AFM force–distance curves during the various steps of this fluid deposition since the original bare substrate.

The observed effect is a direct signal of both solid surface energy and morphology importance in wetting tests. In this latter case in fact the solid substrate presents a high surface energy coupled with a planarity well less than perfect, as on the contrary was postulated for the case A and B. The overall condition, in terms of Fig. 5, may be described as non-flat, non-energetically neither chemically uniform status, due to the unavoidable presence in these sputtered films of some traces of pollutants (as like titanium or nickel coming from the sputtering chamber). Cracking effects of the metallic layer due to autotensions developed after the sputtering deposition step finally determined the conditions for a geometrical highly disturbed surface.

Figure 14A, B and C shows how the metal sputtered surface was looking at the Scanning Polarization Force Microscopy analysis at $t = 0$ and after 10 and 30 sec of oil vapor exposure, while microscopic Fig. 13C has to be intended after 120 sec of the same treatment. The paraffin oil, a low surface tension fluid, initially covered with a very thin and adherent film all the substrate surface. This fact fits with YE statements. The huge energetic difference between the solid and the liquid made the fluid totally spread over the surface.

In a second step indeed, dependent by the amount of vapor oil condensed over the sample, some very early drops begun to grew up starting from the more evident geometric defects. The complete path, from the nanoscopic level up to the microscopic level, is described in [210].

Apart the obvious difference among the wetting behavior of these three unlike samples the striking importance of the solid surface status may be further appreciated in Fig. 15 in which it is reported the completely different behavior of the same oil, deposited by the same condensing technique, over a simply mechanically polished bulk molybdenum surface. As it can be appreciated oil lays over the artificial roughness that shows the same dimensional features of the liquid film becoming a sort of preferential way for driving the fluid displacement far by its natural energetic displacement. This condition is able to create artificial metastable equilibrium states.

Fluid displacing technique assumes therefore a great importance when performing wetting analysis.

Over macroscopic and microscopic dimensions a huge variety of deposition systems, like sessile drop, ADSA or piezo dispensers are available. It has to be

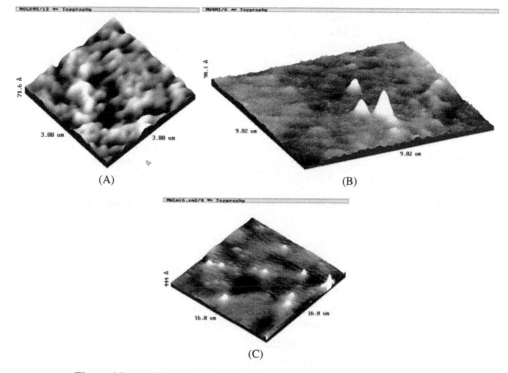

Figure 14. (A) SPFM clean. (B) SPFM 10 seconds. (C) SPFM 30 seconds.

Figure 15. Oil on polished molybdenum surface.

underlined in fact that to be able to fit within pure YE statements it is mandatory that no kinetic effects may influence the fluid natural fashion of disposing over the solid substrate. This may look as a secondary problem but indeed becomes the more important the smaller the dimensions of the evaluated liquid–solid system. The ADSA itself was originally created to be able to provide a better fluid deposition over the sample surface, getting rid of drop falling effects. This system, with its constant improvements [215] greatly amend the way to put a fluid over a substrate even if it still requires a perfectly smooth surface to be able to work within a the lowest possible hysteresis. Indeed some other techniques show a practical interest

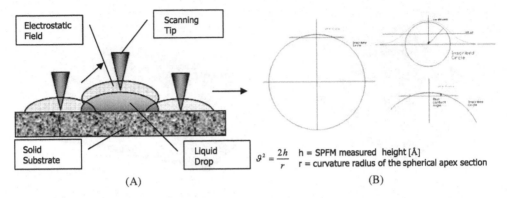

$$\vartheta^2 = \frac{2h}{r}$$

h = SPFM measured height [Å]
r = curvature radius of the spherical apex section

(A)   (B)

Figure 16. (A) SPFM non-contact electrostatic mode. (B) Geometrical approximation.

[202–210] when working at nanoscopic dimension, as like vaporizing-condensing effects or blown-jet removing techniques.

Independently by the evaluated solid–liquid system it is important to finally remark that while at microscopic dimensions experimental techniques as like ESEM may provide a net and clean image of the observed liquid drop, at nanoscopic dimensions this task becomes much difficult.

Noncontact Probe Microscope techniques as like SPFM, further applied in Kelvin Probe double feedback control, are indeed able to give an extremely high scanning resolution, in the order of Angstroms or at least as a function of the best between the tip scanning radius or the scanning distance. This precision is unfortunately much higher over the drop apex while decays over the border of the drop, right the place where one may like it the much. Figure 16A shows the tip-drop working scheme. To overcome the problem and evaluating the total margin of error eventually committing by this procedure Salmeron and co-workers, that promoted this technique, suggested to adopt a geometrical approximation, as described in their papers [203, 206, 207] and briefly reported in Fig. 16B, to get the final contact angle value. This procedure has been found efficient due to the very low contact angle values that normally one has to manage with these tests. At nanodimension, as in the example briefly described, it is quite common to deal with drops lying/coexisting with thin film of their own substance. This situation may be accepted only if one admits that the fluid of the film assumes a 'consistence' different by the one of the normal massive liquid.

This tendency is already known and indicated as *autophobic behaviour* [193, 194]. A fluid, due to particular surface conditions, becomes liophobic to itself. By accepting this fact it can be accepted that a liquid drop stands over a fluid film without loosing its shape. YE holds if one admits that an energetic difference between the film and the drop exists and the substrate layer is considered as a supporting sample for the drop. Figure 17 provides an example of this double-structure recorded by the Author.

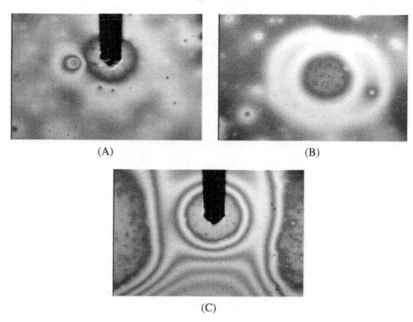

(A)   (B)

(C)

Figure 17. AFM tip opens the bulky liquid oil standing without mixing over the gel-like oil film adherent to the substrate.

The three images, from the original colored pictures, show how, on a microscopic level, it is possible to use the tip of an AFM scanning probe microscope as a spear to 'open' a permanent hole in an amount of already liquid featured fluid condensed over the previous adherent oil film. It can be seen that the very first condensed oil film is still present below the liquid mass. This means that the microscopic liquid amount never mixed within the previous deposited nanoscopic oil film. Being this fluid not polar and neither chemically interactive with the metallic solid surface the observed behavior has to be related to different densities of the oil molecules in the two steric conditions.

I. Conclusions

Nano dimensioned wetting evaluations are showing their great importance in discovering the effective liquid solid interactions to which macroscopic fluid displacements seems to be ascribed. This dependency, already theoretically postulated, is nowadays able to be experimentally observed. YE still holds its theoretical meaning at condition to be correctly applied without misleading simplifications.

J. References

1. Young, T., Philos. Trans. R. Soc., London, 95 (1805) 65.
2. Zisman, W. A., in: Contact Angle, Wettability and Adhesion. F. M. Fowkes, Ed., Advances in Chemistry Series, 43, 1–15. American Chemical Society, Washington, 1964.

3. Fox, H. W. and Zisman, W. A., J. Colloid Sci., 7 (1952) 109.

4. Ellison, A. H. and Zisman, W. A., J. Phys. Chem., 58 (1954) 503.

5. Shafrin, E. G. and Zisman, W. A., J. Phys. Chem., 66 (1962) 740.

6. Bernett, M. K. and Zisman, W. A., J. Phys. Chem., 66 (1962) 1207.

7. De Gennes, P. G., Wetting: statics and dynamics. Review of Modern Physics, 57 (1985) 3 Part I.

8. Joany, J. F. and De Gennes P. G., Role of long range forces in heterogeneous nucleation. Journal of Colloid and Interface Science, 111 (1986) 1.

9. Brochard, F., Motion of droplets on solid surfaces induced by chemical or thermal gradients. Langmuir, 5 (1989) 432–438.

10. Brochard-Wyart, F., di Meglio, J. M., Queré, D. and de Gennes, P. G., Spreading of nonvolatile liquids in a continuum picture. Langmuir, 7 (1991) 335–338.

11. Brochard-Wyart, F., de Gennes, P. G. and Hervet, H., Wetting of stratified solids. Advances in Colloid and Interface Science, 34 (1991) 561–582.

12. Brochard-Wyart, F. and De Gennes, P. G., Dynamics of partial wetting. Advances in Colloid and Interface Science, 39 (1992) 1–11.

13. Cazaba, A. M. and Cohen Stuart, M. A., Dynamics of wetting: effects of surface roughness. J. Phys. Chem., 90 (1986) 5845–5849.

14. Neumann, A. W. and Good, R. J., Thermodynamic of contact angles. Journal of Colloids and Surface, 38 (1972) 2.

15. Neumann, A. W., Good, R. J., Hope, C. J. and Sejpal, M., An equation of state approach to determine surface tension of low energy solids from contact angles. J. Colloid Interface Sci., 49 (1974) 291.

16. Kwok, D. Y. and Neumann, A. W., Contact angle interpretation in terms of solid surface tension. Colloids and Surfaces A: Phys. and Eng. Aspects, 161 (2000) 49–62.

17. Gu, Y. and Li, D., A model for a liquid drop spreading on a solid surface. Colloids and Surfaces A: Phys. and Eng. Aspects, 142 (1998) 243–256.

18. Gu, Y. and Li, D., Liquid drop spreading on solid surfaces at low impact speeds. Colloids and Surfaces A: Phys. and Eng. Aspects, 163 (2000) 239–245.

19. Erickson, D., Blackmore, B. and Li, D., An energy balance approach to modelling the hydrodinamically driven spreading of a liquid drop. Colloids and Surfaces A: Phys. and Eng. Aspects, 182 (2001) 109–122.

20. De Coninck, J., Spreading of chain-like liquid droplets on solids. Colloids and Surfaces A: Phys. and Eng. Aspects, 114 (1996) 155–160.

21. De Coninck, J., De Ruijter, M. and Voué, M., Dynamics of wetting. Current Opinion in Colloid and Interface Science, 6 (2001) 49–53.

22. Blake, T. D., Clarke, A., De Coninck, J., De Ruijter, M. and Voué, M., Droplet spreading: a microscopic approach. Colloids and Surfaces A: Phys. and Eng. Aspects, 149 (1999) 123–130.

23. Blake, T. D., Decamps, C., De Coninck, J., De Ruijter, M. and Voué, M., The dynamic of spreading at the microscopic scale. Colloids and Surfaces A: Phys. and Eng. Aspects, 154 (1999) 5–11.

24. Blake, T. D. and De Coninck, J., The influence of solid–liquid interactions on dynamic wetting. Advances in Colloid and Interface Science, 96 (2002) 21–36.

25. De Ruijter, M., Blake, T. D., Clarke, A. and De Coninck, J., Droplet spreading: a tool to characterize surfaces at the microscopic scale. J. of Petroleum Sci. and Eng., 24 (1999) 189–198.

26. Voué, M. and De Coninck, J., Spreading and wetting at the microscopic scale: recent developments and perspectives. Acta Materialia, 48 (2000) 4405–4417.

27. Shanahan, M. E. R., Spreading of water: condensation effects. Langmuir, 17 (2001) 8229–8235.

28. Shanahan, M. E. R. and Carré, A., Spreading and dynamics of liquid drops involving nanometric deformations on soft substrates. Colloids and Surfaces A: Phys. and Eng. Aspects, 206 (2002) 115–123.

29. Starov, V. M., Kosvintsev, S. R. and Velarde, M. G., Spreading of surfactant solutions over hydrophobic substrates. J. Colloid Interface Sci., 227 (2000) 185–190.

30. Starov, V., Velarde, M. and Radke, C., In Dynamics of Wetting and Spreading. Surfactant Sciences Series, Vol. 138. Taylor & Frances (2007).

31. Lee, K. S., Ivanova, N., Starov, V. M., Hilal, N. and Dutschk, V., Kinetics of wetting and spreading by aqueous surfactant solutions. Advances in Colloid and Interface Science, 144 (2008) 54–65.

32. Wasan, D., Nikolov, A. and Kondiparty, K., The wetting and spreading of nanofluids on solids: Role of the structural disjoining pressure. Current Opinion in Colloid & Interface Science, 16(4) (2011) 344–349.

33. Cazabat, A. M., Gerdes, S., Valignat, A. P. and Villette, S., Dynamics of wetting: from theory to experiment. Interface Science, 5 (1997) 129–139.

34. Burlatsky, S. F., Oshanin, G., Cazabat, A. M., Moreau, M. and Reinhardt, W. P., Spreading of a Thin wetting film: microscopic approach. Physical Review E, 54(4) (1996) 3832–3845.

35. Fanton, X., Cazabat, A. M. and Queré, D., Thickness and shape of films driver by a Marangoni flow. Langmuir, 12 (1996) 24.

36. Chesters, A. K., Elyousfi, A., Cazabat, A. M. and Villette, S., The influence of surfactants on the hydrodynamics of surface wetting. Journal of Petroleum Science and Engineering, 20 (1998) 217–222.

37. Valignat, M. P., Voué, M., Oshanin, G. and Cazabat, A. M., Structure and dynamics of thin liquid films on solid substrates. Colloids and Surfaces A: Phys. and Eng. Aspects, 154 (1999) 25–31.

38. Schneemilch, M. and Cazabat, A. M., Shock separation in wetting films driven by thermal gradients. Langmuir, 16 (2000) 9850–9856.

39. Schneemilch, M. and Cazabat, A. M., Wetting films in thermal gradients. Langmuir, 16 (2000) 8796–8801.

40. Cachile, M., Schneemilch, M., Hamraoui, A. and Cazabat, A. M., Films driven by surface tension gradients. Adv. in Colloid and Interf. Sci., 96 (2002) 59–74.

41. Poulard, C., Benichou, O. and Cazabat, A. M., Freely receding evaporating droplets. Langmuir, 19 (2003) 8828–8834.

42. Guena, G., Poulard, C. and Cazabat, A. M., The leading edge of evaporating droplets. J. of Colloids and Interface Science, 312 (2007) 164–171.

43. Churaev, N. V., Starov, V. M. and Derjaguin, B. V., The shape transition zone between a thin film and bulk liquid and the line tension. Journal of Colloid and Interface Science, 89 (1982) 1.

44. Li, D. and Neumann, A. W., Contact angles on hydrophobic solid surfaces and their interpretation. Journal of Colloids and Interface Science, 141 (1992) 1.

45. Li, D. and Steigmann, D. J., Positive line tension as a requirement of stable equilibrium. Colloids and Surfaces A: Phys. and Eng. Aspects, 116 (1996) 25–30.

46. Kwok, D. Y., Lin, R., Mui, M. and Neumann, A. W., Low rate dynamic and static contact angles and the determination of solid surface tensions. Colloids and Surfaces A: Phys. and Eng. Aspects, 116 (1996) 63–77.

47. Schwartz, W. L. and Garoff, S., Contact angle hysteresis and the shape of the three phase line. Journal of Colloids and Interface Science, 106 (1985) 2.

48. Rusanov, A. I. and Prokhorov, V. A., in: Interfacial Tensiometry, Vol. 3. Möbius, D. and Miller, R., Eds. Amsterdam, Elsevier, 1996.

49. Rusanov, A. I., Classification of line tension. Colloids and Surfaces A: Phys. and Eng. Aspects, 156 (1999) 315–322.

50. Shanahan, M. E. R., Fine structure of perturbed wetting triple line. Colloids and Surfaces A: Phys. and Eng. Aspects, 156 (1999) 71–77.

51. Qu, W., Yang, C. and Li, D. A., Gradient theory approach to line tension of liquid–liquid fluid systems. Colloids and Surfaces A: Phys. and Eng. Aspects, 144 (1998) 275–285.

52. Pismen, L. M., Mesoscopic hydrodynamics of contact line motion. Colloids and Surfaces A: Phys. and Eng. Aspects, 206 (2002) 11–30.

53. Carré, A. and Woehl P., Hydrodynamic behavior at the triple line of spreading liquids and the divergence problem. Langmuir, 18 (2002) 3600–3603.

54. Amirfarzli, A., Keshavartz, A., Zhang, L. and Neumann, A. W., Determination of line tension for systems near wetting. Journal of Colloids and Interface Science, 265 (2003) 152–160.

55. Voué, M., Rioboo, R., Bauthier, C., Conti, J., Charlot, M. and De Coninck, J., Dissipation and moving contact lines on non-rigid substrates. J. of the European Ceramic Society, 23 (2003) 2769–2775.

56. Amirfarzli, A. and Neumann, A. W., Status of the three-phase line tension. Advances in Colloids and Interface Science, 110 (2004) 121–141.

57. Tadmor, R., Line energy and the relation between advancing, receding and Young contact angle. Langmuir, 20 (2004) 7659–7664.

58. Ward, C. A. and Sefiane, K., Adsorption at the solid–liquid interface as the source of contact angle dependence on the curvature of the three-phase line. Advances in Colloid and Interface Science, 161 (2010) 171–180.

59. Sefiane, K., Skilling, J. and MacGillivray, J., Contact line motion and dynamic wetting of nanofluid solutions. Advances in Colloid and Interface Science, 138 (2008) 101–120.

60. Bonn, D., Wetting transitions. Current Opinion in Colloids & Interface Science, 6 (2001) 22–27.

61. Humfeld, K. D. and Garoff, S., Geometry-driven wetting transition. Langmuir, 20 (2004) 9223–9226.

62. Sefiane, K., Shanahan, M. E. R. and Antoni, M., Wetting and phase change: opportunities and challenges. Current Opinion in Colloid & Interface Science, 16(4) (2011) 317–325.

63. Long, J., Hyder, A. M., Huang, R. Y. M. and Chen, P., Thermodynamic modelling of contact angles on rough heterogeneous surfaces. Advances in Colloids and Interface Science, 118 (2005) 173–190.

64. Long, J. and Chen, P., On the role of energy barriers in determining contact angle hysteresis. Advances in Colloids and Interface Science, 127 (2006) 55–66.

65. Chibowsky, E., Surface free energy of a solid from contact angle hysteresis. Advances in Colloids and Interface Science, 103 (2003) 149–172.

66. Chibowsky, E., On some relations between advancing, receding and Young's contact angles. Advances in Colloids and Interface Science, 133 (2007) 51–59.

67. Extrand, C. W. and Kumagai, Y., Contact angles and hysteresis on soft surfaces. Journal of Colloid and Interface Science, 184 (1996) 191–200.

68. Extrand, C. W. and Kumagai, Y., An experimental study of contact angle hysteresis. Journal of Colloid and Interface Science, 191 (1997) 378–383.

69. Extrand, C. W., Contact angles and hysteresis on surfaces with chemically heterogeneous islands. Langmuir, 19 (2003) 3793–3796.

70. Extrand, C. W., A thermodynamic model for wetting free energies from contact angles. Langmuir, 19 (2003) 646–649.

71. Extrand, C. W., Contact angles and their hysteresis as a measure of liquid–solid adhesion. Langmuir, 20 (2004) 4017–4021.

72. Marmur, A., Equilibrium contact angles: theory and experiment. Colloids Surf. A, 116(1/2) (1996) 55–61.

73. Marmur, A., Contact angle hysteresis on heterogeneous smooth surfaces: theoretical comparison of the captive bubble and drop methods. Colloids and Surfaces A: Phys. and Eng. Aspects, 136 (1998) 209–215.

74. Wolansky, G. and Marmur, A., The actual contact angle on a heterogeneous rough surface in three dimensions. Langmuir, 14 (1998) 5292–5297.

75. Marmur, A., Line tension effect on contact angles: axisymmetric and cylindrical systems with rough or heterogeneous solid surface. Colloids and Surfaces A: Phys. and Eng. Aspects, 136 (1998) 81–88.

76. Apel-Paz, M. and Marmur, A., Spreading of liquids on rough surfaces. Colloids and Surfaces A: Phys. and Eng. Aspects, 146 (1999) 273–279.

77. Marmur, A. and Krasovitski, B., Line tension on curved surfaces: liquid drops on solid micro and nanospheres. Langmuir, 18 (2002) 8919–8923.

78. Marmur, A., Wetting on hydrophobic rough surfaces: to be heterogeneous or not to be? Langmuir, 19 (2003) 8343–8348.

79. Drelich, J. and Yu., W., Charge heterogeneity of surfaces: mapping and effects on surface forces. Advances in Colloid and Interface Science, 165(2) (2011) 91–101.

80. Tavana, H., Jehnichen, D., Grundke, K., Hair, M. L. and Neumann, A. W., Contact angle hysteresis on fluoropolymer surfaces. Advances in Colloid and Interface Science, 134/135 (2007) 236–248.

81. Chau, T. T., A review of techniques for measurement of contact angles and their applicability on mineral surfaces. Minerals Engineering, 22 (2009) 213–219.

82. Ramos, S. M. M. and Charlaix, E., Wetting on nanorough surface. Physical Review E, 67 (2003) 031604.

83. Ramos, S. M. M., Charlaix, E. and Benyagoub, A., Contact angle hysteresis on nano-structured surfaces. Surface Science, 540 (2003) 355–362.

84. Ramos-Canut, S., Wetting properties of nano-structured surfaces. Nuclear Instruments and Methods in Physics Research B, 245 (2006) 322–326.

85. Daniel, S. and Chaudhury, M. K., Rectified motion of liquid drops on gradient surfaces induced by vibrations. Langmuir, 18 (2002) 3404–3407.

86. Daniel, S., Sircar, S., Gliem, J. and Chaudhury, M. K., Ratcheting motion of liquid drops on gradient surfaces. Langmuir, 20 (2004) 4085–4092.

87. He, B., Patankar, N. A. and Lee, J., Multiple equilibrium droplet shapes and design criterion for rough hydrophobic surfaces. Langmuir, 19 (2003) 4999–5003.

88. Patankar, N. A., On the modeling of hydrophobic contact angles on rough surfaces. Langmuir, 19 (2003) 1249–1253.

89. Patankar, N. A., Transition between superhydrophobic states on rough surfaces. Langmuir, 20 (2004) 7097–7102.

90. Patankar, N. A., Mimicking the lotus effect: influence of double roughness structures and slender pillars. Langmuir, 20 (2004) 8209–8213.

91. Marmur, A., The lotus effect: superhydrophobicity and metastability. Langmuir, 20 (2004) 3517–3519.

92. Ferrari, M. and Ravera, F., Surfactants and wetting at superhydrophobic surfaces: water solutions and nonaqueous liquids. Advances in Colloid and Interface Science, 161 (2010) 22–28.

93. Krasowska, M., Zawala, J. and Malysa, K., Air at hydrophobic surfaces and kinetics of three phase contact formation. Advances in Colloid and Interface Science, 147–148 (2009) 155–169.

94. Vinogradova, O. I., Slippage of water over hydrophobic surfaces. Int. J. Miner. Process., 56 (1999) 31–60.

95. Christenson, H. G. and Yaminsky, V. V., Is the long range hydrophobic attraction related to the mobility of hydrophobic surface groups? Colloids and Surfaces A: Phys. and Eng. Aspects, 129/130 (1997) 67–74.

96. Yaminsky, V. V. and Ninham, B. W., Surface forces *vs.* surface composition. Colloid science from the Gibbs adsorption perspective. Advances in Colloid and Interface Science, 83 (1999) 227–311.

97. Yaminsky, V. V., The hydrophobic force: the constant volume capillary approximation. Colloids and Surfaces A: Phys. and Eng. Aspects, 159 (1999) 181–195.

98. Von Bahr, M., Tiberg, F. and Yaminsky, V. V., Spreading dynamics of liquids and surfactants solutions on partially wettable hydrophobic substrates. Colloids and Surfaces A: Phys. and Eng. Aspects, 193 (2001) 85–96.

99. Yaminsky, V. V. and Vogler, E. A., Hydrophobic hydration. Current Opinion in Colloids and Interface Science, 6 (2001) 342–349.

100. Shirtcliffe, N. J., McHale, G., Atherton, S. and Newton, M. I., An introduction to superhydrophobicity. Advances in Colloid and Interface Science, 161 (2010) 124–138.

101. Bhushan, B. and Jung, Y. C., Natural and biomimetic artificial surfaces for superhydrophobicity, self-cleaning, low adhesion, and drag reduction. Progress in Materials Science, 56 (2011) 1–108.

102. Nosonovsky, M. and Bhushan, B., Superhydrophobic surfaces and emerging applications: Non-adhesion, energy, green engineering. Current Opinion in Colloid & Interface Science, 14 (2009) 270–280.

103. Sharma, A., Equilibrium contact angle and film thicknesses in the apolar and polar systems: role of intermolecular interactions in coexistence of drops with thin films. Langmuir, 9 (1993) 3580–3586.

104. Sharma A. and Jameel, A. T., Nonlinear stability, rupture and morphological phase separation of thin fluid films on apolar and polar substrates. Journal of Colloids and Interface Science, 161 (1993) 190–208.

105. Jameel, A. T. and Sharma, A., Morphological phase separation in thin liquid films, II-equilibrium contact angles of nanodrops coexisting with thin films. Journal of Colloids and Interface Science, 164 (1994) 416–427.

106. Sharma, A. and Reiter, G., Instability of thin polymer films on coated substrates: rupture, dewetting and drop formation. Journal of Colloids and Interface Science, 178 (1996) 383–399.

107. Khanna, R. and Sharma, A., Pattern formation in spontaneous dewetting in thin apolar films. Journal of Colloids and Interface Science, 195 (1997) 42–50.

108. Sharma, A., Konnur, R. and Kargupta, K., Thin liquid films on chemically heterogeneous substrates: self organization, dynamics and patterns in systems displaying a secondary minimum. Physica A, 318 (2003) 262–278.

109. Manev, E. D. and Nguyen, A. V., Critical thickness of microscopic thin liquid films. Advances in Colloid and Interface Science, 114–115 (2005) 133–146.

110. Saramago, B., Thin liquid wetting films. Current Opinion in Colloid & Interface Science, 15 (2010) 330–340.

111. Bertrand, E., Blake, T. D. and De Coninck, J., Dynamics of dewetting. Colloids and Surfaces A: Physicochemical and Engineering Aspects, 369 (2010) 141–147.

112. Lyklema, J., The surface tension of pure liquids. Thermodynamic Components and corresponding states. Colloids and Surfaces A: Phys. and Eng. Aspects, 156 (1999) 413–421.

113. Lyklema, J., A discussion on surface excess entropies. Colloids and Surfaces A: Phys. and Eng. Aspects, 186 (2001) 11–16.

114. Chen, P., Susnar, S. S. and Neumann, A. W., Thermodynamics of liquid films and film tension measurements, Int. J. Miner. Process., 56 (1999) 75–97.

115. Babak, V. G., Thermodynamic of plane-parallel liquid films. Colloids and Surfaces A: Phys. and Eng. Aspects, 142 (1998) 135–153.

116. Yeh, E. K., Newman, J. and Radke, C. J., Equilibrium configurations of liquid droplets on solid surfaces under the influence of thin film forces. Part I: Thermodynamic. Colloids and Surfaces A: Phys. and Eng. Aspects, 156 (1999) 137–144.

117. Yeh, E. K., Newman, J. and Radke, C. J., Equilibrium configurations of liquid droplets on solid surfaces under the influence of thin film forces. Part II: Shape Calculation. Colloids and Surfaces A: Phys. and Eng. Aspects, 156 (1999) 525–546.

118. Churaev, N. V., Contact angle and surface forces. Advances in Colloid and Interface Science, 58 (1995) 87–118.

119. Churaev, N. V., Setzer, M. J. and Adolphs, J., Influence of surface wettability on adsorption isotherms of water vapour. Journal of Colloid and Interface Science, 197 (1998) 327–333.

120. Starov, V. M. and Churaev, N. V., Wetting film on locally heterogeneous surfaces: hydrophilic surface with hydrophobic spots. Colloids and Surfaces A: Phys. and Eng. Aspects, 156 (1999) 243–248.

121. Churaev, N. V., Surface forces in wetting films. Colloid Journal, 65(3) (2003) 263–274.

122. Churaev, N. V. and Sobolev, V. D., Wetting of low energy surfaces. Advances in Colloid and Interface Science, 2007.

123. Starov, V. M., Surface forces action in a vicinity of three phase contact line and other current problems in kinetics of wetting and spreading. Advances in Colloid and Interface Science, 161 (2010) 139–152.

124. Li, W. and Amirfazli, A., Microtextured superhydrophobic surfaces: A thermodynamic analysis. Advances in Colloid and Interface Science, 132 (2007) 51–68.

125. Boinovich, L. and Emelyanenko, A., Wetting and surface forces. Advances in Colloid and Interface Science, 165(2) (2011) 60–69.

126. Boinovich, L., DLVO forces in thin liquid films beyond the conventional DLVO theory. Current Opinion in Colloid & Interface Science, 15 (2010) 297–302.

127. Toshev, B. V., Thermodynamic theory of thin liquid films including line tension effects. Current Opinion in Colloid & Interface Science, 13 (2008) 100–106.

128. Djikaev, Y. S. and Ruckenstein, E., The variation of the number of hydrogen bonds per water molecule in the vicinity of a hydrophobic surface and its effect on hydrophobic interactions. Current Opinion in Colloid & Interface Science, 16(4) (2011) 272–284.

129. Ramé, E. and Garoff, S., Microscopic and macroscopic dynamic interface shapes and the interpretation of dynamic contact angles. Journal of Colloid and Interface Science, 177 (1996) 234–244.

130. Ramé, E., The interpretation of dynamic contact angles measured by wilhelmy plate method. Journal of Colloid and Interface Science, 185 (1997) 245–251.

131. De Ruijter, M., Kölsch, P., Voué, M., De Coninck, J. and Rabe, J. P., Effects of temperature on the dynamic contact angle. Colloids and Surfaces A: Phys. and Eng. Aspects, 144 (1998) 235–243.

132. Thiele, U., Neuffer, K., Bestehorn, M., Pomeau, Y. and Velarde, M. G., Sliding drops on an inclined plane. Colloids and Surfaces A: Phys. and Eng. Aspects, 206 (2002) 87–104.

133. Della Volpe, C., Contact angle measurements on samples with dissimilar faces by Wilhelmy microbalance. J. of Adhesion Sci. Technol., 10(12) (1994) 1453–1458.

134. Della Volpe, C. and Siboni, S., Analysis of dynamic contact angle on discoidal samples measured by the Wilhelmy method. J. of Adhesion Sci. Technol., 12(2) (1998) 197–224.

135. Della Volpe, C., Maniglio, D., Morra, M. and Siboni, S., The determination of a stable equilibrium contact angle on heterogeneous and rough surfaces. Colloids and Surfaces A: Phys. and Eng. Aspects, 206 (2002) 47–67.

136. Lam, C. N. C., Ko, R. H. Y., Yu, L. M. Y., Ng, A., Li, D., Hair, M. L. and Neumann, A. W., Dynamic cycling contact angle measurements: study of advancing and receding contact angles. Journal of Colloid and Interface Science, 243 (2001) 208–218.

137. Lam, C. N. C., Kim, N., Hui, D., Kwok, D. Y., Hair, M. L. and Neumann, A. W., The effect of liquid properties to contact angle hysteresis. Colloids and Surfaces A: Phys. and Eng. Aspects, 189 (2001) 265–278.

138. Lam, C. N. C., Wu, R., Li, D., Hair, M. L. and Neumann, A. W., Study of the advancing and receding contact angles: liquid sorption as a cause of contact angle hysteresis. Advances in Colloid and Interface Science, 96 (2002) 169–191.

139. Wilhelmy, L., Ann. Phys., (1863) 119–177.

140. Li, D. and Neumann, A. W., in: Applied Surface Thermodynamics. A. W. Neumann & J. K. Spelt Eds. 109–168, Marcel Dekker, NY, 1996.

141. Kwok, D. Y., Lam, C. N. C., Li, A., Leung, A., Wu, R., Mok, E. and Neumann, A. W., Measuring and interpreting contact angles: a complex issue. Colloids and Surfaces A: Phys. and Eng. Aspects, 142 (1998) 219–235.

142. Chen, P., Policova, Z., Pace-Asciak, C. R. and Neumann, A. W., Study of molecular interactions between lipids and proteins using dynamic surface tension measurement: a review. Colloids and Surfaces B: Biointerfaces, 15 (1999) 313–324.

143. Kwok, D. Y. and Neumann, A. W., Contact angle measurement and contact angle interpretation. Advances in Colloids and Interface Science, 81 (1999) 167–249.

144. Kwok, D. Y. and Neumann, A. W., Contact angle interpretation: re-evaluation of existing contact angle data. Colloids and Surfaces A: Phys. and Eng. Aspects, 161 (2000) 49–62.

145. Tavana, H. and Neumann, A. W., Recent progresses in the determination of solid surface tensions from contact angles. Advances in Colloid and Interface Science, 132 (2007) 1–32.

146. Good, R. J. and Koo, M. N., The effect of drop size on contact angle. Journal of Colloids and Interface Science, 71 (1979) 283.

147. Gaydos, J. and Neumann, A. W., The dependence of contact angles on drop size and line tension. Journal of Colloids and Interface Science, 120 (1987) 76.

148. Li, D. and Neumann, A. W., Determination of line tension from the drop size dependence of contact angles. Colloids and Surfaces, 43 (1990) 195.

149. Duncan, D., Li, D., Gaydos, J. and Neumann, A. W., Correlation of line tension and solid–liquid interfacial tension from the measurement of drop size dependence of contact angles. Journal of Colloids and Interface Science, 169 (1995) 256–261.

150. Li, D., Drop size dependence of contact angles and line tension of solid liquid systems. Colloids and Surfaces A: Phys. and Eng. Aspects, 116 (1996) 1–23.

151. Amirfazli, A., Kwok, D. Y., Gaydos, J. and Neumann, A. W., Line tension measurements through drop size dependence of contact angle. Journal of Colloids and Interface Science, 205 (1998) 1–11.

152. Amirfazli, A., Chatain, D. and Neumann, A. W., Drop size dependence of contact angles for liquid tin on silica surface: line tension and its correlation with solid–liquid interfacial tension. Colloids and Surfaces A: Phys. and Eng. Aspects, 142 (1998) 183–188.

153. Gu, Y., Drop size dependence of contact angles of oil drops on a solid surface in water. Colloids and Surfaces A: Phys. and Eng. Aspects, 181 (2001) 215–224.

154. Mac Dowell, L. G., Muller, M. and Binder, K., How do droplets on a surface depend on the system size? Colloids and Surfaces A: Phys. and Eng. Aspects, 206 (2002) 277–291.

155. Brandon, S., Haimovich, N., Yeger, E. and Marmur, A., Partial wetting of chemically patterned surfaces: the effect of drop size. Journal of Colloids and Interface Science, 263 (2003) 237–243.

156. Vafei, S. and Podowski, M. Z., Analysis of the relationship between liquid droplet size and contact angle. Advances in Colloids and Interface Science, 113 (2003) 133–146.

157. Letellier, P., Mayaffre, A. and Turmine, M., Drop size effect on contact angle explained by nonextensive thermodynamics. Young's Equation Revisited. Journal of Colloids and Interface Science, 314(2) (2007) 604–614.

158. Barberis, F. and Capurro, M., Wetting in the nanoscale: a continuum mechanics approach. Journal of Colloid and Interface Science, 326(1) (2008) 201–210.

159. Nikolov, A. D., Wasan, D. T., Chengara, A., Kozco, K., Policello, A. and Kolossvary, I., Superspreading driven by marangoni flow. Advances in Colloids and Interface Science, 96 (2002) 325–338.

160. De Gennes, P. G., On fluid/wall slippage. Langmuir, 18 (2002) 3413–3414.

161. Mao, M., Zhang, J., Yoon, R. H. and Ducker, A., Is there a thin film of air at the interface between water and smooth hydrophobic solids? Langmuir, 20 (2004) 1843–1849.

162. De Gennes, P. G., Polymers at an interface: a simplified view. Advances in Colloid and Interface Science, 27 (1987) 189–209.

163. Good, R. J., in: Adsorption at Interfaces. K. L. Mittal, Ed., ACS Symposium Series No. 8. ACS, Washington, DC, 1975.

164. Tse, J. and Adamson, A. W., Adsorption and contact angle studies. Journal of Colloid and Interface Science, 72 (1979) 3.

165. Gee, M. L., Healy, T. W. and White, R. L., Hydrophobicity effects in the condensation of water films on quartz. Journal of Colloids and Interface Science, 140 (1990) 450–465.

166. Li, D., Xie, M. and Neumann, A. W., Vapour adsorption on hydrophobic solid surfaces. Colloid Polym. Sci., 271 (1993) 573–580.

167. Kwok, D. Y., The usefulness of the Lifshitz–Van der Waals/acid–base approach for surface tension components and interfacial tension. Colloids and Surfaces A: Phys. and Eng. Aspects, 156 (1999) 191–200.

168. Berg, J. C., Ed., Wettability. Dekker, NY (1993).

169. Van Oss, C. J. and Good, R. J., Interfacial Forces in Aqueous Media. Dekker, NY, 1994.

170. Fowkes, F. M., in: Acid–Base Interactions: Relevance to Adhesion Science and Technology. K. L. Mittal and H. R. Anderson, Jr, Eds., 93–115. VSP, Utrecht, 1991.

171. Drago, R. S., Structure and Bonding, 15 73–139. Springer, Berlin, 1973.

172. Van Oss, C. J., Good, R. J. and Chaudhury, M. K., J. Protein Chem., 5 (1986) 385–402.

173. Jacobasch, H. J., Grundke, K., Schneider, S. and Simon, F., J. of Adhesion, 48 (1995) 57–73.

174. Tate, M. L., Kamath, Y. K., Wesson, S. P. and Ruetsch, S. B., Surface energetics of nylon 66 fibers. Journal of Colloids and Interface Science, 177(2) (1996) 579–588.

175. Van Oss, C. J., Chaudhury, M. K., Good, R. J., Monopolar surfaces. Advances in Colloid and Interface Science, 28 (1987) 35–64.

176. Della Volpe, C. and Siboni, S., Some reflections on acid–base solid surface free energy theory. Journal of Colloids and Interface Science, 195 (1997) 121–136.

177. Della Volpe, C. and Siboni, S., Acid–base surface free energies of solids and the definition of scales in the Good–Van Oss–Chaudhury theory. J. of Adhesion Sci. Technol., 14(2) (2000) 235–272.

178. Della Volpe, C. and Siboni, S., Troubleshooting of surface free energy acid–base theory applied to solid surfaces: the case of good, van oss and chaudury theory. Acid–Base Interactions, 2 (2000) 55–90.

179. Della Volpe, C. and Siboni, S., Acid–base surface free energies of solids and the definition of scales in the Good–Van Oss–Chaudury theory. J. Adhesion Sci. Technol., 14(2) (2000) 235–272.

180. Della Volpe, C., Maniglio, D., Brugnara, M., Siboni, S. and Morra, M., The solid surface free energy calculation I. In defense of the multicomponent approach. Journal of Colloids and Interface Science, 271 (2004) 434–453.

181. Siboni, S., Della Volpe, C., Maniglio, D. and Brugnara, M., The solid surface free energy calculations II. The limits of the zisman and of the equation of state approach. Journal of Colloids and Interface Science, 271 (2004) 453–472.

182. Defay, R. and Prigogine, I., Surface Tension and Adsorption. Longmans, London, 1966.

183. Bikerman, J. J., Physical Surfaces. E. M. Loebl Ed. Academic Press, NY & London, 1971.

184. Adamson, A. W. and Gast, A. P., Physical Chemistry of Surfaces, 6th Edn. Wiley & Sons Inc., 1997.

185. Israelachvili, J. N., Intermolecular and Surface Forces, 2nd Edn., Academic Press, 1991.

186. McHale, G., Rowan, S. M., Newton, M. I. and Banerjee, M. K., Evaporation and wetting of a low energy solid surface. J. Phys. Chem. B, 102 (1998) 1964–1967.

187. Erbil, H. Y., McHale, G. and Newton, M. I., Drop evaporation on solid surfaces: constant contact angle mode. Langmuir, 18 (2002) 2626–2641.

188. Brunauer, S., Emmett, P. H. and Teller, E., J. of Am. Chem. Society, 60 (1938) 309.

189. Gregg, S. J. and Singh, K. S. W., Adsorption, Surface Area and Porosity. Academic Press, 1967.

190. Morra, M., Occhiello, E. and Garbassi, F., Contact angle hysteresis in oxygen plasma treated poly(tetrafluoroethylene). Langmuir, 5 (1989) 872–876.

191. Takeda, S., Yamamoto, K., Hayasaka, Y. and Matsumoto, K., Surface OH group governing wettability of commercial glasses. Journal of Non-Crystalline Solids, 249 (1999) 41–46.

192. Takeda, S., Fukawa, M., Hayashi, Y. and Matsumoto, K., Surface OH group governing adsorption properties of metal oxide films. Thin Solid Films, 39 (1999) 220–224.

193. Shanahan, M. E. R., De Gennes, P. G., Start-up of a reactive droplet. C. R. Acad. Sci., Paris, t. 324, Serie II b, (1997) 261–268.

194. De Gennes, P. G., The dynamics of reactive wetting on solid surfaces. Physica A, 249 (1998) 196–205.

195. Searcy, A. W., The dependence of particle shapes on partial free energies of bonding to inert substrates. Scripta Materialia, 40(8) (1999) 979–982.

196. Beruto, D. T., Ferrari, A., Barberis, F. and Giordani, M., Dispersions of micrometric powders of molybdenum and alumina in liquid paraffin: role of the interfacial phenomena on bulk rheological properties. Journal of European Ceramic Society, 22 (2002) 2155–2164.

197. Hartland, S., Surface and Interfacial Tension: Measurement, Theory and Applications. Dekker, NY, 2004.

198. Antonow, G., J. Chem. Phys., 5 (1907) 372.

199. Berthelot, D., Comp. Trend., 126 (1857) 1703–1898.

200. Gu, Y., Li, D. and Cheng, P., Determination of line tension from the shape of axisymmetric liquid–vapor interfaces around a conic cylinder. Journal of Colloids and Interface Science, 180 (1996) 212–217.
201. Lin, F. Y. H. and Li, D., The influence of inclination of a solid surface on contact angles due to the effect of line tension. Colloids and Surfaces, 87 (1994) 93.
202. Hu, J., Xiao, X. D., Ogletree, D. F. and Salmeron, M., Imaging the condensation and evaporation of molecularly thin films of water with nanometer resolution. Science, 268 (1995) 267.
203. Hu, J., Xiao, X.-D. and Salmeron, M., Scanning polarization force microscopy: a technique for imaging liquids and weakly adsorbed layers. Appl. Phys. Lett., 67 (1995) 476.
204. Dai, Q., Hu, J., Freedman, A., Robinson, G. N. and Salmeron, M., Nanoscale imaging of corrosion reaction: sulfuric acid droplets on aluminum surfaces. The Journal of Physical Chemistry, 100 (1996).
205. Salmeron, M., Xu, L., Hu, J. and Dai, Q., High-resolution imaging of liquid structures: wetting and capillary phenomena at the nanometer scale. MRS Bulletin, 22(8) (1997) 36–41.
206. Bluhm, H., Pan, S. H. Xu, L., Inoue, T., Ogletree, D. F. and Salmeron, M., Scanning force microscope and vacuum chamber for the study of ice films: design and first results. Rev. Sci. Instrum., 69 (1998) 1781.
207. Xu, L. and Salmeron, M., Scanning polarization force microscopy study of the condensation and wetting properties of glycerol on mica, J. of Physical Chemistry B, 102(37) (1998) 7210–7215.
208. Rieutord, F. and Salmeron, M., Wetting properties at the submicrometer scale: a scanning polarization force microscopy study. J. of Physical Chemistry B, 102(20) (1998) 3941–3944.
209. Salmeron, M., Nanoscale Wetting and De-Wetting of Lubricants with Scanning Polarization Force Microscopy. Fundamentals of Tribology and Bridging the Gap Between the Macro and Micro-Nanoscales, Vol. 651. Kluwer Academic Publishing, 2001.
210. Barberis, F. and Beruto, D. T., Adsorption of paraffin vapor on oxidized molybdenum substrates at nano and micro-scales. Journal of Colloids and Interface Science, 313 (2007) 592–599.
211. Ruckenstein, E., Microscopic origin of macroscopic wetting. Colloids and Surfaces A: Phys. and Eng. Aspects, 206 (2002) 3–10.
212. Berim, O. G. and Ruckenstein, E., On the shape and stability of a drop on a solid surface. J. Phys. Chem. B, 108 (2004) 19330–19338.
213. Berim, O. G. and Ruckenstein, E., Microcontact and macrocontact angles and the drop stability on a bare surface. J. Phys. Chem. B, 108 (2004) 19339–19347.
214. Berim, O. G. and Ruckenstein, E., Nanodroplets on a planar solid surface: temperature, pressure and size dependence of their density and contact angles. Langmuir, 22 (2006) 1063–1073.
215. Amirfarzli, A., Graham-Eagle, J., Pennell, S., Neumann, A. W., Implementation and examination of a new drop shape analysis algorithm to measure contact angle and surface tension from the diameters of two sessile drops. Colloids and Surfaces A: Phys. and Eng. Aspects, 161 (2000) 63–74.

Review on Kinetics of Spreading and Wetting by Aqueous Surfactant Solutions

J. Radulovic [a], **K. Sefiane** [a], **V. M. Starov** [b], **N. Ivanova** [b], **M. E. R. Shanahan** [c]

[a] Institute for Materials and Processes, The University of Edinburgh, Kings Buildings, Edinburgh, EH9 3JL, United Kingdom. E-mail: ksefiane@ed.ac.uk

[b] Department of Chemical Engineering, Loughborough University, Loughborough, LE11 3TU, United Kingdom

[c] Université de Bordeaux, Institute de Mécanique et d'Ingénierie, CNRS UMR 5295, Bâtiment A4, 351 Cours de la Libération, 33405 Talence Cedex, France

Contents

A. Introduction 37
B. Wetting of Ideal and Real Surfaces 38
C. of Pure Liquids 40
D. Spreading of Surfactant Solutions 41
E. Spreading over Hydrophobic Substrates 43
F. Spreading of Surfactant Solutions over Thin Aqueous Layers: Influence of Solubility and Micelle
    Disintegration 44
G. Instabilities in the Course of Spreading 48
H. Spreading of Surfactant Solutions over Substrates 52
I. superspreading]Superspreading 57
J. References 64

A. Introduction

A great number of industrial processes and practical applications are based on wetting and spreading. Understanding these phenomena is crucial in designing many pharmaceutical, agricultural and bio-medical products, such as herbicides and pesticides, paints and coatings, as well as ink-jet printers and lung-surfactants. Although this interesting topic has been studied for over two centuries, there are still certain underlying mechanisms waiting to be revealed.

Over the years, many eminent scientists have investigated this apparently limitless topic, both theoretically and experimentally. Starting with Young's equation at the beginning of the 19th century, leading to de Gennes' priceless contributions

Drops and Bubbles in Contact with Solid Surfaces
© Koninklijke Brill NV, Leiden, 2012

on the complexity of the phenomenon, up to modern spreading theories, active research is continuing. Many theoretical models have shown good agreement with
experimental results for a number of liquid–solid systems. However, due to a wide
range of hydrophobic surfaces and liquids involved, it is difficult to create a general
model. The use of surfactants as additives further complicates the matter.

To improve wetting, especially on extremely hydrophobic surfaces, surfactants
have been broadly employed for decades. The investigation of spreading of surfactant solutions has led to new approaches and discoveries. As surfactant structure
and performance may greatly differ, proposed models became more complex. This
is particularly the case for trisiloxane surfactants—the 'superspreaders', whose su-
perb behaviour is poorly understood. Another area of interest is spreading on porous
substrates, especially when surfactants are used.

Here we provide a detailed review of significant achievements in the field of
spreading. Acknowledged models are presented, after a comprehensive theoretical
part, which could be very useful for beginners in this scientific area. In addition,
later approaches are discussed for a number of spreading related areas, such as
surfactants on liquid films, on porous substrates and spreading instabilities. As
this fascinating phenomenon is still systematically explored, further discoveries are
soon expected.

B. Wetting of Ideal and Real Surfaces

Wetting is the contact between a liquid and a solid (or another liquid), when the
two are brought into contact. When a liquid drop is placed onto a solid surface, it
will wet the surface to a certain extent, which depends on forces that act on the
three-phase (l–s–g) contact line, the so-called triple line [1, 2], until the equilibrium
is reached. The wetting process is normally characterised by the change in shape
of the drop and, consequently, its geometrical parameters (radius, height, contact
angle). Once the equilibrium, which is defined as the state with the minimal excess
of the total free energy of the system, is reached, the shape of the drop stops varying
and wetting parameters no longer change. Young (1805) was the first to suggest the
constancy of the equilibrium contact angle (θ_{eq}) of a liquid drop on a solid surface,
as a function of the surface free energy, the interfacial free energy and the surface
tension of the liquid. In years to come, Dupré incorporated thermodynamic aspects
and this relationship is nowadays known as the Young–Dupré equation:

$$\cos \theta_{eq} = \frac{\gamma_{sg} - \gamma_{sl}}{\gamma_{lg}}, \tag{1}$$

where γ_{sg}, γ_{sl} and γ_{lg} are interfacial tensions at solid–gas, solid–liquid and liquid–
gas interfaces, respectively (Fig. 1). The first of these is better known as the surface
energy of the solid, while the last is commonly referred to as the surface tension of
the liquid. Surface energy is one of the basic properties of a solid surface and varies
greatly for different materials. The value of the surface energy is the consequence

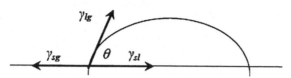

Figure 1. Liquid drop in contact with a solid surface. The contact angle θ depends on three tension forces acting at the three-phase line.

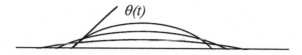

Figure 2. Complete wetting; liquid drop spreads over the surface ($\theta \sim 0°$).

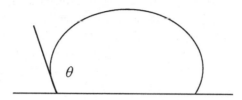

Figure 3. Partial wetting ($0° < \theta < 90°$).

of the type and the intensity of the intermolecular forces acting inside the solid. A general, empirical rule is that liquids with surface tension lower than the surface energy of the solid effectively wet the substrate. Most metals and surfaces similar to silica have high surface energies and are therefore wetted by most liquids, at least when clean [2]. On the other hand, polymers and waxes have low surface energies [3] and to what extent a liquid will wet a low-energy surface primarily depends on the surface tension of the liquid. Hydrophobicity of the substrate is usually characterised in terms of the contact angle that pure water drop exhibits on the particular surface. When a liquid drop is brought in contact with the solid surface, depending on its free energy, three wetting scenarios may occur.

Complete wetting (Fig. 2) happens in the case when the liquid drop spreads over the surface, with the contact angle progressively decreasing to very low values, theoretically zero. Otherwise, a drop forms a finite contact angle with the surface. If the contact angle is between 0° and 90°, partial wetting takes place (Fig. 3). Values of the contact angle bigger than 90° represent what is called non-wetting (Fig. 4) and corresponding surfaces are referred to as hydrophobic, when the liquid is aqueous.

The Young–Dupré equation is based on the equilibrium thermodynamics of an ideal system and is therefore restricted by the following assumptions: line tension contributions can be neglected; the solid surface is inert and flat; and all three interfacial tensions have constant values. Therefore, the Derjaguin–Frumkin equation [4] was proposed as an improved form. In real systems, there are certain irregularities of the solid surface and measured contact angle values vary from the calculated

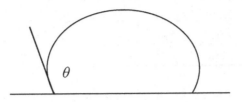

Figure 4. Non-wetting ($\theta > 90°$).

one. These angles fall within a certain range with advanced angles approaching the maximum value ('advancing') and receded angles approaching minimum value ('receding'). The difference between advancing and receding contact angle is called the contact angle hysteresis. Hysteresis phenomena are attributed to surface roughness, mobility and chemical heterogeneity [5]. For inhomogeneous surfaces, any defect will present barriers to the motion of the three-phase contact line and metastable states may occur [6, 7]. Hysteresis effect can also be viewed as an indication of the extent of the change in surface properties caused by wetting [8]. Lam *et al.* have shown that hysteresis depends on the size of liquid molecules and that both advancing and receding contact angles decrease with liquid sorption/retention [9]. Other works investigated the influence of drop size [10] or coverage and distribution of the surfactants [11] on contact angle hysteresis.

Another important parameter for illustration of wetting behaviour is the spreading coefficient, S. The spreading coefficient is defined as the difference in free energies between the bare solid directly in contact with the gaseous phase and the solid covered by a flat liquid layer [1].

$$S = \gamma_{sg} - (\gamma_{sl} + \gamma_{lg}), \tag{2}$$

γ_{sg} here represents the interfacial tension of the 'dry' surface. For $S < 0$, the liquid will form a drop with finite contact angle (partial wetting). For $S \geqslant 0$ drop will flatten forming a thin film spread over the surface (complete wetting with $\theta_{eq} = 0$).

C. Spreading of Pure Liquids

When a liquid drop is placed onto a flat surface, capillary forces drive the interface spontaneously towards equilibrium. At the same time, there is hydrodynamic resistance to spreading that opposes the capillary driving force. Evolution of the drop radius and the contact angle with time are commonly investigated to describe the dynamics of wetting on solid surfaces. For pure liquids, three theoretical approaches have been developed to describe spreading of the drop: the hydrodynamic model [11–14], the molecular-kinetic model [15–17] and the evaporation/condensation model [18–20]. The main difference between the former two descriptions is the dominant energy dissipation channel during drop shape transformation. The dissipation of energy in the hydrodynamic model is due to viscous drag within the spreading droplet, unlike the molecular-kinetic model where the dissipation of energy is assumed to be caused by friction at the three-phase contact line (intermolecular interactions between the solid and the liquid). In the evaporation/condensation model, transfer of liquid at the triple line occurs due non-equilibrium Kelvin pressures. The condensation effect is found to be of greater importance at higher contact angles (e.g., in forced wetting), and the overall features are complementary to the molecular kinetics approach.

The hydrodynamic model was derived from the Navier–Stokes equation by introducing the so-called 'lubrication approximation', in which flow is assumed to be uniquely parallel to the solid surface. It has been shown that the hydrodynamic regime may be characterised asymptotically by the tendency of drop radius and contact angle to follow the power laws: $R(t) \sim t^{1/10}$ and $\theta(t) \sim t^{-3/10}$. Probably the best-known result of the hydrodynamic approach is Voinov–Hoffman–Tanner law [14]. In contrast to the hydrodynamic model, the molecular-kinetic model proposes different dynamic behaviour: $R \sim t^{1/7}$ and $\theta \sim t^{-3/7}$. It originates from the molecular-kinetic theory of Eyring and the behaviour of the wetting line is explained by individual adsorption/desorption molecular displacements. Similar behaviour may be expected with the evaporation/condensation model. While the hydrodynamic model successfully describes complete wetting for viscous systems, the molecular-kinetic model seems to be more appropriate for partial wetting [21]. De Gennes and Brochard-Wyart concluded that hydrodynamics was essential for low angles, while molecular features were important at high velocities and large angles [22].

Ruijter *et al.* [23, 24] presented a combined model to describe the dynamics of a spreading drop, allowing both dissipation by friction against the substrate and hydrodynamic effects. This approach allowed the identification of several stages in the wetting process: a fast, early-time stage (with linear change of the drop radius with time), followed by a kinetic stage, which can be successfully modelled by molecular-kinetic and subsequently hydrodynamic model, and finally an exponential relaxation to the equilibrium state. All the models described above proved to be a powerful tool for portrayal of wetting behaviour of pure liquids, especially when they exhibited partial wetting. Nevertheless, none of them gave a satisfactory fit for wetting by surfactant solutions.

D. Spreading of Surfactant Solutions

Surfactants (*surf*ace *act*ive *agents*) are amphiphilic molecules that consist of hydrophilic (polar) and hydrophobic (non-polar) parts. The hydrophilic part of the molecule is commonly referred to as the head-group and the hydrophobic part as the tail. 'Lipophilic tail' and 'lipophobic head-group' are also commonly used terms. Depending on the nature of the head group, surfactant molecules can be characterised as ionic (anionic, cationic, zwitterionic) or nonionic. The surfactant tail is usually a sufficiently long hydrocarbon chain, although its chemistry may vary (fluorocarbon, siloxane). Due to their dual nature, surfactants tend to pack at interfaces between polar and non-polar regions, allowing enhanced wetting. Therefore, surfactants play a vital role in interface science.

The addition of surfactant can transform a non-wetting aqueous solution into a wetting solution, even on hydrophobic substrates. The term wetting agent is applied to any substance that increases the ability of water or an aqueous solution to wet a solid surface. Altering the wetting ability of water is an important property shown to some degree by all surfactants, although the extent to which they exhibit this phenomenon varies greatly. Surfactants adsorb at relevant interfaces lowering the total free energy of the system. It is often said that surfactants favour expansion of the interface by lowering the surface tension of the liquid. Additionally, micellisation takes place as an alternative mechanism for decreasing the interfacial energy of a surfactant solution by arranging molecules into energetically favourable aggregates. Here we introduce CMC (the *c*ritical *m*icelle *c*oncentration) as the main parameter of a surfactant solution. Below CMC, surfactant molecules are freely solubilised in the solution; above CMC, spontaneous aggregation takes place. Structure of surfactant aggregates mostly depends on the nature and type of the surfactant, but also on the properties of the solvent in which the surfactant is dispersed. Israelachvili [2] showed that geometrical factors dictate the aggregate structure and introduced the surfactant parameter, N_S, (packing ratio) as the most convenient parameter for prediction of the aggregate structure:

$$N_S = \frac{V}{la_o},\tag{3}$$

where V is the volume of the hydrophobic part (the tail), a_o is the area of the surfactant head group and l is the length of the tail. Estimation of these geometrical parameters may be challenging, since the area of the surfactant head and especially the length of the tail are influenced by the type of interactions with the solvent molecules, surfactant concentration and other environmental factors [25]. Very small values of N_S indicate spherical or ellipsoidal micelles; for values of N_S close to unity, vesicles and bilayer structures are expected. When N_S is above 1, inverted micelles are formed [26].

Pure water does not spread spontaneously over hydrophobic surfaces. On hydrophobic substrates, normally characterised by low surface energy, non-wetting occurs with the finite contact angle, $\theta_{eq} > 90°$. However, surfactant solutions can spread spontaneously on hydrophobic substrates and the spreading rate depends on the concentration of the surfactant. In the very beginning of the spreading process, when the surfactant solution is for the first time in contact with the hydrophobic surface, the contact angle, θ, is bigger than 90° and the liquid cannot spread. Transfer of surfactant molecules from the liquid onto all three interfaces may then take place. Surfactant adsorption occurs at (i) the inner solid–liquid interface, which results in a decrease of the solid–liquid interfacial tension, γ_{sl}, (ii) the liquid–gas interface, which results in a decrease of the liquid–air interfacial tension, γ_{lg}, and (iii) transfer from the drop onto the solid–gas interface just in front of the drop.

All three adsorption processes result in a decrease of the excess free energy of the system. However, adsorption processes (i) and (ii) result in a decrease of corresponding interfacial tensions, γ_{sl} and γ_{lg}, but the transfer of surfactant molecules onto the solid–gas interface in front of the drop results in an increase of a local free energy, although the total free energy of the system decreases. That is, surfactant molecule transfer (iii) goes *via* a relatively high potential barrier and, hence, goes considerably slower than adsorption processes (i) and (ii). Therefore, they are 'fast' processes as compared with the third process (iii).

The excess free energy, Φ, of the droplet on a solid substrate is given by (see Fig. 1):

$$\Phi = \gamma_{lg}A + P_e V + \pi R^2(\gamma_{sl} - \gamma_{sg}), \tag{4}$$

where A is the area of the liquid–gas interface; $P_e = P_a - P_l$ is the excess pressure inside the liquid, P_a is the pressure in the ambient air, P_l is the pressure inside the liquid and R is the radius of the droplet base. The last term on the right-hand side of Eq. (4) gives the difference between the energy of the part of the bare surface covered by the liquid drop and the energy of the same solid surface without the droplet. Equation (4) shows that the excess free energy decreases if (i) the liquid–gas interfacial tension, γ_{lg}, decreases, (ii) the solid–liquid interfacial tension, γ_{sl}, decreases, and (iii) the solid–gas interfacial tension, γ_{sg}, increases. This last is a very important conclusion, which is usually overlooked [27].

E. Spreading over Hydrophobic Substrates

Due to their immense importance in numerous practical and industrial applications, enhanced wetting of aqueous surfactant solutions has been the subject of considerable research interest over the past two decades. Scales et al. [28] were one of the first groups to point out the importance of the nature of the solid substrate in the wetting process. They introduced characteristic normalisation of the wetting behaviour as a powerful tool to give both a qualitative and quantitative understanding of adsorption of a series of alkyl aryl polyoxyethylenes. Significance of the substrate hydrophobicity was also shown by Keurentjes et al. [4]. They determined that a surfactant, when adsorbed onto a hydrophobic solid, exposes its polar head groups to the solution, whereas in the case of a hydrophilic solid, a by-layer of surfactants may form. Eriksson et al. [29] gave further understanding why surfactant transfer to the three-phase contact line is the dominant process in wetting. Transfer of surfactant molecules to the interface is widely believed to be the essential element in the overall dynamics of wetting and numerous attempts have been made to quantify this fundamental phenomenon.

Over the years, various wetting and spreading regimes have been revealed, mostly depending on the type of liquid–solid system investigated. Von Bahr *et al.* [30] reported two wetting stages at low surfactant concentrations: a short time regime where spreading occurs rapidly and a long time regime where spreading is slow. This was confirmed by Dutschk *et al.* [31] who investigated wetting behaviour of dilute ionic and nonionic aqueous surfactant solutions over highly hydrophobic and moderately hydrophobic polymer surfaces. They found that nonionic surfactants enhanced spreading on both type of surfaces, whereas ionic surfactants did not spread over highly hydrophobic surfaces. However, they argued that the long time regime goes much slower and concluded that a possible explanation is that adsorption at the expanding solid–liquid interface is slower than the diffusion.

Starov *et al.* [32] described the spreading mechanism of aqueous surfactant solutions over hydrophobic surfaces as a transfer of surfactant molecules on the bare hydrophobic surfaces in front of the moving liquid at the three-phase contact line (Fig. 5). This mechanism was first mentioned elsewhere [33–35], but Starov and co-workers broadened the idea assuming surfactant transfer onto the hydrophobic solid interface to take place only from the liquid–gas interface. As explained before, this increases the solid–gas interfacial tension and 'hydrophilises' the initially hydrophobic solid substrate. Adsorption of surfactant molecules at the liquid–solid and liquid–gas interfaces results in a decrease of the relevant interfacial tensions, and consequently of the total free energy of the system, and the drop spreads. Spontaneous adsorption of surfactant molecules in front of the moving three-phase contact line controls the rate of spreading [36].

The spreading of aqueous solutions of the sodium dodecyl sulphate on two types of hydrophobic substrates, PTFE film and PE wafer, has also been studied [32]. It was found that the evolution of drop radius was in agreement with the theoretical model suggested (Fig. 6). Drop surface coverage was found to be an increasing function of the bulk surfactant concentration inside the drop, and the maximum was reached close to the CMC. Hence, according to the above mechanism, at low surfactant concentrations inside the drop, the characteristic time scale of the surfactant molecular transfer, τ, decreases with increasing concentration, while above the CMC, τ should level off and reach its lowest value. Both of these effects are observed in experimental results [32] (see Fig. 7). The theoretical prediction of τ dependency on surfactant concentration corresponded well to experimental findings and served as a justification for the assumption concerning the transfer mechanism of surfactant molecules.

F. Spreading of Surfactant Solutions over Thin Aqueous Layers: Influence of Solubility and Micelle Disintegration

Thin liquid films can be found in many engineering, geology, and biophysics environments. Their application is significant in many coating processes [37–39] and physiological applications [40]. Presence of non-uniform temperature or surface active compounds across thin liquid films will lead to the formation of shear stresses, also known as Marangoni gradients, at the air–liquid interface. These gradients cause mass transfer on, and in, a liquid layer due to surface tension non-uniformity. Marangoni stresses distribute the liquid from areas of low surface tension to areas

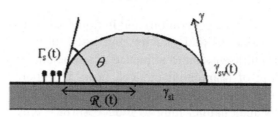

Figure 5. Spreading of surfactant solution over hydrophobic substrate [32]. Adsorption of surfactants on a bare hydrophobic substrate, $\Gamma_s(t)$, is shown in the left hand side. The latter results in the increase of the solid–vapour (gas) interfacial energy, $\gamma_{sv}(t)$.

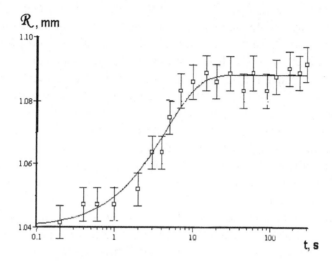

Figure 6. Time evolution of the spreading of a water drop (aqueous SDS solution $c = 0.05\%$, 2.5 ± 0.2 µl volume) over PTFE wafer. Error bars correspond to the error limits of video evaluation of images (pixel size) [32].

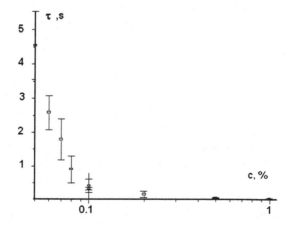

Figure 7. Fitted dependency of τ on SDS surfactant concentration inside the drop (spreading over PTFE wafer). Error bars correspond to the experimental points scattering in different runs; squares are average values [32].

of high surface tension (flow generation) and in doing so, also deform the interface resulting in height variations (deformation and instability of liquid films). In this section, we restrict our discussion to influence of surfactants in thin liquid films.

The understanding of Marangoni induced flows is important, as it can be either beneficial or detrimental in many applications. Surfactants, which are normally present in a healthy mammalian lung to reduce surface tension forces, keep the lung compliant and prevent collapse of the small airways during exhalation. However, most prematurely born babies do not produce an adequate amount of these surfactants, which leads to respiratory distress syndrome. This condition is treated by surfactant replacement therapy where surfactants are introduced into the lungs. These surfactants spread by other forces in the large to medium pulmonary airways. In small airways, surface tension gradients dominate and Marangoni flow distributes the surfactant to the distal regions of the lung [40, 41].

A problem in coating processes where paint films are dried by solvent evaporation is that the non-uniformity of the evaporation leads to Marangoni stresses which cause deformation in the film and hence, permanent defects on the paint surface ('orange peel' effect) [42]. Another example would be in washing latex films displays surfactant non-uniformities (surfactant islands) that leaves permanent indentations in the film [43]. Another application of the Marangoni effect is the use for drying silicon wafers after a wet processing step during the manufacturing of integrated circuits. An alcohol vapour is blown through a moving nozzle over the wet wafer surface and the subsequent Marangoni effect will cause the liquid on the wafer to pull itself off the surface effectively leaving a dry wafer surface [44].

Spreading of surfactant solutions on thin liquid films has been reviewed by Afsar-Siqqiqui *et al.* [45] a few years ago. A moving circular wave front forms after a small droplet of aqueous surfactant solution is deposited on a thin aqueous layer (Fig. 8). The time evolution of the radius of the moving front was monitored [46]. An experimental procedure was designed to investigate the influence of Marangoni force on spreading of surfactant solutions over thin aqueous layers.

Solubility of the surfactants is also believed to have an important influence on wetting. Lee *et al.* [46] considered surfactants of different solubilities at concentrations above CMC, and in all cases observed that spreading considerably depended on the solubility of surfactants: the higher the solubility, the slower the spreading. Von Bahr *et al.* [47] also noticed that the presence of micelles significantly slows surfactant transport, and consequently spreading on the whole. Both groups detected two stages of the front motion: the first fast stage, followed by a slower second stage. High solubility in aqueous solutions, typical of ionic surfactants: SDS and DTAB, slowed down both the first and the second stages [46]; if the solubility is high enough then during the second stage the front reaches some final position and does not move any further. Low soluble nonionic surfactants (Tween® 20, Tergitol® NP10) showed faster wetting during both stages and observed dynamics was successfully modelled by a power law, $R(t) = const \cdot t^n$, where n is the spreading exponent. During the first stage and for low soluble surfactants, n is expected to

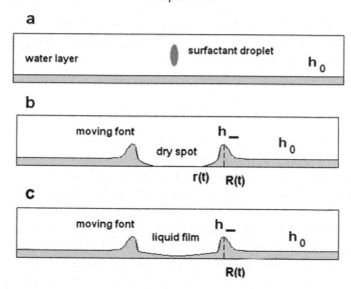

Figure 8. (a) A small droplet of aqueous surfactant solution is deposited on a top of thin aqueous layer of thickness h_0; (b) dry spot formation in the centre: cross section of the system: $R(t)$—radius of a circular moving front, $r(t)$—radius of dry spot in the centre, H—the height of the moving front; (c) the same as in the previous case (b) without dry spot formation in the centre.

reach ~ 0.75. It was also shown in [46], that formation of the dry spot in the centre is determined by the speed of the first stage: the higher this speed, the higher the probability to have dry spot formation. Hence, the dry spot forms in the case of soluble surfactants and does not form in the case of insoluble surfactants.

The observations in [46] differed from the earlier theoretical model [48], thus the influence of surfactant solubility and disintegration of micelles was incorporated to improve on the previous theoretical model. According to theoretical predictions [46], low soluble surfactant produce faster first and second stages. Low soluble surfactant (Tween® 20) produced a power law exponent 0.73 ± 0.01, being close to the maximum attainable spreading rate 0.75 predicted theoretically. For the highly soluble surfactant DTAB, the solubility was most significant during the second stage where the spreading front reached some final position and did not move any further in the agreement with theoretical predictions.

Chan and Borhan [49] deduced that insoluble monolayers enhance the overall spreading rate. They developed a mathematical model for surfactant enhanced spreading that suggests two additional mechanisms which influence the spreading rate: the development of positive surface curvature near the moving contact line, which produces a favourable radial pressure gradient within the drop, and the surfactant convection in a vicinity of the moving contact line. Accumulation of the surfactant at the contact line, due to surface convection, leads to faster spreading of a drop. Seguin *et al.* [50] performed experiments using ionic and non-ionic surfactants in different solvents. For ionic surfactants, effect of a charged head group on micellisation was obvious, while for non-ionic pure ethylene glycol appeared to

be a good alternative to water. These authors concluded the differences in aggregation between these different surfactants and solvent types mirror the changes in the
solvent dielectric constant.

G. Instabilities in the Course of Spreading

Instability at the contact line of drops of surfactant solutions spreading on prewetted hydrophilic surfaces leads to the formation of fingering patterns at the edge
of drops. Fingering at the contact line of drops of surfactant solutions spreading on
solid substrates was first observed by Marmur *et al.* [51]. This fascinating behaviour
has, over the years, been studied theoretically, experimentally and using numerical
simulations [45, 52–74]. It was assumed that a pre-existing water film on the substrate was essential for the growth of the instability. In [57, 70, 71], spreading on
oxidised silicon wafers of solutions of non-ionic surfactants in ethylene glycol or
diethylene glycol are investigated. Using polar solvents other than water allowed
the authors to discriminate between the role of a pre-existing adsorbed film of water, which is known always to be present on a hydrophilic substrate in contact with
the atmosphere, and the role of a possibly thicker film consisting of the same liquid
as the drop.

In [53] an analytical expression was deduced for the growth rate of a disturbance
on a parallel film trapped between two drops based on linear stability analysis. It
was shown that the identified Marangoni effect resulting from the addition of insoluble surfactants dampens the arising instabilities. Under overwhelming presence
of surfactants, the instability becomes insignificant compared to the surfactant-
induced flow along the thin films.

The dynamics of a surfactant influenced thin film on an inclined substrate (gravity driven) were considered [55]. For a constant flux, the formation of a capillary
ridge, with a Marangoni driven fluid 'step' downstream, and a fluid 'hump' up-
stream was observed, while the surfactant concentration reached its maximum at
the capillary ridge. The authors found that the prominent features of the flow, the
ridge, hump, step, and concentration peak become more pronounced *via* an increase
in inclination angle, Marangoni stresses and precursor layer thickness.

Troian *et al.* [63] proposed that the Marangoni effect is responsible for the instability at the edge of spreading surfactant drop. The Marangoni flow in the interface
and the bulk liquid is induced by surfactant concentration gradients along the air-
water interface. In their experiment, drops of aqueous surfactant AOT (sodium
bis-(2-ethyl-hexyl) sulfosuccinate) solution were placed on a hydrophilic surface
covered with a thin water film and immediate spreading, forming fingers advancing
from the contact line, was observed. It was found that the velocity and shape of the
fingers depended on the thickness of the underlying ambient water layer and the
surfactant concentration. On thin water films (0.1 μm), fingers are narrow, sharply
tipped and more branched, than those on a thick (1 μm) film (Fig. 9). The length of

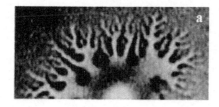

Figure 9. Photographs of spreading drop in 0.5 s from drop deposition. (a) Fingering on a thin water layer; (b) fingering on a thick water layer [63].

fingers during spreading depends on time as t^α, where $\alpha = 0.66$ and 0.7 for the thin and thick films, correspondingly.

Frank and Garoff [65] determined that the formation of fingering patterns depend not only on the surfactant concentration gradients, but on a mechanism of surfactant-surface interactions as well. The same surfactant can exhibit fingering or 'stick-jump' (autophobing) spreading behaviour on the same charged or opposite charged substrates, correspondingly. The spreading behaviour of aqueous surfactants SDS (anionic sodium dodecyl sulphate) and CTAB (cationic cetyltrimethylammonium bromide) solutions on a clean and dry oxidized silicone wafer and a polished sapphire disk has been examined. The SDS solution spreads with the fingering formation on the silicone wafer, while on the sapphire substrate it exhibits an autophobing effect. The spreading behaviour of CTAB solution on these substrates is contrary to the SDS spreading. Authors have deduced that surfactant cannot advance alone; fluid films must be present on the surface. For sufficiently clean surfaces, thin precursor films of fluid move rapidly ahead of the dendrites of the spreading solution. Moreover, on the substrates which are prewetted with a surfactant solution monolayer, no the fingering spreading is observed.

The role of the underlying films on the substrates and viscosity of solvents in the formation of the hydrodynamics instabilities at the contact-line of the spreading drops has been intensively investigated by Cazabat and co-workers [57, 70, 71, 75, 76].

Cachile and Cazabat [70, 71] studied the influence of the ambient relative humidity (RH) and the surfactant concentration on the fingering spreading of solutions of non-ionic $C_{12}E_4$ and $C_{12}E_{10}$ surfactants in ethylene and diethylene glycol on hydrophilic silicon wafers. The results were summarized in a 'phase diagram', as a function of the normalized surfactant concentration, C^*, defined as the ratio of the bulk concentration to the CMC, and the relative humidity, RH (Fig. 10).

In the stable region, at RH $\leqslant 30\%$, there is a capillary spreading regime. The drop spreads out as a uniform circle in the whole concentration range investigated and the radius of the drop R scales as $t^{1/10}$. With increasing RH (unstable region), the spreading process is accelerated and the fingering instabilities are developed. It was found that R^2 grew linearly in time when RH $\approx 60\%$, but at RH $> 85\%$, spreading is faster. At $C^* > 1$, the adsorption of surfactant at the solid–liquid interface creates a

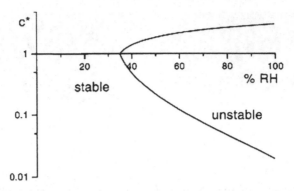

Figure 10. 'Phase diagram' of the behaviour of $C_{12}E_{10}$ in ethylene glycol as a function of the surfactant concentration relative top CMC, and the RH [71].

hydrophobic barrier, which increases with the surfactant concentration and viscous solvent. Because of this barrier, the lateral extension of fingers is limited.

They assumed that the mobility of the surfactant on the solid controls the emergence of a surface-tension gradient at the free liquid surface. Once born, this gradient gives rise to a Marangoni flow, which dominates the spreading process. For a given surfactant, the mobility increases with increasing RH.

This explains why RH controls the start of the Marangoni flow, even on surfaces where adsorbed water is not mobile. This also explains the change in the dendritic pattern with concentration, surfactant or solvent.

Subsequent experiments [31] have been performed to study of the role of the thickness of a precursor wetting film in the amplification of the fingering spreading. For this purpose, clean silicon wafers were deposited with thin films of solution of ethylene glycol in methanol. The surfactant $C_{12}E_{10}$ was chosen at the fixed CMC concentration, the humidity was kept around 60% and the thickness of film was varied from 30 to 250 nm. Under these conditions, the advancing line of the surfactant precursor on the pre-deposited film was clearly visible. It has allowed revealing that instabilities appear behind the contact line of solvent film. When the thickness of pre-deposited film was increased, the branching of fingers and the thickness of their tips were decreased. The spreading dynamics, namely: the radii of fingers R_f from the centre of drop to their tips, and the length of surfactant leading film, R_0, were found to scale as $t^{1/2}$, but the drop radius, R_c, changed with time as t^{α}, where α took on values from 0.3 to 0.45 with increasing thickness of the pre-deposited film.

In [75], the influence of surfactant concentration at the fixed thickness of solvent pre-deposited film was studied. At a concentration below CMC, the thickness and base width of fingers were observed. The profile of the spreading drop is smooth and no boundary between the drop and the fingers was observed. For a concentration near CMC, fingers became more straight. At higher concentrations, fingers disappeared and a hollow part appeared at the tip of the fingers.

Many authors investigated the effect of surfactant solubility on the unstable spreading of aqueous solutions. In the case of sparingly soluble surfactants [75], a drop of AOT solution was deposited on the water layer and it was found that the spreading had stable or fingering behaviour. At surfactant concentrations lower than 0.4 CMC, the spreading process was stable for the all ranges of thicknesses: no fingers are observed. For concentrations in the vicinity of 1 CMC up to 2 CMC, the spreading edge of an AOT drop is uniform for several seconds (from 3 to 10 s), defined by the concentration and the film thickness, then the indistinct, round-tipped and straight fingers appear behind the thickened rim. The fingers become wider when the concentration decreases and the thickness of water layer increases. At the 4 CMC on the 25 μm and 50 μm thick films, spreading is stable and uniform with a distinct surfactant covered disk of liquid in the centre of the drop, which is in agreement with the results of Troian *et al.* [63]. However, on the 100 μm film, fingering spreading behaviour occurs again.

In [45] sparingly and highly soluble anionic surfactants were compared in a wide range of surfactant concentrations with water films ranging from 25 μm to 100 μm in thickness. Sodium dodecyl sulphate (SDS) was used as a soluble surfactant. The spreading behaviour of SDS solutions shows many similarities with some notable differences from the AOT solution [75]. The spreading rates were of the same order of magnitude for both the SDS and AOT, but for the SDS the rate did not vary with the water film thickness. Authors related this fact to fast desorption of surfactant and relatively significant gravity force for the high film thickness. The fingers occurred at the SDS concentration of 0.4 CMC, which is two times less than in the AOT case (0.8 CMC), and almost ten times earlier. The shape of fingers was more pronounced and branched in the case of SDS drop deposition. At the CMC and high film thickness, the fingering was observed in both surfactants. However, at the thick film in the SDS case, the disk remained in the centre after drop deposition, exhibits the protrusion instabilities that extend from the edge of disk, while in the AOT case the disk is stable.

Nikolov *et al.* [77] reported the finger instabilities during the spreading of drop of aqueous trisiloxane solution (Silwet® L-77) on a hydrophobic plate. A droplet of aqueous trisiloxane solution was placed on the cap of a water drop and immediately started to spread at a high rate. In one second, advanced front fingers were generated and became more pronounced during spreading.

Stoebe *et al.* [66] observed that small fingers appeared at the edges of droplets of aqueous trisiloxane solutions spreading on the surfaces of various surface energies. The length of fingers was found to be less than 10% of the dynamic droplet radius, which served as a reason not to consider the fingering in the analysis of their experiment.

Troian *et al.* [63] investigated theoretically the mechanism responsible for the instability during the spreading of surfactant-laden drop. Their physical model represents a hemispherical liquid drop covered with insoluble surfactant spreading on a thin layer of the same liquid.

Considerable efforts were invested in the investigation of instabilities in the course of spreading and progress was achieved in the understanding of the nature of instabilities. However, at the current time it is difficult to extract information from these investigations related to the properties of surfactants.

H. Spreading of Surfactant Solutions over Porous Substrates

Spreading and penetration of liquids on porous media is a fundamental property that affects applications including printing, painting, adhesives, oil recovery, imbibition into soils, health care, and home care products [78–82]. In inkjet printing, the resolution is directly associated to the degree of liquid extension and spreading on the printing media after deposition [82]. The spreading of a drop on a thin permeable medium proceeds in two parts: (a) spreading on the surface of the medium and (b) penetration into the underlying medium. Knowledge of the spreading rate and area covered is critical as drop-to-drop contact would result in unwanted and detrimental effects. Fast penetration of the liquid would limit the time the drop spent on the surface, thereby decreasing coalescence of drops. However, penetration of liquid and medium are usually slow due to the poor wettability by the liquid of the porous medium. Surfactants from this point of view may play a crucial role.

Furthermore, surfactants' role in oil recovery processes is especially important. It is highly desirable to extract oil trapped in the pores of rocks. The injection of surfactants reduces the interfacial tension between the oil and water phases, allowing the extracting of trapped oil in small pores [83]. The importance of such knowledge leads to on-going research on the wetting kinetics of porous media influenced by surfactants.

Recent publications reveal a growing interest in exploring the simultaneous spreading and imbibition processes of aqueous surfactant solutions. However, these studies have so far mostly been restricted to pure liquids simultaneously spreading and imbibing into the porous substrate [15, 84–94].

Clarke *et al.* [95] developed a theoretical model for simultaneous spreading and imbibition by incorporating the molecular kinetic theory of spreading [15] with the modified Lucas–Washburn equation.

Starov and co-workers developed a theoretical model [86, 87] for the case of complete wetting of a drop spreading over a pre-wetted or a dry porous layer. The lubrication theory approximation was used, neglecting gravitational influence, so that only capillary forces are taken into account in the model. They established a system of two differential equations to describe the evolution of the radius of both the drop base ($L(t)$) and wetted region ($l(t)$) inside the porous layer, and an equation describing the dynamic contact angle. Experiments were performed in order to test the theoretical model. Silicon oils were used as the liquid and as porous layers, nitrocellulose membranes with different pore size. By comparing the theoretical model and experimental data on appropriate dimensionless scales, a universal behaviour was observed where experimental data was in good agreement with theoretical prediction [86, 87].

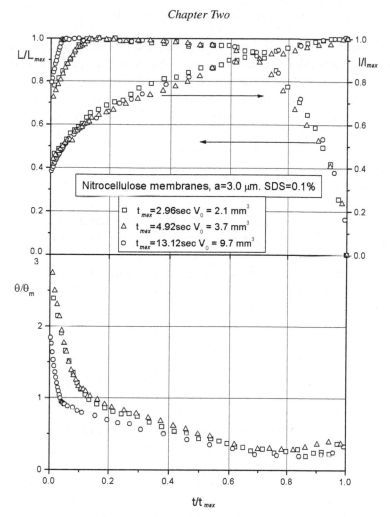

Figure 11. Spreading of droplets of 0.1% SDS solution over Nitrocellulose membrane, average pore size is 3.0 μm. L/L_{max}, dimensionless radius of the drop base; l/l_{max}, radius of the wetted area; θ/θ_m, dynamic contact angle; t/t_{max}, dimensionless time [96].

Starov *et al.* [96] incorporated the previous theoretical model to simulate spreading of surfactant solutions over porous layers. They produced experimental data for the spreading of different concentrations of aqueous SDS solution over different pore sizes of nitrocellulose membranes whilst varying drop volume and properties of porous layers. The dynamic contact angle, radius of the spreading droplet, and the wetted perimeter were monitored for the spreading/penetrating process. The entire process can be divided into three stages (Fig. 11). During the first stage, the drop base spreads until it reaches the maximum position whilst contact angle decreases rapidly. The drop base then remains constant during the second stage, as the contact angle decreases linearly with time. Finally comes the third stage, when the drop base shrinks while the contact angle remains constant until the drop completely disappears. During all three stages, the wetted region continues to expand until an equilibrium value is attained. It was observed that the total spreading duration, the maximum radius of the drop base, and final radius of the wetted region depend considerably on the drop volume, the SDS concentration, the average pore size and the porosity of the membranes used. The overall spreading time decreases when the SDS concentration increases. On dimensionless plots, drops of different volume (over the same porous membrane) showed universal behaviour of the contact angle (Fig. 11). Starov *et al.* derived an equation that predicts the linear dependency of the contact angle during the second stage of spreading/imbibition process.

Contact angle hysteresis data was extracted to show that the difference between advancing and receding contact angles becomes smaller with increasing SDS concentration.

Independent experiments were performed to conclude the constancy of the contact angle during the third stage resulted from the hydrodynamic flow that significantly changes the advancing contact angle and thus hysteresis was not the possible reason to account for the constancy of the contact angle during the third stage.

Using experimental data on the spreading of drops of aqueous SDS solutions over dry porous substrates, the values of advancing, θ_a, and hydrodynamic receding, θ_{rh}, contact angles were extracted as a function of SDS concentration ([96], see an example in Fig. 11). The term "hydrodynamic receding contact angle" and the symbol θ_{rh} are used to distinguish it from the static receding contact angle, which is found equal to zero. The advancing contact angle, θ_a, was defined at the end of the first stage when the drop stopped spreading (the radius of the drop base reached its maximal value). The hydrodynamic receding contact angle, θ_{rh}, was defined for the moment when the drop base started to shrink. In Fig. 12, experimental data on the apparent contact angle hysteresis are summarized [96]. This figure shows that the advancing contact angle, θ_a, decreases with SDS concentration; the hydrodynamic receding contact angle, θ_{rh}, on the contrary, slightly increases with SDS concentration.

Figure 12 shows that the difference between advancing and receding contact angles becomes smaller with the increase in the SDS concentration; the dimensionless time interval when the drop base does not move also decreases with the increase in the SDS concentration.

It is necessary to note that the behaviour of drops of aqueous SDS solutions during the third stage of spreading (partial wetting) is remarkably similar to the behaviour during the second stage of spreading in the case of complete wetting section [86, 87]. Static advancing and static receding contact angles on smooth nonporous nitrocellulose substrate for different SDS concentrations were also measured to compare with those on porous substrate [96].

The static advancing contact angle of pure water on non-porous nitrocellulose substrate was found approximately equal to 70°. The static, advancing, contact angle decreases with the increase of SDS concentration (Fig. 13). This trend continues

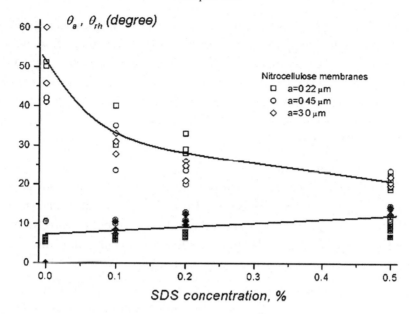

Figure 12. Porous nitrocellulose substrates. Apparent contact angle hysteresis variation with SDS concentration. Nitrocellulose membranes of different average pore sizes. Open symbols correspond to the advancing contact angle, θ_a. The same filled symbols correspond to the hydrodynamic receding contact angle, θ_{rh} [96].

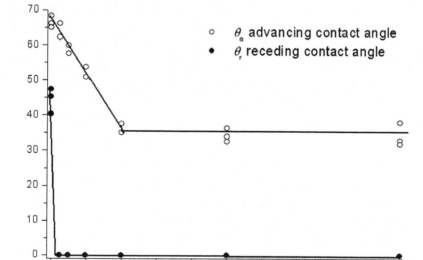

Figure 13. Non-porous nitrocellulose substrate. Advancing and receding contact angles variation with SDS concentration. Open symbols correspond to the static advancing contact angle, θ_a. Filled symbols correspond to the static receding contact angle, θ_r [96].

until the CMC is reached. At concentrations above the CMC, advancing contact angle remains constant and approximately equal to 35°. A non-zero value of the static receding contact angle was found only in the case of pure water droplets. In all other cases (even at the smallest SDS concentrations used 0.025%) the static receding contact angle was found equal to zero in the entire concentration range used: from 0.025% (ten times smaller than CMC) to 1% (five times higher than CMC).

Comparison of Fig. 12 and Fig. 13 shows:

- The advancing contact angle dependence on SDS concentration on porous nitrocellulose substrates is significantly different from the static advancing contact angle dependence on non-porous nitrocellulose substrates. The latter means that in the case of porous substrates, the influence of both hydrodynamic flow caused by imbibition into the porous substrate and the substrate roughness changes significantly the advancing contact angle;

- The hydrodynamic receding contact angle, in the case of the porous substrates, has nothing to do with contact angle hysteresis and is determined completely by the hydrodynamic interactions in a way similar to the complete wetting case.

It is worth mentioning that observed dependencies during imbibition into porous substrates in [86] are very similar to those noticed in droplet spreading and evaporation experiments [97]. When plotted in dimensionless time, changes in droplet radius show surprising resemblance. In both cases, the radius of the drop, L, reaches its maximum value in the very beginning, at time instant t_m, after which it decreases. Compared to the maximum duration of the process, t_{max}, the time needed for complete evaporation of the droplet, and t_p^*, the time for imbibition into the porous substrate, it was found that $t_m \sim 0.06 t_{max}$ and $t_m \sim 0.08 t_p^*$, in the cases of evaporation and imbibition, respectively (Fig. 14).

Experimental data on a spreading/evaporation curve showed the same trend for liquids with very different volatility: alkanes, water and silicone oils. Although theoretical models have been successfully developed, it is still necessary to explain why mass loss in the case of sessile drops manifests itself in an identical manner.

Daniel and Berg [98] developed a model to predict the simultaneous spreading and penetration of surfactant solutions, based on energy considerations of the system. Comparison with [86] showed that the energy-based model is functionally equivalent. The derived energy-based model was tested against experimental data for spreading of commercial surfactants over a variety of papers relevant to thermal ink-jet printing.

It is necessary to emphasise that, in spite of enormous industrial importance, the kinetics of spreading/imbibition of surfactant solutions into porous media is far less investigated than it deserves.

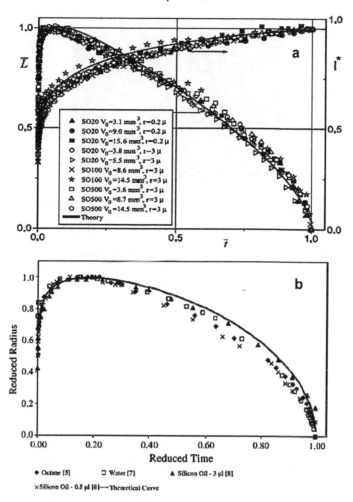

Figure 14. (a) Measured dependencies of dimensionless radii of the drop base, $\bar{L} = L/L_m$, and radii of the wetted region inside the porous layer, $\bar{l} = l/l^*$, over dimensionless time, $\bar{t} = t/t_p^*$, where t_p^* is the total time of imbibition into porous substrate [86]. (b) Dimensionless radius, $\bar{L} = L/L_m$, against dimensionless time, $\bar{t} = t/t_{max}$, curve for the behaviour of different liquids spreading/evaporating on solid substrates [97].

I. Superspreading

Since the early 90s, silicone surfactants have been the subject of substantial scientific interest. Due to their exceptional wetting abilities, trisiloxanes, in particular, have drawn significant attention recently. Trisiloxane surfactants are commonly denoted as $M(D'EO_nR)M$, where M represents the trimethylsiloxy group, $M = (CH_3)_3–SiO–$, the term $D' = –Si(CH_3)(R')–$, where $R' = (CH_2)_3$ onto which is polyoxyethylene group, $EO_n = (CH_2CH_2O)_n$, attached, with n as the average number of polyoxyethylene groups, and R stands for an end-capping group, usually $–H$,

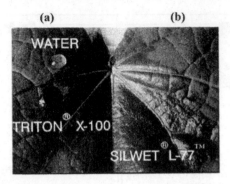

Figure 15. Photograph [77] depicting the spreading of: (a) a water drop and a drop of 0.25% Triton®
X-100 solution, (b) 0.1% Silwet® L-77 solution on a velvetleaf surface.

$-CH_3$, or $-Ac$; $Ac = -C(O)CH_3$. Until recently, a trisiloxane molecule with eight
poly(oxyethylene) groups was generally referred to as E_8.

However, a large selection of commercial trisiloxane products is available on the
world market nowadays and these are called correspondingly by their registered
trade names. Trisiloxanes possess an unusual ability to induce highly efficient wetting properties even on extremely hydrophobic surfaces. It has been reported that
the addition of trisiloxane surfactants can transform a small spherical drop into an
'infinitely' thin liquid film—the phenomenon popularly known as 'superspreading'
[99]. Thus, trisiloxane surfactants are widely recognized as 'the superspreaders'.
Due to its unique characteristics and colossal practical employment, the superspreading phenomenon has attracted much attention, especially from the theoretical
point of view. In spite of considerable research interest, there is still a lack of
explanation of the underlying mechanism of superspreaders' behaviour and understanding of the necessary conditions for its realisation. Superiority of trisiloxanes
over conventional surfactant types has been confirmed in numerous publications
over the years [67, 77, 99–107].

Silwet® L-77™, a commercially available trisiloxane surfactant (polydispersed
with an average of 7.5 ethoxylate groups), has been widely used since the late eighties. Aqueous drops with sufficiently high bulk concentration of trisiloxanes spread
rapidly, even when placed on hydrophobic surfaces, and completely wet the solid
with no measurable final contact angle at the contact line [99, 108].

It is believed that the overall wetted area achievable by an aqueous droplet containing trisiloxane surfactant can be as much as 50 times greater than pure water,
and 25 times more effective than a conventional surfactant [77]. Figure 15 illustrates this point by comparing the relative spreading properties of water with 0.25%
Triton® X-100 (polyoxyethylene (10)-octylphenyl ether), and with aqueous solution of 0.1% Silwet® L-77 surfactant on a velvetleaf (*Abutilion theophrasti*). Water
alone on velvetleaf makes a contact angle bigger than 90°; aqueous Triton® X-100
solution slightly reduces the contact angle due to a decrease in surface tension and
gives a bigger spreading area. However, Silwet® L-77 provides a most significant

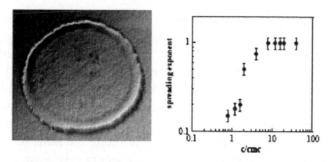

Figure 16. Left: Photograph of the spreading trisiloxane solution droplet at the highest surfactant concentration used. Right: power n from the evolution of the radius $R = C \cdot t^n$ for the 'superspreading' of the trisiloxane solutions *versus* concentration [112].

increase in spreading area and decrease in contact angle making it an exceptionally effective wetting agent. Nevertheless, the debate continues about the nature of the driving force for superspreading and interpretation of the observed dynamics.

Svitova *et al.* [109] established the essential parameters of trisiloxane wetting: the critical wetting concentration, CWC, and the critical aggregation concentration, CAC; the latter being very similar to CMC. CWC is considered to be the concentration above which spreading occurs over a (moderately hydrophobic) solid and liquid substrates. Hence, it is associated with the beginning of superspreading. CWC, which is reported to be independent of the substrate surface energy, represents the maximal spreading potential of a trisiloxane surfactant [109, 110]. The occurrence of the transition from partial wetting to complete spreading at CWC was first mentioned in [111]. Recently, three regimes of spreading were identified: (i) complete non-wetting during the spreading process at low concentrations, (ii) a transition from initial non-wetting to partial wetting at the end of the spreading process at intermediate concentrations, and (iii) partial wetting both at the beginning and at the end of the spreading process at higher concentrations. Transitions between different regimes flawlessly correspond to relevant critical concentrations: CAC or CMC, and CWC [100].

Both the extent and the dynamics of trisiloxane spreading attracted significant attention over the years. Rafai and Bonn [112, 113] suggested a $R \propto t^n$ power law to describe trisiloxane spreading with a power n larger than 0.1 for concentrations both above and below CMC (Fig. 16). For very high bulk concentrations, a linear relation $R \propto t$ ($n = 1$) was reported.

In [111], three regimes of the spreading dynamics were observed: early stages where wetting diameter is proportional to t^n with n in the range 0.12–0.22; during the second stage the exponent increases to 0.38–0.58; during the last stage of spreading the surface roughness and local tension gradients lead to an asymmetric drop shape and formation of fingers and dendrites. Nikolov *et al.* [77] also reported finger instabilities during spreading of the drop of aqueous Silwet® L-77 solution on a hydrophobic plate. Stoebe *et al.* [66] observed the small fingers appeared at the edges of droplets of aqueous trisiloxane solutions, but found the length of fingers was less than 10% of the dynamic droplet radius, which served as a reason not to consider the fingering in the analysis of their experiment. Recently, Radulovic *et al.* [114] argued that the power law was not a suitable model to describe trisiloxane dynamics and proposed an exponential relationship, $R \propto \exp(-t/\tau)$, for solutions above CWC. Their model gave a satisfactory fit for both the evolution of the contact angle and drop radius with time, introducing the time constant, τ, as the main parameter of trisiloxane wetting which is directly related to the underlying diffusion of superspreader molecules. Values of the time constant were found to decrease with increasing concentration; hence, the faster wetting. Ivanova *et al.* [100] also dismissed the power law and suggested that partial wetting of trisiloxanes proceeds in two stages: the fast short first stage, which is followed by a much slower second stage. They found that the characteristic time for the second stage decreased with increasing concentrations. Certain discrepancies from the proposed model were observed during the first wetting stage, which is understandable as the characteristic time scale for the first stage is an order of magnitude smaller than that for the second stage.

Initial attempts to shed light on the wetting ability of trisiloxanes related the phenomenon to its ability to adsorb at the interface and significantly reduce tensions at the moving liquid front, creating a positive spreading coefficient. However, the positive value of the spreading coefficient during spreading could not be either proved or theoretically explained, because the adsorption cannot keep pace with the expansion of the drop due to diluted solutions at the perimeter and the surface concentrations becomes reduced at the contact line. The main drawback of the latter explanation is that it can be equally applied to any aqueous surfactant solution and unfortunately is not specific to trisiloxanes. Subsequent theories included solution turbidity, the presence of a disperse phase, as a vital parameter in superspreading. It was first mentioned in [115, 116] that at concentrations < 0.1 wt%, trisiloxane solutions contain small aggregates, such as vesicles. These vesicles and/or other aggregates disintegrate efficiently to transfer surfactant molecules to the contact surfaces and enhance spreading. Zhu *et al.* [108] confirmed that superspreading occurs only when dispersed particles (vesicles) are present. Ruckenstein [117] argued bilayer adsorption on hydrophobic surfaces and the Marangoni effect it triggers are responsible for superwetting. Stoebe *et al.* [67] also concluded that the aggregate disintegration plays an important part as it results in sudden increases in surface tension gradients and a corresponding increase in Marangoni flow. Svitova *et al.* [111] gave further insight into surfactant self-assembly and aggregate structure as main factor responsible for transition between partial to complete wetting.

Nikolov *et al.* [77] suggested that although aggregates are present in the solution, they do not play a crucial role in initiating the Marangoni effect which they believed to contribute to the rapid spreading. The surface concentrations at the drop apex are assumed to remain high compared to those at the perimeter, resulting in the drop being pulled out by the higher tension at the perimeter than at the apex. To maintain a high apex concentration, surfactant adsorption must exceed the rate of interfacial dilation at the apex due to the outward flow. Chengara *et al.* [118] also provided considerable evidence of surface tension gradient being an important driving force in superspreading. Kumar *et al.* [119] pointed out that bilayer adsorption may account for the ability of superspreaders to maintain the low apex concentration necessary to drive Marangoni spreading, or a high concentration at the contact line to maintain a low contact angle and force spreading.

A number of hypotheses are based on the formation of a precursor film on a solid surface. Zhu *et al.* [108] noted the influence of the water vapour pressure in the surrounding atmosphere on the trisiloxane wetting as superspreading was observed only in saturated or supersaturated water vapour. This has given rise to the suggestion that fast spreading may be caused by surface flow of a thin precursor film formed from the vapour phase. Churaev *et al.* [120] showed that fast spreading and climbing of trisiloxane surfactant solutions over hydrophobic surfaces, are caused by formation of extremely thick wetting films, stabilized by mutual repulsion of vesicles, and emphasized the important role of disjoining pressure. Other attempts to explain trisiloxane wetting included the idea of the surfactant molecule transfer on a bare hydrophilic solid in front of the moving contact line—the autophilic phenomenon [32, 78]. This theory was used to explain the second slow stage of wetting found by Ivanova *et al.* [100], while the first fast stage was related to disintegration of aggregates in the vicinity of the three-phase contact line.

Ananthapadmanabhan *et al.* [101] first postulated that the rapid spreading is a consequence of the unusual shape of the hydrophobe group. However, Hill claimed that 'T' ('umbrella-like') shape of superspreader molecule is not the explanation for unusual wetting properties and emphasised the importance of the end-capping groups [116]. Kumar *et al.* [119] compared superspreaders to *n*-alkyl polyoxyethylene surfactants, which have *n*-alkyl chain instead of trisiloxane backbone. The hydrophobicity of the alkyl chain and the trisiloxane backbone is approximately the same, but their size is very different, which served as a reason for the different spreading power of polyethoxylates with alkyl groups and trisiloxanes. Radulovic *et al.* [105, 106] confirmed the superiority of trisiloxanes over conventional nonionic surfactant. They compared Silwet® L-77 with Triton® X-100, surfactants which have the same chemical composition and similar length of poly(ethyleneoxide) hydrophilic tails, but very different structure, size and hydrophobicity of the hydrophobic head (Fig. 17). It was concluded that the larger area of the trisiloxane hydrophobic head undoubtedly contributes towards a more efficient (and quicker) decrease in interfacial tensions; hence, the higher wetting potential. Svitova *et al.* [102] showed that the rate of trisiloxane spreading depends on surfactant nature, its structure, and concentration and the subphase nature. Generally, it was concluded that E_8 has one of the best wetting properties. The increase in surfactant tail length ($n = 12, 16$) is known to suppress the superspreading ability of siloxane surfactants [108].

Trisiloxane molecules with short chain lengths also exhibited poorer wetting ability than those with medium lengths [67]. This is in agreement with the recent

Figure 17. Molecular structures of the alkyl polyethoxylates (left) and trisiloxane surfactants (right). Grey atoms are oxygen; black atoms are carbon; small white atoms are hydrogen; and large white atoms are silicon [105].

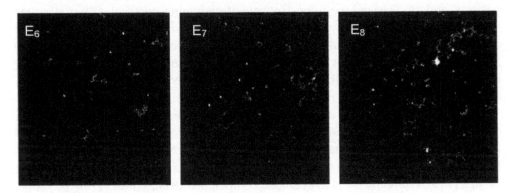

Figure 18. BAM images for trisiloxanes E_6, E_7, E_8 at $C = 2$ mmol/m^{-3} [121].

studies which confirmed slightly higher contact angles and surface tension values for E_4 and E_5 [110, 121]. Optimal wetting abilities were found for chain lengths in range of 6–9 ethoxy groups. Silwet® L-77 in particular, and a number of similar commercial products, with an average of $n = 7.5$ were found to have the most advanced wetting properties [105].

Recently, Ritacco *et al.* [121, 122] have measured dynamic surface tensions for trisiloxane surfactants for the short time (<1 s). They discovered that in the case of trisiloxanes (with relatively long EO$_n$ chains, $n = 6$–9) possessing superspreading character, two inflection points were detectable on dynamic surface tension curves [121]. Using Brewster Angle Microscopy, the authors directly observed aggregates appearing on the solution interface for those trisiloxanes in a certain range of concentrations, Fig. 18. This allows us to suggest that the surfactant molecules are present at the liquid/air interface in two states: as monomers and as surface aggregates [121, 122]. The aggregates could act as reservoirs of surfactant monomers in the course of spreading, completely confirming the suggestion made by Kumar

Figure 19. Spontaneous increase in the wetting angle of aged 0.1 wt% Silwet® L-77 aqueous solution. Day 1—53°; day 6—72°; day 8—91°; and day 10—103° [126].

et al. [119] about direct adsorption surfactant aggregates on the liquid/air interface supporting superspreading.

Other attempts to reveal the true nature of the superspreading phenomenon have involved simulations. Shen *et al.* [123] studied the influence of surfactant structure on spreading using molecular dynamics simulations. They compared the T-shaped structure of trisiloxanes to the linear chain surfactant, treating surfactant molecules as united atom structures, with all interactions given by Lennard–Jones potentials, while modelling the solvent as a monatomic liquid and the substrate as a particle lattice. The results from this study were somewhat coarse and authors admitted improvement of the model was necessary. Other studies included molecular dynamics simulations using all-atom force fields by Halverston *et al.* [124]. Spherical water nanodroplets with surfactant molecules adsorbed at the liquid–vapour interface were placed in the vicinity of a graphite substrate and allowed to spread freely at room temperature. The authors disclosed that their simulations of superspreading produced results that did not match the experimental observations, probably because the drop was too small, or the time interval studied was too short. Another possible reason could be the insufficient length of the hydrophilic tail since they used E_4 trisiloxane surfactant in their model.

Possibly the biggest drawback of trisiloxanes is their hydrolytical stability. It was shown that practically instantaneous hydrolysation of the trisiloxane head takes place at low pH [106], leading to complete loss of their wetting ability. At more moderate pH values, somewhat longer shelf-life is expected [125]. However, a gradual loss of trisiloxanes' interfacial activity was noted after several days even in aqueous environment (Fig. 19), as a consequence of spontaneous hydrolysation of Si–O bond [126].

This has led to synthesis of the alternative, more hydrolytically stable, trisiloxane structures. Zhang *et al.* tested the spreading behaviour of new glucosamide-based trisiloxane surfactants on wheat and cabbage leaves [127].

Their latest study on spreading of the same surfactants on polystyrene surfaces proved that these molecules exhibit superspreading, but not to the extent ethylene oxide trisiloxanes do [128]. Wetting enhancement trisiloxanes exhibit is undoubtedly superior to any other type of surface-active agents. Impressive and unique superspreading phenomenon is still one of the mysteries of the interfacial science and further research is necessary until light is shed on the underlying activity of superspreaders.

J. References

1. Joanny, F. J. and de Gennes, P. G., A model for contact angle hysteresis. J. Chem. Phys, 81(1) (1984) 552–562.
2. Isrealchvilli, J., Intermolecular and Surface Forces, 2nd edn. Academic Press, San Diego, 1991.
3. Brandrup, J., et al., Polymer Handbook, 4th edn., Vol. I. John Wiley & Sons, Inc., New York, 1999.
4. Keurentjes, J. T. F., et al., Surfactant-induced wetting transitions: Role of surface hydrophobicity and effect on oil permeability of ultrafiltration membranes. Colloids and Surfaces, 51 (1990) 189–205.
5. Shaw, D. J., Introduction to Colloid and Surface Chemistry, 2nd edn. Butterworths, London, 1970.
6. Wang, J.-H., et al., Dynamic contact angles and contact angle hysteresis of plasma polymers. Langmuir, 10 (1994) 3887–3897.
7. Raphaël, E. and de Gennes, P. G., Dynamics of wetting with nonideal surfaces. The single defect problem. J. Chem. Phys., 90(12) (1989) 7577–7584.
8. Lam, C. N. C., et al., Study of the advancing and receding contact angles: liquid sorption as a cause of contact angle hysteresis. Advances in Coll. and Interface Sci., 96 (2002) 169–191.
9. Drelich, J., Miller, J. D. and Good, R. J., The effect of drop (bubble) size on advancing and receding conract angle for heterogeneous and rough surfaces as observed with sessile-drop and captive bubbles techniques. J. Colloid Interface Sci., 179 (1995) 37–50.
10. Schwartz, L. W. and Garoff, S., Contact angle hysteresis on heterogeneous surfaces. Langmuir, 1 (1985) 279–230.
11. Cox, R. G., The dynamics of the spreading of liquids on a solid surface. Part 2. Surfactants. J. Fluid Mech., 168 (1986) 194–220.
12. Hoffmann, R. L., A study of advancing interface. I. Inetrface shape in liquid–gas systems. J. Colloid Interface Sci., 50(2) (1975) 228–241.
13. Voinov, O. V., Hydrodynamics of wetting. Izv. Akad. Nauk SSSR, Mekh. Zhidk. i Gaza, 76(5) (1976).
14. Tanner, L. H., The preading of silicone oil drops on horizontal surfaces. J. Phys. D: Appl. Phys., 12 (1979) 1473–1484.
15. Blake, T. D., et al., Contact angle relaxation during droplet spreading: comparison between molecular kinetic theory and molecular dynamics. Langmuir, 13 (1997) 2164–2166.
16. Hayes, R. A. and Ralston, J., The molecular-kinetic theory of wetting. Langmuir, 10 (1994) 340–342.
17. Lin, S. H., Ziv, A. R. and Eyring, H., Quantun statistical mechanical theory of diffusion and reaction on solid subsrates. Proc. Natl. Acad. Sci. USA, 78(7) (1981) 3989–3992.
18. Wayner Jr, P. C., The effect of interfacial mass transport on flow in thin liquid films. Colloids and Surfaces, 52 (1991) 71–84.
19. Shanahan, M. E. R., Spreading of water: condensation effects. Langmuir, 17(26) (2001) 8229–8235.
20. Shanahan, M. E. R., Condensation transport in dynamic wetting. Langmuir, 17(13) (2001) 3997–4002.
21. De Ruijter, M. J., et al., Contact angle relaxation during the spreading of partially wetting drops. Langmuir, 13 (1997) 7293–7298.
22. Brochard-Wyart, F. and de Gennes, P. G., Dynamics of partial wetting. Advances in Colloid and Interface Science, 39 (1992) 1–11.

23. De Ruijter, M. J., De Coninck, J. and Oshanin, G., Droplet spreading: partial wetting regime revisited. Langmuir, 15 (1999) 2209–2216.
24. De Ruijter, M. J., et al., Experimental evidence of several time scales in drop spreading. Langmuir, 16 (2000) 2363–2368.
25. Myers, D., Surfactant science and technology, 3rd edn. Wiley-Interscience, John Wiley & Sons, Inc, Hoboken, New Jersey, 2006.
26. Nagarajan, R., Molecular packing parameter and surfactant self-assembly: the neglected role of the surfactant tail. Langmuir, 18(1) (2002) 31–38.
27. Lee, K. S., et al., Kinetics of wetting and spreading by aqueous surfactant solutions. Advances in Colloid and Interface Science, 144(1–2) (2008) 54–65.
28. Scales, P. J., et al., Contact angle changes for hydrophobic and hydrophilic surfaces induced by nonionic surfactants. Colloids and Surfaces, 21 (1986) 55–68.
29. Eriksson, J., Tiberg, F. and Zhmud, B., Wetting effects due to surfactant carryover through the three-phase contact line. Langmuir, 17(23) 2001 7274–7279.
30. von Bahr, M., Tiberg, F. and Yaminsky, V., Spreading dynamics of liquids and surfactant solutions on partially wettable hydrophobic substrates. Colloids and Surfaces A: Physicochemical and Engineering Aspects, 193(1–3) (2001) 85–96.
31. Dutschk, V., et al., Unusual wetting dynamics of aqueous surfactant solutions on polymer surfaces. Journal of Colloid and Interface Science, 267(2) (2003) 456–462.
32. Starov, V. M., Kosvintsev, S. R. and Velarde, M. G., Spreading of Surfactant Solutions over Hydrophobic Substrates. Journal of Colloid and Interface Science, 227(1) (2000) 185–190.
33. Churaev, N. V. and Zorin, Z. M., Penetration of aqueous surfactant solutions into thin hydrophobized capillaries. Colloids and Surfaces A: Physicochemical and Engineering Aspects, 100 (1995) 131–138.
34. Zolotarev, P. P., Starov, V. M. and Churaev, N. V., Colloid. J. USSR Academy of Sciences, English Translation, 38 (1976) 895.
35. Churaev, N. V., et al., Some features of capillary imbibition of surfactant solutions. Colloid & Polymer Sci., 259(7) (1981) 747–752.
36. Starov, V., Ivanova, N. and Rubio, R. G., Why do aqueous surfactant solutions spread over hydrophobic substrates? Advances in Colloid and Interface Science, 161(1–2) (2010) 153–162.
37. Schwartz, L. W., Weidner, D. E. and Eley, R. R., Proceedings of the ACS Division of Polymeric Materials Science and Engineering, 73 (1995) 490.
38. Patzer, J., Fuchs, J. and Hoffer, E. P., Proc. SPIE-Int. Soc. Opt. Eng., 167 (1995) 2413.
39. Le, H. P., J. Imaging Sci. Tech., 42 (1998) 49.
40. Shapiro, D. L., In Surfactant Replacement Therapy. AR Liss, New York, 1989.
41. Tsai, W.-t. and Liu, L.-Y., Transport of exogenous surfactants on a thin viscous film within an axisymmetric airway. Colloids and Surfaces A: Physicochemical and Engineering Aspects, 234(1–3) (2004) 51–62.
42. Birwagen, G. P., Prog. Org. Coat., 19 (1991) 59.
43. Gundabala, V. R. and Routh, A. F., Thinning of drying latex films due to surfactant. Journal of Colloid and Interface Science, 303(1) (2006) 306–314.
44. Chang, I.-S. and Jae-Hyung, K., J. Cleaner Production, 9 (2001) 227.
45. Afsar-Siddiqui, A. B., Luckham, P. F. and Matar, O. K., The spreading of surfactant solutions on thin liquid films. Advances in Colloid and Interface Science, 106(1–3) (2003) 183–236.
46. Lee, K. S. and Starov, V. M., Spreading of surfactant solutions over thin aqueous layers: Influence of solubility and micelles disintegration. Journal of Colloid and Interface Science, 314(2) (2007) 631–642.

47. von Bahr, M., Tiberg, F. and Zhmud, B. V., Spreading dynamics of surfactant Solutions. Langmuir, 15(20) (1999) 7069–7075.
48. Starov, V. M., de Ryck, A. and Velarde, M. G., On the spreading of an insoluble surfactant over a thin viscous liquid layer. Journal of Colloid and Interface Science, 190(1) (1997) 104–113.
49. Chan, K. Y. and Borhan, A., Surfactant-assisted spreading of a liquid drop on a smooth solid surface. Journal of Colloid and Interface Science, 287(1) (2005) 233–248.
50. Seguin, C., et al., SANS studies of the effects of surfactant head group on aggregation properties in water/glycol and pure glycol systems. Journal of Colloid and Interface Science, 315(2) (2007) 714–720.
51. Marmur, A. and Lelah, M. D., Chem. Eng. Commun., 13 (1981) 133.
52. Gaver, D. P. and Grotberg, J. B., The dynamics of a localized surfactant on a thin film. Journal of Fluid Mechanics Digital Archive, 213(−1) (2006) 127–148.
53. Yeo, L. Y. and Matar, O. K., Hydrodynamic instability of a thin viscous film between two drops. Journal of Colloid and Interface Science, 261(2) (2003) 575–579.
54. Matar, O. K. and Lawrence, C. J., The flow of a thin conducting film over a spinning disc in the presence of an electric field. Chemical Engineering Science, 61(12) (2006) 3838–3849.
55. Edmonstone, B. D., Matar, O. K. and Craster, R. V., Surfactant-induced fingering phenomena in thin film flow down an inclined plane. Physica D: Nonlinear Phenomena, 209(1–4) (2005) 62–79.
56. Edmonstone, B. D. and Matar, O. K., Simultaneous thermal and surfactant-induced Marangoni effects in thin liquid films. Journal of Colloid and Interface Science, 274(1) (2004) 183–199.
57. Cachile, M., et al., Films driven by surface tension gradients. Advances in Colloid and Interface Science, 96(1–3) (2002) 59–74.
58. Borgas, M. S. and Grotberg, J. B., Monolayer flow on a thin film. Journal of Fluid Mechanics Digital Archive, 193(−1) (2006) 151–170.
59. Jensen, O. E. and Grotberg, J. B., Insoluble surfactant spreading on a thin viscous film: shock evolution and film rupture. Journal of Fluid Mechanics Digital Archive, 240(−1) (2006) 259–288.
60. Troian, S. M., Herbolzheimer, E. and Safran, S. A., Model for the fingering instability of spreading surfactant drops. Physical Review Letters, 65(3) (1990) 333.
61. Jensen, O. E., The spreading of insoluble surfactant at the free surface of a deep fluid layer. Journal of Fluid Mechanics Digital Archive, 293(-1) (2006) 349–378.
62. Princen, H. M., et al., Instabilities during wetting processes: Wetting by tensioactive liquids. Journal of Colloid and Interface Science, 126(1) 1988 84–92.
63. Troian, S. M., Wu, X. L. and Safran, S. A., Fingering instability in thin wetting films. Physical Review Letters, 62(13) (1989) 1496.
64. Lin, Z., et al., Determination of wetting velocities of surfactant superspreaders with the quartz crystal microbalance. Langmuir, 10(11) (1994) 4060–4068.
65. Frank, B. and Garoff, S., Origins of complex motion of advancing surfactant solutions. Langmuir, 11 (1995) 87–93.
66. Stoebe, T., et al., Surfactant-enhanced spreading. Langmuir, 12(2) (1996) 337–344.
67. Stoebe, T., et al., Superspreading of aqueous films containing trisiloxane surfactant on mineral oil. Langmuir, 13(26) (1997) 7282–7286.
68. Stoebe, T., et al., Enhanced spreading of aqueous films containing ionic surfactants on solid substrates. Langmuir, 13(26) (1997) 7276–7281.
69. Stoebe, T., et al., Enhanced spreading of aqueous films containing ethoxylated alcohol surfactants on solid substrates. Langmuir, 13(26) (1997) 7270–7275.

70. Cachile, M., et al., Spontaneous spreading of surfactant solutions on hydrophilic surfaces. Colloids and Surfaces A: Physicochemical and Engineering Aspects, 159(1) (1999) 47–56.

71. Cachile, M. and Cazabat, A. M., Spontaneous spreading of surfactant solutions on hydrophilic surfaces: CnEm in ethylene and diethylene glycol. Langmuir, 15(4) (1999) 1515–1521.

72. Matar, O. K. and Troian, S. M., The development of transient fingering patterns during the spreading of surfactant coated films. Physics of Fluids, 11(11) (1999) 3232–3246.

73. Matar, O. K. and Troian, S. M., Growth of non-modal transient structures during the spreading of surfactant coated films. Physics of Fluids, 10(5) (1998) 1234–1236.

74. Matar, O. K. and Troian, S. M., Linear stability analysis of an insoluble surfactant monolayer spreading on a thin liquid film. Physics of Fluids, 9(12) (1997) 3645–3657.

75. Hamraoui, A., et al., Fingering phenomena during spreading of surfactant solutions. Colloids and Surfaces A: Physicochemical and Engineering Aspects, 250(1–3) (2004) 215–221.

76. Cachile, M., et al., Contact-line instabilities in liquids spreading on solid substrates. Physica A: Statistical Mechanics and its Applications, 329(1–2) (2003) 7–13.

77. Nikolov, A. D., et al., Superspreading driven by Marangoni flow. Advances in Colloid and Interface Science, 96(1–3) (2002) 325–338.

78. Starov, V. M., Velarde, M. G. and Radke, C. J., Wetting and Spreading Dynamics. Surfactant Sciences Series, Vol. 138. Taylor & Frances, 2007.

79. Kumar, S. M. and Deshpande, A. P., Dynamics of drop spreading on fibrous porous media. Colloids and Surfaces A: Physicochemical and Engineering Aspects, 277(1–3) (2006) 157–163.

80. Maiti, R. N., et al., The liquid spreading on porous solids: Dual action of pores. Chemical Engineering Science, 60(22) (2005) 6235–6239.

81. Maiti, R. N., et al., Enhanced liquid spreading due to porosity. Chemical Engineering Science, 59(13) (2004) 2817–2820.

82. Holman, R. K., et al., Spreading and infiltration of inkjet-printed polymer solution droplets on a porous substrate. Journal of Colloid and Interface Science, 249(2) (2002) 432–440.

83. Lake, L. W., Enhanced Oil Recovery. Prentice-Hall, Englewood Cliffs, NJ, 1996.

84. Alleborn, N. and Raszillier, H., Spreading and sorption of droplets on layered porous substrates. Journal of Colloid and Interface Science, 280(2) (2004) 449–464.

85. Alleborn, N. and Raszillier, H., Spreading and sorption of a droplet on a porous substrate. Chemical Engineering Science, 59(10) (2004) 2071–2088.

86. Starov, V. M., et al., Spreading of liquid drops over dry porous layers: complete wetting case. Journal of Colloid and Interface Science, 252(2) (2002) 397–408.

87. Starov, V. M., et al., Spreading of liquid drops over porous substrates. Advances in Colloid and Interface Science, 104(1–3) (2003) 123–158.

88. Davis, S. H. and Hocking, L. M., Spreading and imbibition of viscous liquid on a porous base. Physics of Fluids, 11(1) (1999) 48–57.

89. Aradian, A., Raphaël, E. and de Gennes, P. G., Dewetting on porous media with aspiration. Eur. Phys. J. E. 2 Soft Matter, (2000) 367–376.

90. Bacri, L. and Brochard-Wyart, F., Droplet suction on porous media. Eur. Phys. J. E. 3 Soft Matter, (2000) 87–97.

91. Modaressi, H. and Garnier, G., Mechanism of wetting and absorption of water droplets on sized paper: effects of chemical and physical heterogeneity. Langmuir, 18(3) (2002) 642–649.

92. Heilmann, J. and Lindqvist, U., Effect of drop size on the print quality in continuous ink jet printing. J. Imaging Sci. Technol., 44(6) (2000) 491–494.

93. Heilmann, J. and Lindqvist, U., Significance of paper properties on print quality in continuous ink jet printing. J. Imaging Sci. Technol., 44(6) (2000) 495–499.

94. Denesuk, M., et al., Dynamics of incomplete wetting on porous materials. Journal of Colloid and Interface Science, 168(1) (1994) 142–151.

95. Clarke, A., et al., Spreading and imbibition of liquid droplets on porous surfaces. Langmuir, 18(8) (2002) 2980–2984.

96. Zhdanov, S. A., et al., Spreading of aqueous SDS solutions over nitrocellulose membranes. Journal of Colloid and Interface Science, 264(2) (2003) 481–489.

97. Lee, K. S., et al., Spreading and evaporation of sessile droplets: universal behaviour in the case of complete wetting. Colloids and Surfaces A: Physicochemical and Engineering Aspects, 323(1–3) (2008) 63–72.

98. Daniel, R. C. and Berg, J. C., Spreading on and penetration into thin, permeable print media: Application to ink-jet printing. Advances in Colloid and Interface Science, 123–126 (2006) 439–469.

99. Hill, R. M., Superspreading. Current Opinion in Colloid & Interface Science, 3(3) (1998) 247–254.

100. Ivanova, N., et al., Spreading of Aqueous Solutions of Trisiloxanes and Conventional Surfactants over PTFE AF Coated Silicone Wafers. Langmuir, 25(6) (2009) 3564–3570.

101. Ananthapadmanabhan, K. P., Goddard, E. D. and Chandar, P., A study of the solution, interfacial and wetting properties of silicone surfactants. Colloids and Surfaces, 44 (1990) 281–297.

102. Svitova, T., Hoffmann, H. and Hill, R. M., Trisiloxane surfactants: surface/interfacial tension dynamics and spreading on hydrophobic surfaces. Langmuir, 12(7) (1996) 1712–1721.

103. Hill, R. M., ed. Silicone Surfactants. Surfactant Science Series, Vol. 086. Marcel Dekker, 1999.

104. Hill, R. M., Silicone surfactants—new developments. Current Opinion in Colloid & Interface Science, 7(5–6) (2002) 255–261.

105. Radulovic, J., Sefiane, K. and Shanahan, M. E. R., Spreading and wetting behaviour of trisiloxanes. Journal of Bionic Engineering, 6(4) (2009) 341–349.

106. Radulovic, J., Sefiane, K. and Shanahan, M. E. R., On the effect of pH on spreading of surfactant solutions on hydrophobic surfaces. Journal of Colloid and Interface Science, 332(2) (2009) 497–504.

107. Rosen, M. J. and Song, L. D., Superspreading, skein wetting, and dynamic surface tension. Langmuir, 12(20) (1996) 4945–4949.

108. Zhu, S., et al., Superspreading of water–silicone surfactant on hydrophobic surfaces. Colloids and Surfaces A: Physicochemical and Engineering Aspects, 90(1) (1994) 63–78.

109. Svitova, T., et al., Wetting and interfacial transitions in dilute solutions of trisiloxane surfactants. Langmuir, 14(18) (1998) 5023–5031.

110. Ivanova, N., et al., Critical wetting concentrations of trisiloxane surfactants. Colloids and Surfaces A: Physicochemical and Engineering Aspects, 354(1–3) (2010) 143–148.

111. Svitova, T., Hill, R. M. and Radke, C. J., Adsorption layer structures and spreading behavior of aqueous non-ionic surfactants on graphite. Colloids and Surfaces A: Physicochemical and Engineering Aspects, 183–185 (2001) 607–620.

112. Rafai, S. and Bonn, D., Spreading of non-Newtonian fluids and surfactant solutions on solid surfaces. Physica A: Statistical Mechanics and Its Applications, 358(1) (2005) 58–67.

113. Rafai, S., et al., Superspreading: aqueous surfactant drops spreading on hydrophobic surfaces. Langmuir, 18(26) (2002) 10486–10488.

114. Radulovic, J., Sefiane, K. and Shanahan, M. E. R., Dynamics of trisiloxane wetting: effects of diffusion and surface hydrophobicity. The Journal of Physical Chemistry C, 114(32) (2010) 13620–13629.

115. He, M., et al., Phase behavior and microstructure of polyoxyethylene trisiloxane surfactants in aqueous solution. J. Phys. Chem., 97(34) (1993) 8820–8834.

116. Hill, R. M., et al., Comparison of the liquid crystal phase behavior of four trisiloxane superwetter surfactants. Langmuir, 10(6) (1994) 1724–1734.

117. Ruckenstein, E., Effect of short-range interactions on spreading. Journal of Colloid and Interface Science, 179(1) (1996) 136–142.

118. Chengara, A., Nikolov, A. and Wasan, D., Surface tension gradient driven spreading of trisiloxane surfactant solution on hydrophobic solid. Colloids and Surfaces A: Physicochemical and Engineering Aspects, 206(1–3) (2002) 31–39.

119. Kumar, N., Couzis, A. and Maldarelli, C., Measurement of the kinetic rate constants for the adsorption of superspreading trisiloxanes to an air/aqueous interface and the relevance of these measurements to the mechanism of superspreading. Journal of Colloid and Interface Science, 267(2) (2003) 272–285.

120. Churaev, N. V., et al., The superspreading effect of trisiloxane surfactant solutions. Langmuir, 17(5) (2001) 1338–1348.

121. Ritacco, H. A., et al., Equilibrium and dynamic surface properties of trisiloxane aqueous solutions: Part 1. Experimental results. Colloids and Surfaces A: Physicochemical and Engineering Aspects, 365(1–3) (2010) 199–203.

122. Ritacco, H. A., et al., Equilibrium and dynamic surface properties of trisiloxane aqueous solutions. Part 2. Theory and comparison with experiment. Colloids and Surfaces A: Physicochemical and Engineering Aspects, 365(1–3) (2010) 204–209.

123. Shen, Y., et al., Molecular dynamics study of the influence of surfactant structure on surfactant-facilitated spreading of droplets on solid surfaces. Langmuir, 21(26) (2005) 12160–12170.

124. Halverson, J. D., et al., Wetting of hydrophobic substrates by nanodroplets of aqueous trisiloxane and alkyl polyethoxylate surfactant solutions. Chemical Engineering Science, 64(22) (2009) 4657–4667.

125. Knoche, M., Tamura, H. and Bukovac, M. J., Performance and stability of the organosilicon surfactant L-77: effect of pH, concentration, and temperature. Journal of Agricultural and Food Chemistry, 39(1) (2002) 202–206.

126. Radulovic, J., Sefiane, K. and Shanahan, M. E. R., Ageing of trisiloxane solutions. Chemical Engineering Science, 65(18) (2010) 5251–5255.

127. Zhang, Y., Zhang, G. and Han, F., The spreading and superspeading behavior of new glucosamide-based trisiloxane surfactants on hydrophobic foliage. Colloids and Surfaces A: Physicochemical and Engineering Aspects, 276(1–3) (2006) 100–106.

128. Zhang, Y. and Han, F., The spreading behaviour and spreading mechanism of new glucosamide-based trisiloxane on polystyrene surfaces. Journal of Colloid and Interface Science, 337(1) (2009) 211–217.

Wetting Dynamics of Aqueous Surfactant Solutions on Polymer Surfaces

Victoria Dutschk [a,b], **Alfredo Calvimontes** [b] **and Manfred Stamm** [b]

[a] Engineering of Fibrous Smart Materials (EFSM), Faculty for Engineering Technology (CTW), University of Twente, P.O. Box 217, 7500 AE Enschede, The Netherlands

[b] Leibniz-Institut für Polymerforschung Dresden e.V., Hohe Strasse 6, 01069 Dresden, Deutschland

Contents

A. Basics of Wetting 71
    1. Wetting and Wettability 72
    2. Three Phase Contact (TPC) Zone 72
    3. Contact Angle Hysteresis 73
B. Wetting on Rough Surfaces 74
    1. Surface Roughness of Solids 75
    2. Different Wetting Regimes 76
C. Wetting Dynamics 79
    1. Spreading Theories 79
    2. Spreading as a Rate Process 80
    3. Spreading of Aqueous Surfactant Solutions 82
    4. Spreading in Technological Applications 83
D. Summary 90
E. Acknowledgements 90
F. References 90

A. Basics of Wetting

The physical bases of wetting are molecular interactions within a solid or liquid or across the interface between a liquid and a solid. The wetting behavior of liquids on solid surfaces is determined by surface tensions (strictly speaking by interfacial tensions) of solid or liquid and liquid–solid interfacial tensions. Curvature and roughness of contact surface are the two other critical factors for wetting phenomena.

Drops and Bubbles in Contact with Solid Surfaces
© Koninklijke Brill NV, Leiden, 2012

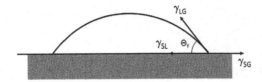

Figure 1. Liquid drop on a solid surface: Θ_Y is the Young contact angle; γ_{LG} is the interfacial tension liquid–gas; γ_{SV} and γ_{SL} are interfacial tensions solid–gas and solid–liquid.

1. Wetting and Wettability

The term "wetting" describes a displacement of a solid–gas (air) interface with a solid–liquid interface. That is a process in which free energy decreases in a system consisting of three contacting phases. The term "wettability" describes the ability of a surface to maintain a contact with a liquid. The degree of wetting and wettability is determined by a force balance between adhesive and cohesive molecular forces. The forces in equilibrium at a solid–liquid boundary are commonly described by the Young equation shown in Fig. 1a as a quasi-static wetting situation:

$$\gamma_{SG} = \gamma_{SL} + \gamma_{LG} \cdot \cos \Theta_Y, \tag{1}$$

where γ_{SG}, γ_{SL} and γ_{LG} are interfacial solid–gas, solid–liquid and liquid–gas tensions, respectively, and Θ_Y is the equilibrium contact angle. Since the quantities γ_{SG} and γ_{SL} are generally inaccessible to experiments, in contrast to γ_{LG}, the Young equation is often used for solving the inverse problem, namely to determine the difference $(\gamma_{SG} - \gamma_{SL})$, which is referred to as wetting tension or *adhesive tension*, by means of experimental values for static or quasi-static contact angle and interfacial tension γ_{LG}. The magnitude of the contact angle depends on the strength of molecular interactions between liquid molecules inside the drop as well as between a liquid and a solid surface.

The surface tension of a liquid is a measure of its resistance to increase its surface area. Liquids with relative large intermolecular forces such as water tend to have relatively high surface tensions, that is, high resistances to increasing their surfaces areas. When a small droplet of a pure liquid is put in contact with a flat surface, two distinct equilibrium regimes may be found: *partial wetting* with a finite contact angle, or *complete wetting* with a zero contact angle [1]. However, if the contact angle is larger than 90°, the liquid does not wet the surface and the situation is referred to as *non-wetting*. A more detailed description of advancing, receding, Young's contact angles as well as problems of experimental and theoretical verification of equilibrium contact angle is recently provided in [2].

2. Three Phase Contact (TPC) Zone

The three-phase contact (TPC) zone is the region where three immiscible coexisting phases meet, unless one of them wets the interface between the other two. The intersection of the three interfaces forms the three phase contact zone, the so-called TPC line as shown in Fig. 2.

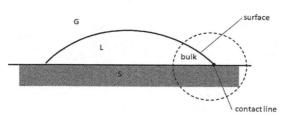

Figure 2. Liquid drop on a solid surface: G, L and S stand for gas, liquid and solid, respectively.

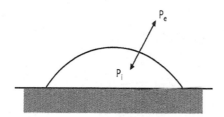

Figure 3. Shape of a liquid drop on a solid surface: $\Delta P = P_i - P_e$.

For a sufficiently small drop of a partial wetting or non-wetting liquid placed on a planar surface, gravity effects can be neglected. For such a drop, hydrostatic pressure inside the drop equilibrates and the drop adopts a shape to conform to the Laplace law:

$$\Delta P = \gamma_{LG}\left(\frac{1}{R_1} + \frac{1}{R_2}\right), \tag{2}$$

where ΔP is the pressure difference between two sides of a curved interface characterized by the principal radii of curvature R_1 and R_2. The drop shape would be spherical as shown in Fig. 3. For complete wetting of a flat surface, this pressure can be reduced towards zero by simultaneously increasing both R_1 and R_2 conserving the volume of the liquid.

The Young equation (1) describes the mechanical balance at the TPC line for an ideal surface. However, the equilibrium contact angle Θ_Y in the equation can be obtained only on a perfectly smooth and homogeneous solid surface. In case of the non-ideality of *real* solid surfaces, i.e., physical roughness and heterogeneity of surface energy densities across the surface, a given solid–liquid system experimentally produced different contact angles depending on how the experiment was performed. In particular, careful experiments have shown two relatively reproducible values of the contact angle—one as the TPC line advances across the surface and one as it recedes. Therefore, there is a fundamental issue of how liquid displacement at a solid surface is to be understood.

3. *Contact Angle Hysteresis*

Upon causing rapid displacement of a TPC line over a solid surface to a new location, experiments have shown static contact angles depend on the *direction* of

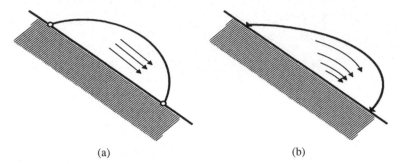

(a)   (b)

Figure 4. Different wetting regimes: (a) ideal Young regime, barrierless TPC line and (b) contact angle hysteresis, pinned TPC line.

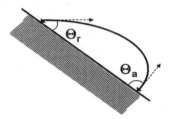

Figure 5. Advancing Θ_a and receding Θ_r contact angles.

recent movement of the TPC line. In Fig. 4, two different wetting regimes—ideal and real—are shown.

The difference between the so-called advancing Θ_a and receding Θ_r contact angles is called contact angle hysteresis: $\Theta_Y = \Theta_a - \Theta_r$. In Fig. 5, the advancing and receding contact angles are illustrated.

The difference can depend on the time interval between the movement and measurement, on contamination, and on many aspects of the solid surface state. The work of adhesion while receding is larger. It is essential to measure both contact angles and report the contact angle hysteresis to fully characterize a surface. The hysteresis can be classified in thermodynamic and kinetic terms. Roughness and heterogeneity of the surface are sources for hysteresis in thermodynamic sense. Kinetic hysteresis is characterized by time-dependent changes in contact angle which depends on deformation, reorientation and mobility of the surface as well as liquid penetration. Advancing and receding contact angle measurements are possible force-driven if the drop volume will be increased or decreased.

B. Wetting on Rough Surfaces

Surface roughness can be viewed as surface energy fluctuations [1], which act as barriers for the TPC line propagation. As shown in Fig. 6, the TPC line is anything but straight or smooth and its motion is scarcely studied. From the viewpoint of fluid mechanics, there is *no-slip* boundary condition [3]. In wetting experiments, suitable precautions have to be used to separate dynamic effects from static hysteresis. For

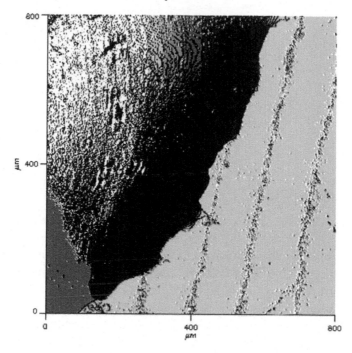

Figure 6. Microscopic image of a water droplet on a steel surface [4].

this purpose, a thorough topographic characterisation of a surface to be evaluated with respect to its wetting properties is absolutely essential.

1. Surface Roughness of Solids

Surfaces as physical entities possess many attributes, geometry being one of them. Surface geometry of real materials by nature is three-dimensional and its detailed features are termed topography. In engineering field, topography represents the main external features of a surface which is determined by the description of its morphology and topometry. In practice, the notion of a surface extends to sublayers of solid boundaries and the surface assumes certain internal features [5], e.g., hardness, residual stress, deformation, chemical composition and reactions, microstructure, capillary, hydrophobicity, that are often of foremost concern in an application. Surface topography often interrelates with these features in complicated manners and in three dimensions to define certain engineering properties. Surface topography is, therefore, significant for surface performance and the importance of its measurement by means of functional analysis and prediction is obvious.

Engineering surfaces are produced in various ways, typically by machining, surface treatment and coating. Surface topography modification is therefore performed by material removal, transformation or addition. Combinations of various machining, treatment and coating operations are employed to produce surfaces with desired characteristics for a particular application. Surface topography, therefore, contains 'signatures' of the surface generation process and as such can be used to

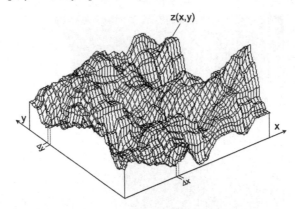

Figure 7. Coordinate system used for surface topography representation.

diagnose, monitor and control the manufacturing process. From an engineering perspective, the ultimate objective of a surface topography measurement, as a mean of control and knowledge, is to establish a correlation between an engineering surface transformation (e.g., wear, chatter, soiling, cleanability, permeability, etc.) and its topographical characteristics—waviness, roughness, porosity, fractal dimension, etc. Surface topography measurement, therefore, serves as a link between manufacturing, functional performance analysis and prediction, and surface design.

The main topographic characteristics are surface morphology and surface topometry. Surface morphology qualitatively describes the form and structure of a surface disregarding of fine details. Surface topometry means the exact coordinates of each single point. A "topography measurement" is the determination of x-, y- and z-coordinates for a set of representative points of the surface, as shown in Fig 7.

The topography analysis allows filtering measured total profiles of a surface to split them into two analytical representations for displaying surface features—roughness and waviness. The first information represents the shorter spatial wavelengths, whereas the second one represents the longer wavelength features of the surface. Both filtered profiles as well as the schematic of a complex fabric structure are illustrated in Fig. 8.

Some measurements possibilities—both conventional and modern—as well as topogaphic characterisation on different length scales are described in detail in [6].

2. Different Wetting Regimes

In general, surfaces can be either *randomly* rough or *periodically* rough. A liquid drop placed on a rough surface can sit on peaks or wet grooves depending on the geometry of surface roughness as well as on surface tension of a liquid.

The apparent static contact angle θ on a rough surface depends on the intrinsic Young contact angle Θ_Y. Existing theories for pure liquids can be taken into account in case of different wetting regimes (i) homogeneous surfaces according to Wenzel [7]; (ii) heterogeneous surfaces according to Cassie and Baxter [8]; (iii) composite surfaces according to Dettre and Johnson [9]; (iv) TPC line approach

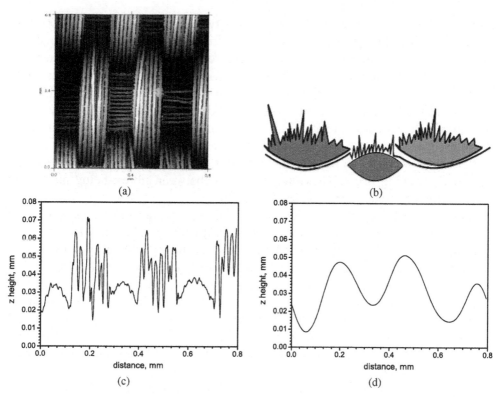

Figure 8. An optical image of a plain fabric surface (a); schematic of a complex fabric structure: the total profile contains both roughness and waviness information (b); filtered roughness profile of a plain fabric surface as an example (c); filtered waviness profile of a plain fabric surface as an example (d). Reprinted from *Textile Research Journal 78, Hasan M. M. B., Calvimontes A., Synytska A., Dutschk V., pp. 996–1003, Copyright (2008),* with permission from SAGE publications.

according to Extrand and Kumagai [10]. A textured solid can be considered as a two-dimensional porous material in which liquid can be absorbed by surface wicking [11]. This wetting regime is intermediate between spreading and imbibitions. When the contact angle is smaller than a critical value θ_{cr}, a film propagates from a deposited droplet, a small amount of liquid is sucked into the texture and remaining drop sits on a patchwork of a solid and liquid. It is very common that fibrous materials encounter roughness on surfaces and walls of pores. The driving force for such surface wicking depends on the geometry of the grooves, the surface tension of the liquid and the free-energies of the solid–gas and solid–liquid interfaces. An excellent overview on different wetting regimes for rough surfaces is given in [12].

New topographic concepts for a mechanistic understanding of wetting phenomena were presented by the authors in [13–16]. Moreover, in [17] was shown that there are significant differences between the soiling behaviour and cleanability of polyester textile materials with different topographic structures despite the similarity of their chemical nature. Toward a better mechanistic understanding soil removal effects as well as a quantitative correlations between them and topographical non-identity of fabrics, roughness and dynamic wetting results [17, 18], were analysed in respect of liquid transport driven into a porous system by capillary forces. In general, wetting of a fibrous assembly, such as a fabric, is a complex process. Particularly, capillary flow is not determined by a constant advancing contact angle, as frequently assumed [11], but it depends on a dynamic contact angle corresponding to the instantaneous velocity of the moving meniscus. In our studies was found that measured dynamic contact angle of water, as a consequence of wetting, is less expressive in respect to changes in the topographic structure of textile surfaces. Moreover, wicking occurring within the spaces between fibres, being previously wetted, strongly controls the fabrics cleanability.

As most textile processes, including their detergency, are time-limited, the rate and direction of wicking is therefore important. It is known that the wicking rate is not solely governed by interfacial tensions and the fibres wettability but it is also depends on capillary substrate dimensions and the liquid viscosity [11, 19].

On the basis of macroscopic water drop base changes, the wetting behaviour of the water drop can be divided into three different regimes (Fig. 9): dynamic wetting, defined as growing of the drop diameter depending on time (also know as spreading), quasi-static wetting, here the drop diameter remains approximately constant and penetration, which is marked by liquid drop absorption into fabrics depending on time.

As periodic surfaces, all textile materials show horizontal and vertical repetitive unities. For this reason, different length scales have to be taken into account by interpreting topographic data measured [14]. The topographical study of textile materials using scales permits to characterize surfaces considering their specific morphologies due to the type of weave, yarn and filament/fibres separately. The use of a scales concept to characterize textile surfaces aims to be a new skill to correlate textile parameters, topography and topographical changes with interface

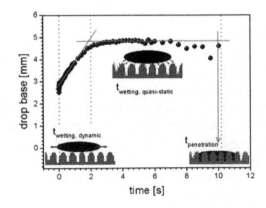

Figure 9. Three different wetting regimes on a textile surface. Reprinted from *Textile Research Journal 80, Calvimontes A., Dutschk V., Stamm M., pp. 1004–1015, Copyright (2010), with permission from SAGE publications.*

phenomena such as spreading, wetting, capillary and soil release. Three-scale concept includes (i) macro-morphological irregularities of textiles such as folds and wrinkles; (b) meso-scale of textile materials describing the surface topography produced by the type of weave and yarn used, without attending previous defined macro-topographic irregularities and details corresponding to fibres or filaments and (iii) micro-length scale revealing the influence of filaments and fibres characteristics on the resulting topography. Profile, fineness and natural or machined texture of these elements or distances between them, are only some of the possible characteristics that as a whole define the resulting morphology and topometry at the micro-length scale. Dimensional changes (relaxation/shrinkage) of fabrics at macro scale influence their meso- and micro-topography due to the modification of repetitive unit dimensions and therefore the distances between yarns, filaments and fibres.

To reach the macroscopic details of the liquid wetting behaviour in fibrous media, various computer simulation techniques have been applied in this filed to accommodate more complexity so as to investigate more realistic systems, and to better understand and explain experimental results.

C. Wetting Dynamics

Contact angle as a thermodynamic equilibrium property, virtually all the published data for which reproducibility is claimed, are measurements of advancing contact angle within a minute of TPC line displacement. The second category is that of truly dynamic contact angles. If the TPC line as a phase boundary liquid–solid simultaneously moves relative to the adjacent solid surface, a dynamic contact angle will be observed. Dynamic contact angle means the contact angle as a function of time, which can significantly differs from the static contact angle.

1. Spreading Theories

The dynamic behaviour of a pure liquid on an ideal solid surface can be successfully mathematically described by the equilibrium contact angle Θ, the dynamic (time-dependent) contact angle $\theta(t)$ as well as the spreading velocity dr/dt, where r is the base radius of a spreading drop. The *spreading velocity* or *spreading rate* as a time-dependent drop radius variation is often an important criterion on which basis the efficiency of surface-active substances (surfactants) can be estimated.

In the hydrodynamic consideration disturbed capillary equilibrium leads to the spreading force $\gamma_{LG}(\cos\theta_0 - \cos\theta(t))$, where θ_0 is initial contact angle. The most popular hydrodynamic models gave a successful interpretation of experimental data on complete wetting and propagation at high capillary numbers [3, 20–22]. Work is necessary to expanding the solid–liquid interface, and energy will dissipate due to viscose shear in the liquid. Such a consideration assumes a slippage of the liquid with respect to the solid in the TPC line vicinity. This theory is applicable to the description of a slow spreading near equilibrium. The molecular-kinetic theory [23, 24] assumes, however, particular displacements on a molecular level at the

TPC line as a possible reason for the spreading force; it is suitable for describing high spreading velocities far from equilibrium. Although this theory, in contrast to the hydrodynamic theory, includes surface effects, its application to predict the spreading velocity is rather problematic since the molecular parameters such as the density of the adsorption centres and the distance between them on real surfaces is unknown und generally inaccessible to experiments. By completely neglecting the viscous drag, a theoretical dependence of the dynamic contact angle $\theta(t)$ on the TPC line velocity is determined from the balance of the driving force and the friction force in the TPC zone. This approach is based on Eyring theory for transport phenomena [25]. There are some attempts for simultaneous analysis of the bulk flow and the TPC line friction [1, 21, 26, 27].

2. *Spreading as a Rate Process*

Spreading of evaporating droplets is determined by the spreading rate law dr/dt and evaporation rate law $V(t)$, where r is the base radius of a spreading drop and V is the drop volume. In many cases, the spreading rate law was found to be bi-exponential [28], while the evaporation has a well documented proportionality to the TPC line length [29–31]. If these two laws are in force, the spreading kinetics of a liquid drop, i.e., the dependence of the base radius r and contact angle θ on time, can be predicted. A more general step-mechanism of the TPC line motion was recently proposed [32]. Here, surface energy fluctuations serve as energy barrier of the spreading process, described as a nucleation process driven by capillary waves at liquid surfaces. It was shown, that such a mechanism leads to a spreading rate law similar to one obtained by molecular-kinetic treatment, without being based on adsorption/desorption mechanism of spreading [23, 24, 28]. The well-documented exponential law of spreading [28, 33–35] can also be interpreted by this mechanism.

Spreading rate law dr/dt can be physically determined by bulk friction or by friction in the vicinity of the TPC zone [1, 28] as schematically shown in Fig. 10.

In other words, spreading of droplets is a hydrodynamic problem with slip boundary conditions at the TPC line [36] such as bulk properties, existing in the hydrodynamic equations, or TPC zone properties, occurring in their boundary condition, that are the decisive factors for spreading. In the following, the case, where spreading rate is determined mainly by the TPC line region properties, is assumed.

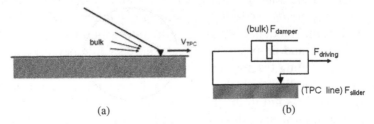

(a)   (b)

Figure 10. Schematic of a global spreading resistance (a); a parallel damper-slider model.

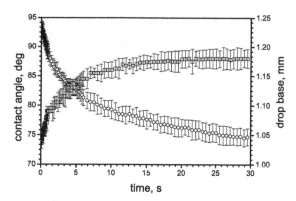

Figure 11. Contact angle and drop base of a water drop, containing a non-ionic surfactant on a hydrophobic surface: (○) contact angle; (□) drop diameter.

A general form of the spreading rate law in such case is given by Eyring's bi-exponential form [25]

$$\frac{dr}{dt} = A_A e^{-b_A \cos\theta} - A_R e^{b_R \cos\theta}, \qquad (3)$$

where θ is dynamic contact angle, i.e., the contact angle as a function of time $\theta = \theta(t)$. The indices A and R describe the parameters of advancing and receding movement of the TPC line. For $\theta > \Theta$, the TPC-line is advancing, e.g., in the case of spreading of aqueous surfactant solutions; for $\theta < \Theta$, the TPC-line is receding, e.g. during evaporation of a water drop. The spreading law (3) reflects the barrier character of the TPC line movement [26], which was confirmed for the adsorption-desorption mechanism of the TPC movement [37, 38]. Most of wetting characteristics such as mobility and immobility of the TPC line—pinning effect [39], quasi-static advancing and receding contact angles, characteristic spreading velocity—can be interpreted with the help of equation (3). The parameters $A_{A,R}$ and $b_{A,R}$ depend on a particular mechanism, but in the case of equilibrium $dr/dt = 0$, the following relationship is valid

$$\cos\Theta = \frac{1}{b_A + b_R} \ln \frac{A_A}{A_R}. \qquad (4)$$

The cosine of Θ is defined by the Young equation (1) and does not depend on any mechanisms. From a simple geometrical consideration [40, 41], the base radius r for a small drop can be expressed as

$$r^3 = \frac{3}{\pi} \frac{(1 + \cos\theta)^{3/2}}{(1 - \cos\theta)^{1/2}(2 + \cos\theta)} V, \qquad (5)$$

where V is the drop volume, which is supposed to remain constant, and θ is the contact angle changing with time. In Fig. 11, the contact angle and base radius of an evaporating drop (water) are illustrated.

In summary, the cosine of dynamic contact angle $\theta(t)$ is determined by the spreading law (3), drop geometry (5) and evaporation rate law $V(t)$. The problem of the determination of $\cos\theta(t)$ and $r(t)$, assuming that drop volume remains constant, was solved numerically [42], where the exponential factors b_A und b_R were assumed to be identical. In this case, the spreading law of Eq. (3) reduces to

$$\frac{dr}{dt} = A \sinh(b(\cos\Theta - \cos\theta)). \tag{6}$$

3. Spreading of Aqueous Surfactant Solutions

If a surface-active substance and a real (i.e., rough, inhomogeneous) solid surface is of interest, some attempts to describe the wetting behaviour theoretically are rather of a speculative nature. Aqueous surfactant solutions differ from pure liquids by the fact that their surface tension γ_{LV} and solid–liquid interfacial tension γ_{SL} are functions of time, and the molecular orientation influences surfactant-solid surface interactions. In the study of aqueous surfactant solutions, along with the solid surface state (chemical and morphologic nature), additional factors such as the solution concentration, the chemical nature of a surfactant (non-ionic, anionic, cationic, amphoteric) have to be taken into account. In the last decade, various authors, using dynamic contact angle measurements, found that the spreading velocity of aqueous surfactant solutions is strongly affected by the solid surface energy.

An adequate interpretation of the results of contact angle measurements is additionally complicated because a solid surface is able to adsorb water vapour from humid air. So, the surface energy value γ_{SV} depends on the thickness of the adsorption film. Disjoining pressure isotherms in the presence of surfactants are well investigated in the case of free liquid films [43], much less is known in the case of liquid films on solid substrates [44]. At the present, an answer to the question *how surfactant molecules are transferred in the TPC line vicinity* is not given. In the case of aqueous surfactant solutions, the knowledge of the transition zone behaviour from meniscus to thin films in front is very limited as referred in [45].

Hitherto, either only the solid–vapour interfacial tension γ_{SV} or only the surface tension γ_{LV} was assumed in the literature, though variations of the solid–liquid interfacial tension γ_{SL} was not excluded in this case. Depending on these assumptions, two fundamentally different *spreading mechanisms* of aqueous surfactant solutions were proposed for hydrophobic surfaces [45]. Following von Bahr *et al.* ideas [46] the surfactant molecules adsorb at the freshly formed solid–liquid interface *behind* the advancing wetting front. The transfer of molecules from the liquid–vapour interface occurs very quickly. The replenishment of this interface happens by diffusion of surfactant molecules from the volume phase and depends linearly on the root of time (*diffusion-controlled*). In contrast with it, Starov *et al.* [40] assumed that the surfactant molecules adsorb onto the solid–liquid interface *before* the wetting front. As this takes place, interfacial tension γ_{SV} increases. The transfer of molecules from the liquid–vapour interface occurs very slowly. The reason for the spreading force is the difference $(\Gamma_S(t) - \Gamma_e)$, where $\Gamma_S(t)$ is the current value of the adsorption quantity on the solid surface and Γ_e is the corresponding equilibrium value.

While many monographs and reviews are devoted to capillarity and wetting by pure liquids, the wetting by surfactant solution is not described in monographs, with the exception for [45, 47, 48]. Interest in wetting dynamics processes has immensely increased during the past 15 years. In many industrial and medical applications, some strategies to control drop spreading on solid surfaces are being developed. One possibility is that a surfactant, a surface active polymer or polyelectrolyte is added to a liquid. A reduction of water surface tension by adsorption of surfactant molecules on a water vapour interface and adsorption of surfactant molecules on solid–liquid and solid vapour interfaces alter non-wetting behaviour of aqueous solutions on hydrophobic substrates into partial or complete wetting behaviour.

The same conclusion can be drawn in the case of water penetration into hydrophobic porous media. Aqueous surfactant solutions can spontaneously penetrate into hydrophobic porous substrates and the penetration rate depends on both the surfactant type and its concentration. Both the liquid–vapour interfacial tension γ_{LV} and the contact angle of moving meniscus θ_a (advancing contact angle) are concentration dependent. Despite the enormous technical importance of spreading of aqueous surfactant solutions over solid surfaces, information on possible spreading mechanisms is limited in literature. Disjoining pressure isotherms in the presence of surfactants are well investigated in the case of free liquid films [49], much less is known in the case of liquid films on solid substrates [44]. At the present, we are not able to give an answer how surfactant molecules are transferred in the TPC line vicinity. In the case of aqueous surfactant solutions our knowledge of the transition zone behaviour from meniscus to thin films in front is very limited. A systematic representation of the main areas of this broad topic can be found in a review [45] recently published.

4. Spreading in Technological Applications

For many applications, surfactants are introduced into the aqueous phase to increase the rate and uniformity of wetting. Despite their enormous technical importance, there is a lack of data in the literature about the spreading dynamics of aqueous surfactant solutions on surfaces of practical interest, e.g., rough, inhomogeneous. The knowledge of how surfactant adsorption at the surfaces involved affects the spreading mechanism and dynamics is also limited. Dynamic wetting measurements allow studying surfactants, polyelectrolytes or other surface-active substances as well as their mixtures and engineered surfaces. A methodology for screening of technical surfactant solutions applicable for the wetting dynamic measurement on a wide range of solid surfaces was developed and is outlined below.

On the one hand, the screening method has to make possible a selection of a surfactant suitable for certain purpose. On the other hand, new application fields for known surfactants could be opened up. By a screening method, a methodol-

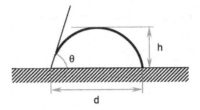

Figure 12. Sessile drop method: θ—contact angle, d—base and h—height of the drop.

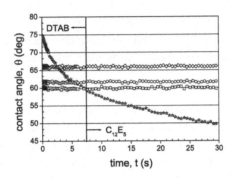

Figure 13. Contact angle of aqueous surfactant solutions on polypropylene surface as function of time ($c = $ cmc): $\bigcirc$ SDS; $\diamond$ DTAS; $\square$ DTAB; $\star$ $C_{12}E_5$.

ogy for analysis of wetting properties of surfactant solutions is meant including a suitable measuring method as well as a number of appropriate model surfaces with manufacturing and cleaning instructions.

The sessile drop method was used to investigate wetting of aqueous surfactant solutions on solid surfaces. The measurement data are the contact angle, base and height of the drop as shown in Fig. 12. For surfactant solutions, the contact angle as a function of time may be significant as illustrated in Fig. 13.

In the time interval of 0 to 7 s, the most effective, i.e., the lowest contact angle, surfactant for highly hydrophobic polypropylene (PP) surface is cationic dodecyltrimethylammonium bromide (DTAB). Beginning at approximately 7 s, a non-ionic pentaethylene glycol monododecyl ether ($C_{12}E_5$) becomes more effective than DTAB, cationic dodecyltrimethylammonium sulfate (DTAS) and anionic sodium dodecyl sulfate (SDS).

Dynamic wetting measurements were carried out with a FibroDAT 1122 dynamic contact angle tester (Fibro Systems AB, Sweden) equipped with a video camera which allows collecting up to 1000 images per second. The device has some advantages over other contact angle measuring systems, which are especially important for performing the dynamic wetting measurements with aqueous surfactant solutions:

(i) Adjustment of a specified (small) drop volumes using a liquid delivery system;

(ii) tuning the deposition parameters such as the distance between a drop and a solid surface, the time difference between drop formation on the syringe tip and drop deposition on the surface, as well as the intensity of a very short stroke from an electromagnet to release the drop from the tip;

(iii) checking the surface tension of surfactant solutions before wetting measurements.

Some surfactant solutions have a very long relaxation towards equilibrium. Dynamic surface tension is the change in surface tension before equilibrium conditions are obtained. Hence, in addition to static surface tension measurements, dynamic surface tension measurements were performed. To obtain the static surface tension, the pendant drop method was used (Fibro DAT1122). The shape of the drop is a measure of the surface tension. The dynamic surface tension was measured using the maximum bubble pressure method (SINTERFACE, Golm, Germany) as described in [50].

It is well known that the state of a surface affects crucially its wettability. Therefore, controlling the surface properties such as roughness, surface structure and chemical composition is of great importance. At first, a set of polymer samples with the same surface chemical composition but different prehistory (sheets, foils and spin-coated films) were used. Sample surface topography was examined using a scanning force microscope (Digital Instruments, USA). To obtain the degree of surface hydrophobicity, advancing contact angles of water drops were measured by the sessile drop method using a Krüss DSA 10 drop shape analyser (Germany). The chemical surface composition was investigated by means of X-ray spectroscopic analysis. For the appreciation of the chosen polymer samples on their ability to be model surfaces for screening method, the values for water and SDS contact angles as well as their surface roughness were compared. The comparison between industrial polymer samples (sheets and foils) and spin-coated polymer films proved that:

(i) the wetting results for each set of surfaces are in good agreement;

(ii) due to chemical heterogeneity connected with manufacturing process, the water contact angles for sheets are lower than that for foils and films (Fig. 14);

(iii) due to considerable surface roughness of the polymer sheets, the relative error in the measured contact angles for sheets was larger than those for films and foils as illustrated in Fig. 13.

Consequently, spin-coated polymer films and industrially manufactured polymer foils were chosen as model surfaces for screening.

Anionic SDS, cationic DTAB and DTAS and non-ionic $C_{12}E_5$ (all with the same alkyl chain length) were used as model surfactants, as mentioned above. In addition, technical surfactants anionic n-C_{12}–C_{13}-alkylbenzene sulphonate Na salt (Marlon ARL) and non-ionic C_{12}–C_{14}-fatty alcohol polyethylene glycol ether (E_6, Marlipal 24/60) were used to verify the screening method. For all surfactants used, their

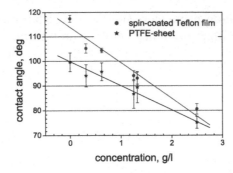

Figure 14. Water contact angles for spin-coated Teflon film and industrial PTFE sheets.

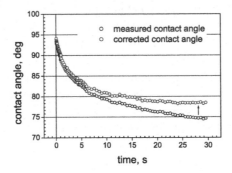

Figure 15. Measured and corrected contact angles for a $C_{12}E_5$ drop on Teflon AF (c = cmc).

critical concentration values (cmc) were experimentally determined. Aqueous solutions of these surfactants with desired concentrations in the range 0.0625 to 2 part of the corresponding cmc were prepared by dilution from the corresponding stock solutions with reagent-grade water.

In a typical experiment, the drop volume decreases linearly with time beginning at approximately 5 s. Proceeding from the assumption that the drop volume losses occur due to evaporation caused by relatively low humidity during measurements (40%), the measured contact angles were corrected by assuming volume constancy [41] shown in Fig. 15.

The amorphous Teflon AF (a copolymer of tetrafluoroethylene and 2,2-bis(trifluoroethylene)-4,5-difluoro-1,3-dioxole) which is a product of Du Pont, represents a new class of fluoropolymers. The stability of the surface properties of Teflon AF films were described in detail [51]. The main conclusion drawn was that the use of untreated Teflon AF as a model experimental object during long contact with water is limited by the processes of its swelling and expansion.

Effectiveness of aqueous surfactant solutions at the cmc for model surfactants used is presented in Fig. 16. Apparently from these experimental data, the following conclusions arise:

(i) the effectiveness of surfactants strongly depends on the polymer surface used;

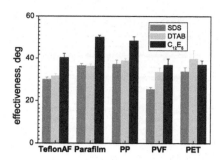

Figure 16. Effectiveness of surfactant solutions: a difference between the water contact angle and contact angle of corresponding surfactant solution at the cmc; PVF: polyvinyl fluoride, PET: polyethylene terephthalate.

(ii) the more hydrophilic the surface, the less effective is the respective surfactant;

(iii) with regard to the changes in wettability, the largest effects were obtained with the non-ionic surfactant, especially on highly hydrophobic polymer surfaces.

From the experiment as shown in Fig. 17, polymer surfaces may be divided into classes with reference to the spreading behaviour of aqueous surfactant solutions:

– highly hydrophobic surfaces such as Teflon AF, Parafilm and PP with the surface free energy of 11.8, 18.4 and 23.2 mJ/m^2, respectively;

– moderately hydrophobic surface such as PVF and PET with the surface free energy of 36.2 and 36.7 mJ/m^2, respectively.

For the highly hydrophobic surfaces, the contact angles for SDS, DTAB and DTAS solutions do not change with time at any concentrations investigated. In contrast to the highly hydrophobic surfaces, the contact angles of the ionic surfactant solutions on the moderately hydrophobic surfaces strongly depend on time. Similarly to the behaviour observed for ionic surfactants on the moderately hydrophobic surfaces, non-ionic $C_{12}E_5$ spread well over both moderately and highly hydrophobic surfaces, but in the last case, the effect is less pronounced.

For the highly hydrophobic surfaces, the contact angles for SDS, DTAB and DTAS solutions do not change with time at any concentrations investigated. In contrast to the highly hydrophobic surfaces, the contact angles of the ionic surfactant solutions on the moderately hydrophobic surfaces strongly depend on time. Similarly to the behaviour observed for ionic surfactants on the moderately hydrophobic surfaces, non-ionic $C_{12}E_5$ spread well over both moderately and highly hydrophobic surfaces, but in the last case, the effect is less pronounced.

An important parameter to compare the wetting behaviour of surfactant solutions is the wetting activity (or wetting power), calculated from the derivative of the equilibrium contact angle with respect to relative concentration (c/cmc). This quantity relies not only on the surface activity of a surfactant at the liquid–vapour interface, but it also depends on molecular interactions between the surfactant and a solid

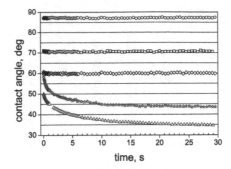

Figure 17. Contact angle of aqueous DTAB solution at the cmc on different polymer surfaces. ○ Teflon AF; □ Parafilm; ◇ PP; ⋆ PVF; △ PET.

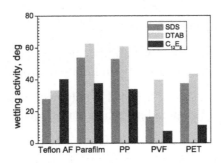

Figure 18. Wetting activity of surfactants under study for polymer surface used.

surface. Using the wetting activity parameter, a suitable surfactant can be selected for a given surface. The larger the value of the wetting activity, the more active is surfactant. In our case, the most suitable surfactant for very hydrophobic Teflon AF might be $C_{12}E_5$ as shown in Fig. 18. For other surface under investigation, DTAB may be used as the 'best' surfactant with respect to the wetting activity, apart from other decision criteria.

Another parameter, the spreading rate, is of great significance in order to estimate the spreading behaviour of a given surfactant on a given surface (cf. Section C.2). Our results indicate that spreading, if it occurs, may be divided into two regimes: I—the short time regime (fast spreading). In this regime, until approximately 1 s, the base radius depends linearly on time; II—the long time regime (slow spreading). In the long time regime, from 1 s to the equilibrium state, the radius values may be fitted by a power function. Exponent value of this power function serves as a measure of the spreading extent with respect to time. The larger the exponent value, the faster the spreading as shown in Fig. 19.

The analysis of time dependencies of the drop base radius reveals that the slow wetting dynamics observed for both ionic and non-ionic surfactants on hydrophobic surfaces can be explained neither by surfactant diffusion from the bulk of the drop to the expanding liquid–vapour interface nor in terms of viscous spreading. Two different explanations were provided [52]: First, a slow rearrangement of sur-

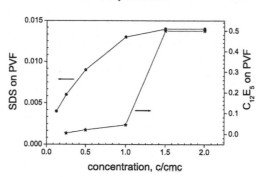

Figure 19. Exponent values of the power function for SDS and $C_{12}E_5$ on PVF.

factant molecules adsorbed at the solid–liquid interface occurs inside of the drop, probably caused by the bi-layer formation due to low monomer solubility which, in turn, may be affected by low surrounding humidity leading to water evaporation of the drop. Second, there is other fairly realistic possibility should be considered when analyzing the reasons for observed low spreading rates—the transfer of surfactant molecules onto the bare hydrophobic surfaces which is a relatively slow but a spontaneous process. The characteristic time scale of the surfactant molecules transfer onto the hydrophobic surface decreases with increasing surfactant concentration for all surfactant/polymer systems studied as predicted in [40]. Experimental data in [41, 52, 53] indicate that it is valid for both highly and moderately hydrophobic surfaces. Moreover, it was found that the characteristic time scale for transfer of individual ethoxylated alcohol surfactants $E_m E_5$ estimated for hydrophobic polypropylene surface increases with increasing the hydrocarbon chain length, probably indicating steric limitations with increasing molecular size. As concluded in [52], based on experimental data the characteristic scale time transfer which can be act as a spreading characteristic sensibly responds to changes both the surfactant nature (ionic, non-ionic) and surface free energy.

Dynamic contact angle measurements of aqueous solutions of technical surfactants were carried out on Parafilm surface. The equilibrium contact angles are good reproducible within 1 to 2% (Fig. 20). Considering the spreading behaviour, they spread faster on Parafilm surface than similar model surfactants. Especially at relatively low concentrations 0.0625·cmc and 0.125·cmc, their aqueous solutions already spread on highly hydrophobic Parafilm.

A methodology to analyze and compare aqueous solutions of different surfactants with respect to their wetting and spreading on polymer surfaces is provided. In order to better distinguish between very similar surfactants, highly hydrophobic polymer samples with very smooth and homogeneous surfaces are recommended to use as model surfaces. In many respects (costs, repeated use, reproducibility), the best model sample seems to be the surface of Parafilm, a product of Pechiney Plastic Packaging, USA. The developed method was verified by two selected technical surfactants, both products of Sasol Germany (Marl).

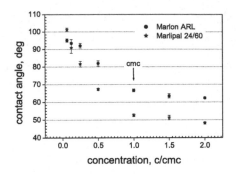

Figure 20. Equilibrium contact angles of aqueous solutions of technical surfactants on Parafilm.

D. Summary

Basics of wetting including wetting on rough surfaces was presented. Spreading as a rate process was considered and described. Wetting dynamics of aqueous surfactant solutions over different surfaces was outlined. Both model surfaces and surfaces of technical relevance as well as pure and technical surfactants were taken into account. Two most relevant models from literature explaining the spreading behaviour of aqueous surfactant solutions over hydrophobic surfaces were displayed. Some practical tools for technological applications based on the wetting dynamics of surfactant solutions were illustrated.

E. Acknowledgements

Victoria Dutschk is indebted to Boryan Radoev (Sofia University, Bulgaria) for the careful attention, fruitful discussions about a possible description of spreading as a rate process according to Eyring's approach. The author is grateful to Radomir Slavchov (Sofia University, Bulgaria) for performing several numerical studies. This research was supported by Sasol Germany (Marl). Victoria Dutschk is obliged to Burkhard Breitzke, Martin Stolz, Wulf Ruback and Herbert Koch for proving technical surfactants, solid materials of technological relevance, giving many useful advises in the field of surfactants. The authors are thankful to Victor Starov (Loughborough University, UK) for many useful discussions.

F. References

1. De Gennes, P. G., Wetting and static dynamics, Review of Modern Physics, 57 (1985) 827.
2. Chibowski, E., Adv. Colloid Interface Sci., 133 (2007) 51.
3. Huh, C. and Scriven, L. E., J. Colloid Interface Sci., 35 (1971) 85.
4. Radoev, B., unpublished results.
5. Stout, K. J., Sullivan, P. J., Dong, W. P., Mainsah, E., Luo, N., Mathia, T. and Zahouani, H., The development of methods for the characterisation of roughness in three dimensions, Brussels–Luxemburg: Commission of the European Communities, Brussels–Luxembourg, 1993.

6. Calvimontes, A., Topographic characterization of polymer materials at different length scales and the mechanistic understanding of wetting phenomena, PhD thesis, Technische Universität Dresden, 2009.

7. Wenzel, R. N., Ind. Eng. Chem., 28 (1936) 98.

8. Cassie, A. B. D. and Baxter, S., Trans. Faraday Soc., 40 (1944) 546.

9. Dettre, R. H. and Johnson, R. E., in: Contact angle, Wettability and Adhesion. Advances in Chemistry Series 43. ACS, Washington DC, 1964.

10. Extrand, C. W. and Kumagai, Y., J. Colloid Interface Sci., 170 (1995) 515.

11. Kissa, E., Tex. Res. J., 66 (1996) 660.

12. Rengasamy, R. S., Wetting phenomena in fibrous materials. Thermal and moisture transport in fibrous materials, Woodhead Publishing, Cambridge, England, 2006.

13. Hasan Badrul, M. M., Calvimontes, A. and Dutschk, V., J. Surfact. Deterg., 12 (2009) 285.

14. Calvimontes, A., Dutschk, V. and Stamm, M., Textile Res. J., 80 (2010) 1004.

15. Calvimontes, A., Hasan Badrul, M. M. and Dutschk, V., Effects of topographic structure on wettability of differently woven fabrics. Woven Fabric Engineering, SCIYO, 2010.

16. Calvimontes, A., Saha, R. and Dutschk, V., AUTEX Research J., 11 (2011) 24.

17. Calvimontes, A., Dutschk, V., Koch, H. and Voit, B., Tenside Surf. Det., 42 (2005) 210.

18. Calvimontes, A., Dutschk, V., Breitzke, B., Offermann, P. and Voit, B., Tenside Surf. Det., 42 (2005) 17.

19. Chwastiak, S., J. Colloid Interface Sci., 42 (1973) 298.

20. Cox, R. G., J. Fluid Mech., 168 (1986) 169.

21. Voinov, O. V., Fluid Dyn., 11 (1976) 714.

22. Shikhmurzaev, Y. D., Capillary Flows with Forming Interfaces. Chapman & Hall/CRC, Roca Raton, 2008.

23. Blake, T. D. and Haynes, J. M., J. Colloid Interface Sci., 30 (1969) 421.

24. Ruckenstein, E., Langmuir, 8 (1992) 3038.

25. Glasstone, S., Laidler, K. J. and Eyring, H. J., The Theory of Rate Processes. McGraw-Hill, New York, 1941.

26. Petrov, J. and Radoev, B., Colloid Polym. Sci. 259 (1981) 753.

27. Petrov, P. G. and Petrov, J., Langmuir 8 (1992) 1762.

28. Blake, T. D., in: Wettability (John, C. B., Ed.), Surfactant Science Series, 49 (1993) 291–309.

29. Deegan, R. D., Bakajin, O., Dupont, T. F., Huber, G., Nagel, S. R. and Witten, T. A., Phys. Rev. E, 62 (2000) 756.

30. Hu, H. and Larson, R. G., J. Phys. Chem. B, 106 (2002) 1334.

31. Guéna, G., Allancon, P., Cazabat, A. M., Colloids Surfaces A, 300 (2007) 307.

32. Slavchov, R., Dutschk, V., Heinrich, G. and Radoev, B., Colloids Surfaces A, 354 (2010) 252.

33. Scheemilch, M., Hayes, R. A., Petrov, J. G. and Ralston, J., Langmuir, 14 (1998) 7047.

34. Petrov, J. G., Ralston, J., Scheemilch, M. and Hayes, R. A., J. Phys. Chem. B, 107 (2003) 1634.

35. Petrov, J. G. and Petrov, P. G., Colloid Surfactants A, 64 (1992) 143.

36. Dussan, E. B. V., Ann. Rev. Fluid Mech., 11 (1979) 371.

37. Hayes, R. A. and Ralston, J., Colloids Surfaces A, 93 (1994) 15.

38. Ruckenstein, E. and Dunn, C. S., J. Colloid Interface Sci., 59 (1977) 135.

39. Moffat, J. R., Sefiane, K. and Shanahan, M. E. R., J. Nano Res., 7 (2009) 75.

40. Starov, V. M., Kosvintsev, S. R. and Velarde, M. G., J. Colloid Interface Sci., 227 (2000) 185.

41. Dutschk, V., Sabbatovskiy, K. G., Stolz, M., Grundke, K. and Rudoy, V. M., J. Colloid Interface Sci., 267 (2003) 456.

42. Semal, S., Blake, T. D., Geskin, V. and de Ruijter, M. J., Gastelein, G., de Coninck, J., Langmuir, 15 (1999) 8765.
43. Exerowa, D. and Krugliakov, P., in: Foam and Foam Films: Theory, Experiment, Application. Studies in Interface Science, Vol. 5. Elsevier, New York, 1988.
44. Churaev, N. and Zorin, Z., Adv. Colloid Interface Sci., 40 (1992) 109.
45. Lee, K. S., Ivanova, N., Starov, V. M., Hilal, N. and Dutschk, V., Adv. Colloid Interface Sci., 144 (2008) 54.
46. von Bahr, M., Tiberg, F. and Yaminsky, V., Colloid Surfaces A, 193 (2001) 85.
47. Starov, V., Velarde, M. and Radke, C., in: Dynamics of Wetting and Spreading. Surfactant Sciences Series, Vol. 138. Taylor & Frances, 2007.
48. de Gennes, P. G., Brochard-Wyart, F. and Quere, D., Capillarity and Wetting Phenomena. Springer, 2003.
49. Exerowa, D. and Krugliakov, P., in: Foam and Foam Films: Theory, Experiment, Application. Studies in Interface Science, Vol. 5, Elsevier, New York, 1988.
50. Dutschk, V., Wetting dynamics of aqueous solutions on solid surfaces, in: Progress in Colloid and Interface Science, Vol. 2, Bubble and Drop Interfaces, 2009.
51. Sabbatovskii, K. G., Dutschk, V., Nitschke, M., Simon, F. and Grundke, K., Colloid J., 66 (2004) 239.
52. Dutschk, V. and Breitzke, B., Tenside Surfact Det., 42 (2005) 82.
53. Dutschk, V., Breitzke, B. and Grundke, K., Tenside Surfact Det., 40 (2003) 250.

Hydrodynamic Interactions Between Solid Particles at a Fluid–Gas Interface

B. Cichocki [a], **M. L. Ekiel-Jeżewska** [b], **G. Nägele** [c] **and E. Wajnryb** [b]

[a] Institute of Theoretical Physics, University of Warsaw, Hoża 69, 00-681 Warsaw, Poland
[b] Institute of Fundamental Technological Research, Polish Academy of Sciences, Świętokrzyska 21, 00-049 Warsaw, Poland
[c] Institut für Festkörperforschung, Soft Matter Division, Forschungszentrum Jülich, D-52425 Jülich, Germany

Abstract

We analyze the hydrodynamic interactions between spherical solid particles which are suspended in a quiescent fluid and in contact with a planar fluid–gas interface. Stick boundary conditions are assumed on the sphere surfaces, and the free surface boundary conditions are accounted for by the method of images. The one-sphere hydrodynamic resistance operator of such a quasi-two-dimensional system is calculated numerically. Using a spherical multipole expansion with symmetry-adjusted basis functions, explicit results are derived for the long-distance terms of the two-sphere mobility tensor up to cubic order in the inverse interparticle distance. The point particle model is also constructed, taking into account the constraint forces necessary to keep the point-particles at a fixed 'radius' a apart from the interface. The accuracy of both far field approximations is discussed by comparing them with the precise many-sphere mobility, evaluated by the multipole expansion.

Contents

A. Introduction 94
B. Theoretical Method 96
C. Single-Sphere Mobilities 99
D. Many-Sphere Mobility Tensor for Large Interparticle Distances 100
E. Point-Particle Mobility 101
F. Discussion 102
    1. Comparison with the Three-Dimensional Unbounded Fluid 102
    2. Accuracy of far-field]Far-Field Approximations 102
G. Acknowledgements 103
H. References 103

Drops and Bubbles in Contact with Solid Surfaces
© Koninklijke Brill NV, Leiden, 2012

A. Introduction

In the past, a substantial amount of research in the soft condensed matter field has been done on the dynamics of model suspensions of spherical colloidal particles. The dynamics of colloidal particles is strongly influenced, in addition to direct interparticle and Brownian forces, by complicated hydrodynamics interactions (HI) mediated by the embedding solvent [1]. While research was mainly focused on the dynamics of three-dimensional bulk dispersions, the center of intense interest has shifted meanwhile to static and dynamic properties of colloids near boundaries like a single hard wall [2–5]. Of considerable interest are, in particular, so-called quasi-two-dimensional (Q2D) dispersions, where the particles are spatially confined to form a planar or curved monolayer. In such systems the particle dynamics is more complicated as in unbounded bulk dispersions, since additional fluid boundary conditions need to be satisfied, and the particle interactions are usually strongly different from the bulk ones. A prominent example of such a Q2D system are macroscopically extended monolayers of colloidal spheres confined between two narrow parallel plates. For this system, diffusion properties like lateral particle mean-squared displacements [6, 7], relative pair mobilities [8–11], and real-space van Hove and wave-number dependent hydrodynamic functions [7, 10] have been studied in great detail both in experiment, theory and simulation. Another class of Q2D systems relevant to emulsion technology, food science and biomedicine are particles trapped at liquid–liquid interfaces such as in water and oil mixtures [12–16], and in liquid–gas interfaces [12, 14] such as in foams [17]. In many of these systems, the colloidal particles or proteins in the case of biological systems [12], become trapped at the interface by the effects of surface tension or image charges. The presence of particles in the interface of water-in-oil emulsions can stabilize the droplets against coalescence even in the absence of emulsifiers, giving rise to so-called Pickering emulsions [18]. The work on particles at liquid interfaces noted above deals in particular with interfacial rheological features and the formation of aggregated structures. Hydrodynamic interactions between colloidal particles close to a clean or surfactant-covered planar fluid–fluid interface have been discussed in [19, 20] and [21], respectively.

A particularly clean-cut and well-studied Q2D model system of particles confined to a fluid–gas interface is given by micron-sized super-paramagnetic colloidal spheres suspended in water next to a planar water-air interface [22–27]. By means of an experimental hanging-drop geometry, the spheres are gravitationally confined to lateral motion along the interface without capillary forces, since the they are completely wetted by the water. The spheres repel each other through long-range dipolar magnetic forces induced by an external magnet field pointing perpendicular to the interface. This model system is of particular interest both from the theoretical and experimental point of view since the direct magnetic interaction are very well characterized. The strength of the dipolar interactions can be precisely tuned by the magnetic field strength, and the two-dimensional particle trajectories can be monitored over an extended range of time using video imaging. This method has allowed to study in great detail phenomena like the hydrodynamic enhancement of self- and collective diffusion [22–26], two-dimensional melting- and freezing transitions *via* an intermediate hexatic phase of quasi-long range orientational order [27], elastic properties and the overdamped phonon band structure of Q2D colloidal crystals [28, 29], and the partial clustering [30] and glass formation [31, 32] in binary magnetic mixtures. To describe quantitatively the lateral particle diffusion properties, as well as the overdamped lattice dynamics (see, e.g., [33–36]), requires to know precisely the lateral hydrodynamic mobility tensors of the spheres in presence of the liquid–gas interface. The mobility tensors are thus an essential input in theoretical and dynamic computer simulation studies of in-plane diffusion and non-equilibrium microstructures. Most likely, also the vitrification dynamics of dense magnetic monolayers is affected by the form of the hydrodynamic mobilities, even though a comparison of experimental data with recent two-dimensional mode coupling theory calculations suggests the hydrodynamic influence to be rather small [37].

In typical monolayers of super-paramagnetic spheres the long-range magnetic dipole repulsions render particle distances shorter than $r < 6a$ very unlikely. Therefore, the long-distance form of the Q2D mobility matrix of two spheres should suffice to describe the hydrodynamics of the translational (and rotationally constrained) motion of spheres along the interface. In this article, we explain how such a far-field expression is constructed and we estimate its precision by comparing it with the precise values of the mobility coefficients, evaluated by the multipole expansion.

The present review is organized as follows. In Section B, we describe how to use the spherical multipole method to determine the low-Reynolds-number HI between colloidal spheres of radius a, confined to lateral motion along a planar fluid–gas interface, and with the sphere surfaces permanently touching the interface. In particular, we show how to construct the many-sphere Q2D mobility tensors using the method of images [38], in combination with an irreducible multipole expansion method [39], adapted to the present system symmetry. The mobility matrix is constructed in terms of a multiple scattering series [40–43]. In Section C, explicit numerical results are presented for the translational and rotational single-sphere mobilities. In Section D, we provide explicit results for the long-distance contributions to the self- and distinct Q2D hydrodynamic mobility tensors of two spheres up to order $1/r^3$ in the interparticle distance r. An alternative far-field expression, corresponding to a particle approximation, is presented in Section E. To keep the point-particles at a fixed distance from the interface, additional constraint forces are introduced. In Section F, the two-sphere far-field Q2D mobility is shown to differ significantly from the corresponding 3D Rotne-Prager expression for an unbounded fluid. Furthermore, both versions of the long-distance asymptotic mobilities are compared with the accurate numerical values, evaluated by the multipole expansion for test configurations of two spherical particles.

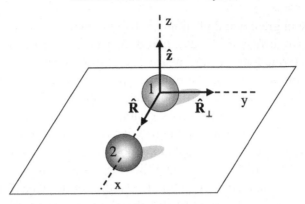

Figure 1. The system of spheres embedded in a semi-infinite fluid ($z > 0$). The spheres are in single-point contact with a planar fluid–gas interface at $z = 0$.

B. Theoretical Method

We consider N identical spherical particles immersed in a low-Reynolds number fluid of shear viscosity η, which occupies the region $z > 0$. The particles touch the planar free interface located at $z = 0$, so that during their motion, the sphere centers remain within the plane $z = a$ (cf. Fig. 1). To satisfy the free surface boundary conditions of zero normal fluid velocity and zero tangential-normal components of the fluid stress tensor, we apply the method of images [38]. Within this method, the fluid is mentally extended to fill the whole space, by reflecting the N spheres and the incident flow field $\mathbf{v}_0$ at $z = 0$ into the mirror region $z < 0$. For creeping flow as described by the Stokes equations [1], the fluid velocity field, $\mathbf{v}(\mathbf{r})$, can be written as [44],

$$\mathbf{v}(\mathbf{r}) - \mathbf{v}_0(\mathbf{r}) = \sum_{j=1}^{N} \int \mathbf{T}_F(\mathbf{r}, \bar{\mathbf{r}}) \cdot \mathbf{f}_j(\bar{\mathbf{r}}) \, d^3\bar{\mathbf{r}}, \tag{1}$$

with the integral kernel, $\mathbf{T}_F(\mathbf{r}, \bar{\mathbf{r}})$, equal to [45]

$$\mathbf{T}_F(\mathbf{r}, \bar{\mathbf{r}}) = \mathbf{T}_0(\mathbf{r} - \bar{\mathbf{r}}) + \mathbf{T}_0(\mathbf{r} - \bar{\mathbf{r}}') \cdot \mathcal{R}_F. \tag{2}$$

Here, $\mathbf{T}_0(\mathbf{r}) = (\mathbf{I} + \hat{\mathbf{r}}\hat{\mathbf{r}})/(8\pi\eta r)$ is the Oseen tensor, and $\mathcal{R}_F = \mathbf{1} - 2\hat{\mathbf{z}}\hat{\mathbf{z}}$ is the reflection operator with respect to $z = 0$. The reflection of a position vector $\bar{\mathbf{r}} = (\bar{x}, \bar{y}, \bar{z})$ is given by $\bar{\mathbf{r}}' = \mathcal{R}_F \bar{\mathbf{r}} = (\bar{x}, \bar{y}, -\bar{z})$. The subscript F is used to indicate a system with a planar free interface. The induced force density, $\mathbf{f}_i(\mathbf{r})$, exerted by the sphere i on the fluid and located on the surface S_i of the sphere i, is determined by the stick boundary conditions $\mathbf{v}(\mathbf{r}) = \mathbf{w}_i(\mathbf{r}) \equiv \mathbf{U}_i + \mathbf{\Omega}_i \times (\mathbf{r} - \mathbf{r}_i)$, for $\mathbf{r} \in S_i$, where $\mathbf{U}_i$ and $\mathbf{\Omega}_i$ are the translational and angular velocities of sphere i.

To determine the force densities $\{\mathbf{f}_i\}$, the l.h.s. of Eq. (1) is evaluated on the sphere surfaces using the stick boundary conditions. This leads to a set of boundary integral equations, written in a compact notation as

$$\mathbf{w}_i - \mathbf{v}_0 = \mathbf{Z}_0^{-1}(i)\mathbf{f}_i + \sum_{j=1}^{N} \mathbf{G}_F(ij)\mathbf{f}_j, \quad \text{for } \mathbf{r} \in S_i, \tag{3}$$

with the single-sphere resistance operator, $\mathbf{Z}_0(i)$, of an unbounded fluid, and the interparticle Green propagator, which contains a part diagonal and a part off-diagonal in the particle indices i and j,

$$\mathbf{G}_F(ij) = \delta_{ij}\mathbf{G}_0(ii')\mathcal{R}_F + (1 - \delta_{ij})\tilde{\mathbf{G}}(ij), \tag{4}$$

where the Green operator for an unbounded fluid with Oseen kernel $\mathbf{T}_0$ is denoted as $\mathbf{G}_0$. The off-diagonal operator,

$$\tilde{\mathbf{G}}(ij) = \mathbf{G}_0(ij) + \mathbf{G}_0(ij')\mathcal{R}_F, \tag{5}$$

accounts for the velocity fields generated by the force densities $\mathbf{f}_j$ of all the other particles $j \neq i$, incident on S_i. The non-zero diagonal part of the Green propagator $\mathbf{G}_F(ii) \neq 0$, absent in the unbounded fluid, now does not vanish owing to the presence of the free surface. Moreover, the off-diagonal part of $\mathbf{G}_F(ij)$, $i \neq j$, is also modified with respect to the Oseen propagator for the unbounded fluid. Both diagonal and off-diagonal extra terms, caused by the free surface, are generated by the image spheres i' and j'.

The Q2D boundary integral equation (3) has the same form as in infinite space, but with the modified Q2D single-sphere operator $[\mathbf{Z}_0^{-1}(i) + \mathbf{G}(ii')]$ rather than $\mathbf{Z}_0^{-1}(i)$, and the modified Q2D propagator $\tilde{\mathbf{G}}(ij)$ rather than $\mathbf{G}_0(ij)$,

$$\mathbf{w}_i - \mathbf{v}_0 = [\mathbf{Z}_0^{-1}(i) + \mathbf{G}_0(ii')]\mathbf{f}_i + \sum_{j \neq i} \tilde{\mathbf{G}}(ij)\mathbf{f}_j, \quad \text{for } \mathbf{r} \in S_i, \tag{6}$$

In the following, we will briefly describe the procedure of solving Eq. (6) for the force density, and evaluating the $6N \times 6N$ friction $\boldsymbol{\zeta}$ and mobility $\boldsymbol{\mu}$ matrices. The focus of this paper is on the Q2D mobility matrix, which relates the hydrodynamic forces and torques acting on the N spheres to the resulting translational and rotational velocities [1]. To determine the leading far-field expression, we aim towards constructing $\boldsymbol{\mu}$ as a series of scattering sequences (reflections), which involve the Q2D single-sphere friction operators and the Q2D propagators between pairs of spheres.

HI between many spheres in a fluid bounded by a planar free surface have been analyzed in Ref. [38]. In that work, the multiple expansion has been carried out with respect to a basis set of multipole functions identical to the set used for an unbounded fluid. However, for a genuine Q2D system where all spheres touch the interface at $z = 0$, this basis set should be used with care. Indeed, the multipole projection of the boundary integral equation (6), gives the degenerate matrix of the multipole elements of the single-particle operator $[\mathbf{Z}_0^{-1}(i) + \mathbf{G}_0(ii')\mathcal{R}_F]$. Inverting it would lead to spurious divergences of most of the elements. To solve the problem, Eq. (6) has been first projected onto those multipole functions, which are symmetric with respect to reflections in the plane $z = 0$. The corresponding projection operator has the form,

$$\mathbf{Pv(r)} = \frac{1}{2}[\mathbf{v(r)} + \mathcal{R}_F \mathbf{v}(\mathcal{R}_F \mathbf{r})]. \tag{7}$$

In this way the boundary conditions at the free surface are automatically satisfied. The multipole matrix elements of $\mathbf{P}(i)$ are determined from displacement theorems [46]. We refer to [41, 42] for a derivation of the explicit formulas for these elements.

The projection operator $\mathbf{P}$ is used to define the Q2D one-sphere resistance operator $\mathbf{Z}_F(i)$, as the inversion in the projected subspace of the symmetric multipole functions,

$$\mathbf{Z}_F(i) = \{\mathbf{P}^T(i)[\mathbf{Z}_0^{-1}(i) + \mathbf{G}_0(ii')\mathcal{R}_F]\mathbf{P}(i)\}^{-1}, \tag{8}$$

where the superscript T denotes the operation of transposition. The multipole elements of the re-defined one-sphere resistance operator, $\mathbf{Z}_F(i)$, are finite even when sphere i is in contact with the interface.

For the Q2D off-diagonal propagator, the following relation has been proved [43],

$$\tilde{\mathbf{G}}(ij) = 2\mathbf{P}^T(i)\mathbf{G}_0(ij)\mathbf{P}(j). \tag{9}$$

The physical implication of the spurious divergences noted above is that, once sphere i touches the interface, lubrication effects forbid its translational motion perpendicular to the interface, i.e., $U_{iz} = 0$, and its rotation along an axis parallel to the interface, i.e., $\Omega_{ix} = \Omega_{iy} = 0$.[1] For more information, we refer to [41]. For zero incident flow, the translational and rotational velocities, $\bar{\mathbf{U}} \equiv (U_{1x}, U_{1y}, \Omega_{1z}, \ldots, U_{Nx}, U_{Ny}, \Omega_{Nz})$ depend linearly on the horizontal components of the external forces and the vertical components of the external torques, $\bar{\mathcal{F}} \equiv (\mathcal{F}_{1x}, \mathcal{F}_{1y}, \mathcal{T}_{1z}, \ldots, \mathcal{F}_{Nx}, \mathcal{F}_{Ny}, \mathcal{T}_{Nz})$,

$$\bar{\mathcal{F}} = \bar{\xi} \cdot \bar{\mathbf{U}}, \tag{10}$$

with the N-particle friction tensor,

$$\bar{\xi} = \mathcal{P}_F[\mathbf{Z}_F^{-1} + 2\mathbf{G}_0]^{-1}\mathcal{P}_F \equiv \mathcal{P}_F[\mathbf{Z}_F - \mathbf{Z}_F 2\mathbf{G}_0\mathbf{Z}_F + \cdots]\mathcal{P}_F, \tag{11}$$

where the operator $\mathcal{P}_F$ selects those spherical multipoles, which correspond to the components of forces and torques (and to translational and angular particle velocities), symmetric under reflection at the interface, and the inverse, limited to the

---

[1] In reality, the surfaces of spheres are not perfectly smooth, the free surface may be curved rather than flat in vicinity of a touching particle and, moreover, a liquid–gas interface always shows thermally induced undulations. Therefore, it would be of interest to check experimentally to what extent the constrained-motion predictions of lubrication theory are met in real quasi-two-dimensional systems and how accurately the model analyzed in this paper approximates the real geometry of the system.

subspace of the symmetric multipole functions, stands for the multiple scattering series.

Our goal is to evaluate the quasi-two-dimensional mobility $\bar{\mu}$, which is a $3N \times 3N$ matrix depending on the configuration of the sphere centers in the plane $z = a$,

$$\bar{\mu} = \bar{\xi}^{-1}. \tag{12}$$

Finally, after some additional algebraic steps (cf. [41] for details), the Q2D mobility is obtained in form of a multiple scattering series,

$$\bar{\mu} = \mu_F + 2\mu_F \mathbf{Z}_F \frac{1}{1 + 2\mathbf{G}_0 \hat{\mathbf{Z}}_F} \mathbf{G}_0 \mathbf{Z}_F \mu_F, \tag{13}$$

where the quasi-two-dimensional operator $\hat{\mathbf{Z}}_F$ is defined as

$$\hat{\mathbf{Z}}_F = \mathbf{Z}_F - \mathbf{Z}_F \mu_F \mathbf{Z}_F. \tag{14}$$

Here, μ_F is the Q2D mobility operator of a single sphere touching the free surface, i.e.,

$$\mu_F = [\mathcal{P}_F \mathbf{Z}_F \mathcal{P}_F]^{-1}. \tag{15}$$

C. Single-Sphere Mobilities

From evaluating the multipole matrix elements of $\mathbf{Z}_F$ and transforming them to Cartesian coordinates, we obtain the Q2D single-sphere mobility tensor μ_F for a sphere touching the planar free surface. By definition, its translation-translational, μ_F^{tt}, and rotation-rotational, μ_F^{rr}, components relate the sphere translational, $\mathbf{U}$, and angular, $\mathbf{\Omega}$, velocities to the external force, $\mathcal{F}$, and torque, $\mathcal{T}$,

$$\mathbf{U} = \mu_F^{tt} \cdot \mathcal{F}, \tag{16}$$
$$\mathbf{\Omega} = \mu_F^{rr} \cdot \mathcal{T}. \tag{17}$$

The single sphere mobility has the form,

$$\mu_F^{tt} = \mu_F^{tt} [\hat{\mathbf{R}}\hat{\mathbf{R}} + \hat{\mathbf{R}}_\perp \hat{\mathbf{R}}_\perp], \tag{18}$$
$$\mu_F^{tt} = \mu_F^{rr} \hat{\mathbf{z}}\hat{\mathbf{z}}. \tag{19}$$

Here μ_F^{tt} and μ_F^{rr} can be interpreted as 2×2, and 1×1 tensors, the components of the reduced mobility, which describes only the relevent degrees of freedom, or, alternatively, as the standard 3×3 tensors with a number of zeros, which eliminate the unphysical motions. The scalar coefficients are,

$$\mu_F^{tt}/\mu_0^{tt} = 1.3799554, \tag{20}$$
$$\mu_F^{rr}/\mu_0^{rr} = 1.10920983. \tag{21}$$

Here, $\mu_0^{tt} = [6\pi\eta a]^{-1}$ and $\mu_0^{rr} = [8\pi\eta a^3]^{-1}$ are the translation-translational and rotation-rotational mobilities, respectively, of a single sphere in an unbounded fluid.

In Ref. [41], we have obtained these numerical values from a truncated multipole expansion of order $L = 80$. Our numerical values agree with earlier calculations of the reduced inverse drag and turn coefficients of two touching spheres [47], which are given by 1.3801 and $4/(3\zeta(3)) \approx 1.1092098301$, respectively. The single-sphere mobilities in presence of the interface are larger than in the bulk since the particle experiences less solvent friction, in particular regarding its rotational motion. Note that the method developed in this work allows to evaluate with high numerical precision additional hydrodynamic coefficients which characterize the response of a single sphere to an incident fluid flow.

D. Many-Sphere Mobility Tensor for Large Interparticle Distances

The Q2D mobility tensor for two spheres can be expressed in dyadic notation by employing the unit vectors $\hat{\mathbf{z}}$, $\hat{\mathbf{R}} = (\mathbf{r}_2 - \mathbf{r}_1)/r$ and $\hat{\mathbf{R}}_\perp = \hat{\mathbf{z}} \times \hat{\mathbf{R}}$ depicted in Fig. 1. The basic mobility relation (10) acquires here the form,

$$\mathbf{U}_i = \sum_{j=1}^{2} \boldsymbol{\mu}_{ij}^{tt} \cdot \boldsymbol{\mathcal{F}}_j + \boldsymbol{\mu}_{ij}^{tr} \cdot \boldsymbol{\mathcal{T}}_j, \tag{22}$$

$$\boldsymbol{\Omega}_i = \sum_{j=1}^{2} \boldsymbol{\mu}_{ij}^{rt} \cdot \boldsymbol{\mathcal{F}}_j + \boldsymbol{\mu}_{ij}^{rr} \cdot \boldsymbol{\mathcal{T}}_j. \tag{23}$$

The corresponding translation-translational, translation-rotational, rotation-translational and rotation-rotational mobilities can be interpreted likewise as 2×2, 2×1, 1×2 and 1×1 tensors, the components of the reduced matrix $\bar{\boldsymbol{\mu}}$, which describes only the relevent degrees of freedom, or, alternatively, as the standard 3×3 tensors $\boldsymbol{\mu}$ with a number of zeros as elements, which eliminate the unphysical motions. These tensors are specified by 4 scalar functions of the dimensionless interparticle distance $R = r/(2a)$, with $r = |\mathbf{r}_2 - \mathbf{r}_1|$.

The long-distance approximation of the Q2D mobility to $\mathcal{O}(1/R^3)$, is constructed from Eq. (13) by considering only terms in the scattering expansions with no more than a single propagator $\mathbf{G}_0$ (for details, see [41]). This yields the translation-translational long-distance part of the Q2D mobility matrix as

$$\boldsymbol{\mu}_{11}^{tt}/\mu_0^{tt} = 1.3799554[\hat{\mathbf{R}}\hat{\mathbf{R}} + \hat{\mathbf{R}}_\perp\hat{\mathbf{R}}_\perp], \tag{24}$$

$$\boldsymbol{\mu}_{12}^{tt}/\mu_0^{tt} = \frac{3}{2R}\left[\hat{\mathbf{R}}\hat{\mathbf{R}} + \frac{1}{2}\hat{\mathbf{R}}_\perp\hat{\mathbf{R}}_\perp\right] - \frac{1}{R^3}[1.159862\hat{\mathbf{R}}\hat{\mathbf{R}} + 0.111686\hat{\mathbf{R}}_\perp\hat{\mathbf{R}}_\perp]. \tag{25}$$

Likewise, the long-distance asymptotic translation-rotational, rotation-translational and rotation-rotational mobilities are obtained to $\mathcal{O}(1/R^3)$ as

$$\boldsymbol{\mu}_{11}^{tr}/\mu_0^{rt} = 0, \qquad\qquad \boldsymbol{\mu}_{12}^{tr}/\mu_0^{rt} = -\hat{\mathbf{R}}_\perp\hat{\mathbf{z}}/(4R^2), \tag{26}$$

$$\boldsymbol{\mu}_{11}^{rt}/\mu_0^{rt} = 0, \qquad\qquad \boldsymbol{\mu}_{12}^{rt}/\mu_0^{rt} = \hat{\mathbf{z}}\hat{\mathbf{R}}_\perp/(4R^2), \tag{27}$$

$$\boldsymbol{\mu}_{11}^{rr}/\mu_0^{rr} = 1.10920983\hat{\mathbf{z}}\hat{\mathbf{z}}, \qquad \boldsymbol{\mu}_{12}^{rr}/\mu_0^{rr} = -\hat{\mathbf{z}}\hat{\mathbf{z}}/(8R^3), \tag{28}$$

where $\mu_0^{rt} = (4\pi\eta a^2)^{-1}$.

Finally, let us now discuss the far-field approximation of the many-sphere mobility. In general, the mobility matrix $\bar{\mu}(1 \cdots N)$ for N spheres is not pairwise-additive, i.e., $\bar{\mu}_{11}(1 \cdots N) \neq \bar{\mu}_{11}(1)$ and $\bar{\mu}_{12}(1 \cdots N) \neq \bar{\mu}_{12}(12)$. However, the far-field approximation of the N-sphere mobility contains only a single propagator G_0, and therefore it is the superposition of the two-sphere mobilities specified in Eqs (24)–(28),

$$\bar{\mu}_{11}(1 \cdots N) = \bar{\mu}_{11}(1) \equiv \mu_F, \tag{29}$$

$$\bar{\mu}_{12}(1 \cdots N) = \bar{\mu}_{12}(12). \tag{30}$$

E. Point-Particle Mobility

In discussing the far-field approximation of the Q2D mobility for a system of many spheres, it is interesting to investigate also the point-particle model, where the spheres touching a flat interface are approximated by points located at their centers. The image points are now constructed by reflection with respect to the free surface, and, according to the method of images, [1], the original system is thus replaced by a doubled number of points and their images placed in an unbounded fluid. The subtlety of the Q2D point-particle model is that, in addition to the horizontal radial and transversal external forces $\mathcal{F}_{ix} = \mathcal{F}_{i'x}$ and $\mathcal{F}_{iy} = \mathcal{F}_{i'y}$, additional constraint forces perpendicular to the interface $\mathcal{F}_{iz}$ are required to keep the point particles i at the fixed distance $z = a$ from the interface (and $\mathcal{F}_{i'z} = -\mathcal{F}_{iz}$ to keep the images i' at $z = -a$). Using the condition $U_{iz} = U_{i'z} = 0$, the constraint forces $\mathcal{F}_{iz}$ can be eliminated, leading to the Q2D N-point mobilities [42]. Unlike in case of spheres, for points the mobility matrix contains only the translation-translational tensors. Indeed, for point particles of zero extension, it does not make sense to specify angular velocities and torques. Therefore, for two points the mobility relation has the form,

$$\mathbf{U}_i = \sum_{j=1}^{2} \mathcal{M}_{ij}^{tt} \cdot \mathcal{F}_j, \quad i = 1, 2, \tag{31}$$

and the corresponding mobility tensors in the dyadic notation are expressed as

$$\mathcal{M}_{1i}^{tt} = \mathcal{M}_{1i}^{R} \hat{\mathbf{R}}\hat{\mathbf{R}} + \mathcal{M}_{1i}^{T} \hat{\mathbf{R}}_{\perp}\hat{\mathbf{R}}_{\perp}, \quad i = 1, 2, \tag{32}$$

with the radial and transversal Q2D mobility coefficients for two points given by,

$$\mathcal{M}_{11}^{R} = \left[\frac{11}{8} - \frac{3RA}{8(R^2 + 1)^{3/2}}\right]\mu_0^{tt}, \quad \mathcal{M}_{11}^{T} = \frac{11}{8}\mu_0^{tt}, \tag{33}$$

$$\mathcal{M}_{12}^{R} = \frac{3}{8}\left(\frac{2}{R} + \frac{2R^2 + 1 + RAB}{(R^2 + 1)^{3/2}}\right)\mu_0^{tt},$$

$$\mathcal{M}_{12}^{T} = \frac{3}{8}\left(\frac{1}{R} + \frac{1}{\sqrt{R^2 + 1}}\right)\mu_0^{tt}, \tag{34}$$

with the following functions A and B of the interparticle distance R,

$$A = \frac{3R}{2(R^2 + 1)^{3/2}(1 - B^2)}, \qquad B = -\frac{3}{2}\left(\frac{1}{R} - \frac{R^2 + 2}{(R^2 + 1)^{3/2}}\right). \qquad (35)$$

As the consequence of the constraints, the two-point Q2D mobilities, restricted to the Q2D degrees of freedom, are different from the Oseen tensors, and the N-point Q2D mobilities are not pairwise additive, unlike in the 3D unbounded fluid with 6N degrees of freedom. These features, paradoxically, make the concept and evaluation of the Q2D many-point mobility significantly more complicated than the concept and evaluation of the far-field Q2D many-sphere mobility, s since the latter is constructed by superposition of the pairwise mobilities (24)–(28).

F. Discussion

1. Comparison with the Three-Dimensional Unbounded Fluid

The Q2D mobility, given in Eq. (25), will be now compared with the corresponding 3D Rotne–Prager tensor [48],

$$\mu_{12}^{tt}/\mu_0^{tt} = \frac{3}{4R}\left[\hat{\mathbf{R}}\hat{\mathbf{R}} + \frac{1}{2}(\mathbf{1} - \hat{\mathbf{R}}\hat{\mathbf{R}})\right] - \frac{1}{8R^3}\left[\hat{\mathbf{R}}\hat{\mathbf{R}} - \frac{1}{2}(\mathbf{1} - \hat{\mathbf{R}}\hat{\mathbf{R}})\right], \qquad (36)$$

which is the asymptotic form of the three-dimensional trans lational-translational mobility of two spheres moving without constraints in an unbounded fluid. The leading order terms in the inverse distance expansion of the rotational-rotational ($\sim 1/R^3$), rotational-translational ($\sim 1/R^2$) and translational-translational ($\sim 1/R$) parts of the Q2D mobility μ_{12} are twice as large as their three-dimensional counterparts (cf. [1] and Eq. (25)). This observation is easily understood qualitatively within the method of images. To the lowest order in the $1/R$ expansion, the position of sphere 2 as "seen" from the distant sphere 1, coincides with the position of its image $2'$. Therefore the presence of the image effectively doubles the forces and torques acting on sphere 2, and also the resulting translational and angular velocities of sphere 1. The next order correction ($\sim 1/R^3$) to the translational-translational quasi-two-dimensional mobility μ_{12}^{tt} in Eq. (25) cannot be interpreted in a simple way.

2. Accuracy of Far-Field Approximations

The accuracy of the far-field approximations described above was extensively discussed in Ref. [42], where the two-particle translational-translational mobilities μ_{11}^{tt} and μ_{12}^{tt} were evaluated very precisely by the multipole method with a very high order of truncation $L = 20$, see Ref. [42] for the definition of L, resulting in two radial coefficients: $\mathbf{R} \cdot \bar{\mu}_{1i}^{tt} \cdot \mathbf{R}$, self ($i = 1$) and distinct ($i = 2$), and two transversal ones, $\mathbf{R}_\perp \cdot \bar{\mu}_{1i}^{tt} \cdot \mathbf{R}_\perp$, also self ($i = 1$) and distinct ($i = 2$). These values were used as a reference for testing accuracy of the far field approximation for spheres, see

Eq. (25), and for point-particles, see Eqs (32)–(35). Typically, for all the approximations, in the range $2.5 \leqslant R \leqslant 5$, the worst relative accuracy was observed for the radial distinct mobility. For example, at a typical [24–27] interparticle distance $R = 3$, the relative error of the $\mathcal{O}(1/R)$ approximation of the radial distinct mobility coefficient for spheres, equivalent to the point-particles with no constraints, was as large as 8%. A significantly smaller 2.5% relative error of the mobility was achieved for point-particles with the constraints, and the best 1% accuracy was reached for the $\mathcal{O}(1/R^3)$ pair-wise approximation for spheres. The last expression is therefore the far-field approximation of choice, because of the high precision and because of the pairwise additivity.

G. Acknowledgements

Financial support of the COST Action P21 Physics of droplets and of the Polish Ministry of Science and Higher Education grant No. COST/116/2007 are kindly acknowledged by M. L. E.-J. and E. W.

H. References

1. Kim, S. and Karrila, S. J., Microhydrodynamics: Principles and Selected Applications. Butterworth-Heinemann, London, 1991.
2. Aderogba, K. and Blake, J. R., Bull. Aust. Math. Soc., 18 (1978) 345.
3. Jones, R. B. and Cichocki, B., Physica A, 258 (1998) 273.
4. Jones, R. B. and Kutteh, R., J. Chem. Phys., 112 (2000) 11080.
5. Carbajal-Tinoco, M. D., Lopez-Fernandez, R. and Arauz-Lara, J. L., Phys. Rev. Lett., 99 (2007) 138303.
6. Lobry, L. and Ostrowsky, N., Phys. Rev. B, 53 (1995) 12050.
7. Pesché, R., Kollmann, M. and Nägele, G., Phys. Rev. E, 62 (2000) 5432.
8. Cui, B., Diamant, H., Lin, B. and Rice, S. A., Phys. Rev. Lett., 92 (2004) 258301.
9. Diamant, H., Cui, B. and Rice, S. A., J. Phys.: Condens. Matter, 17 (2005) S2787.
10. Santana-Solano, J., Ramirez-Saito, A. and Arauz-Lara, J. L., Phys. Rev. Lett., 95 (2005) 198301.
11. Valley, D. T., Rice, S. A., Cui, B., Ho, H. M., Diamant, H. and Lin, B., J. Chem. Phys., 126 (2007) 134908.
12. Cicuta, P., Stancik, E. J. and Fuller, G. G., Phys. Rev. Lett., 90 (2003) 236101.
13. Reynaert, S., Moldenaers, P. and Vermant, J., Langmuir, 22 (2006) 4936.
14. Reynaert, S., Moldenaers, P. and Vermant, J., Phys. Chem. Chem. Phys., 9 (2007) 6463.
15. Leunissen, M. E., van Blaaderen, A., Hollingsworth, A. D., Sullivan, M. T. and Chaikin, P. M., Proc. Natl. Acad. Sci. U.S.A., 104 (2007) 2585.
16. Leunissen, M. E., Zwanikken, J., van Roij, R., Chaikin, P. M. and van Blaaderen, A., Phys. Chem. Chem. Phys., 9 (2007) 6405.
17. Sethumadhavan, G. N., Nikolov, A. D. and Wasan, D. T., J. Colloid Interface Sci., 240 (2001) 105.
18. Pickering, S. U., J. Chem. Soc., 91 (2001) 1907.
19. Jones, R. B., Felderhof, B. U. and Deutch, J. M., Macromolecules, 8 (1975) 680.
20. Lee, S. H., Chadwick, R. S. and Leal, G., J. Fluid Mech, 93 (1979) 705.
21. Bławzdziewicz, J., Cristini, V. and Loewenberg, M., Phys. Fluids, 11 (1999) 251.
22. Pesché, R., Kollmann, M. and Nägele, G., J. Chem. Phys., 114 (2001) 114.

23. Nägele, G., Kollmann, M., Pesché, R. and Banchio, A., Mol. Phys, 100 (2002) 2921.
24. Zahn, K., Mendez-Alcaraz, J. M. and Maret, G., Phys. Rev. Lett., 1997 (1997) 175.
25. Rinn, B., Zahn, K., Maass, P. and Maret, G., Europhys. Lett., 46 (1999) 537.
26. Kollmann, M., Hund, R., Rinn, B., Nägele, G., Zahn, K., König, H., Maret, G., Klein, R. and Dhont, J. K. G., Europhys. Lett., 58 (2002) 919.
27. Zahn, K. and Maret, G., Phys. Rev. Lett., 85 (2000) 3656.
28. Keim, P., Maret, G., Herz, U. and von Grünberg, H. H., Phys. Rev. Lett., 92 (2004) 215504.
29. von Grünberg, H. H., Keim, P., Zahn, K. and Maret, G., Phys. Rev. Lett., 93 (2004) 255703.
30. Hoffmann, N., Ebert, F., Likos, C. N., Löwen, H. and Maret, G., Phys. Rev. Lett., 97 (2006) 078301.
31. König, H., Zahn, K. and Maret, G., Glass transition in a two-dimensional system of magnetic colloids, in: Tokuyama, M. and Oppenheim, I., Eds. AIP Conference Proceedings: Slow Dynamics in Complex Systems. Butterworth-Heinemann, London, 2004.
32. König, H., Hund, R., Zahn, K. and Maret, G., Eur. Phys. J. E, 18 (2005) 287.
33. Hurd, A. J., Clark, N. A., Mockler, R. C. and O'Sullivan, W. J., Phys. Rev. A, 26 (1982) 2869.
34. Felderhof, B. U. and Jones, R. B., Faraday Discuss. Chem. Soc., 83 (1987) 69.
35. Tata, B. V. R., Mohanty, P. S., Valsakumar, M. C. and Yamanaka, J., Phys. Rev. Lett., 93 (2004) 268303.
36. von Grünberg, H. H. and Baumgartl, J., Phys. Rev. E, 75 (2007) 051406.
37. Bayer, M., Brader, J. M., Ebert, F., Fuchs, M., Lange, E., Maret, G., Schilling, R., Sperl, M. and Wittmer, J. P., Phys. Rev. E, 76 (2007) 011508.
38. Cichocki, B., Jones, R. B., Kutteh, R. and Wajnryb, E., J. Chem. Phys., 112 (2000) 2548.
39. Schmitz, R. and Felderhof, B. U., Physica A, 113 (1982) 90.
40. Felderhof, B. U., Physica A, 151 (1988) 1.
41. Cichocki, B., Ekiel-Jezewska, M. L., Nägele, G. and Wajnryb, E., J. Chem. Phys., 121 (2004) 2305.
42. Cichocki, B., Ekiel-Jezewska, M. L., Nägele, G. and Wajnryb, E., Europhys. Lett., 67 (2004) 383.
43. Cichocki, B., Ekiel-Jezewska, M. L. and Wajnryb, E., J. Chem. Phys., 126 (2007) 184704.
44. Pozrikidis, C., Boundary Integral and Singularity Methods for Linearized Viscous Flow. Cambrigde Uiversity Press, Cambrigde, 1992.
45. Perkins, G. and Jones, R. B., Physica A, 171 (1991) 575.
46. Felderhof, B. U. and Jones, R. B., J. Math. Phys., 30 (1989) 339.
47. Jeffrey, D. J. and Onishi, Y., J. Fluid Mech., 139 (1984) 261.
48. Rotne, J. and Prager, S., J. Chem. Phys., 50 (1969) 4831.

Microdrops Evaporating from Deformable or Soluble Polymer Surfaces

Elmar Bonaccurso [a,c], **Ramon Pericet-Camara** [b,c], **Hans-Jürgen Butt** [c]

[a] Center of Smart Interfaces, Technical University of Darmstadt, Petersenstr. 32, 64287 Darmstadt, Germany. E-mail: bonaccurso@csi.tu-darmstadt.de
[b] Laboratory of Intelligent Systems, Ecole Polytechnique Fédérale de Lausanne, 1015 Lausanne, Switzerland. E-mail: ramon.pericet@epfl.ch
[c] Max-Planck-Institute for Polymer Research, Ackermannweg 10, 55128 Mainz, Germany. E-mail: butt@mpip-mainz.mpg.de

Keywords
Microdrop evaporation, surface tension, capillary pressure, Young's equation, polymer surface, elastic solid, inkjet etching, solvent, swelling, diffusion, precipitation, three phase contact line, pinning, fluid flow, skin formation

List of Abbreviations and Symbols

std. dev.: Standard deviation

dpi: Dots per inch

NPT: Normal pressure and temperature

RH: Relative humidity

AFM: Atomic force microscope

QCM: Quartz crystal microbalance

TPCL: Three phase contact line

CCR: Constant contact radius

CCA: Constant contact angle

SEM: Scanning electron microscope

PS: Polystyrene

PEMA: Polyethylmethacrylate

PVP: Polyvinylphenol

AP: Acetophenone

EA: Ethyl acetate

BSA: Bovine serum albumine

PDMS: Polydimethysiloxane

$\Delta\sigma$: Surface stress [N m^{-1}]

Θ: Drop contact angle [°]

γ: Liquid surface tension [N m^{-1}]

τ: Evaporation time [s]

η: Viscosity [mPa s]

ρ: Density [g cm^3]

α: Thermal expansion coefficient [K^{-1}]

v: Poisson's ratio

a: Drop contact radius [m]

m: Drop mass [kg]

g: Gravitational acceleration [$g = 9.8$ kg m^{-1} s^{-2}]

f: Frequency [Hz]

k: Cantilever spring constant [N/m]

d: Cantilever thickness [m]

w: Cantilever width [m]

l_0: Cantilever length [m]

V: Drop volume [m^3]

K: Capillary length [m]

ΔP: Laplace-, or capillary pressure [Pa]

P_0: Vapour pressure [Pa]

D: Diffusion coefifcient [cm^2 s^{-1}]

T: Temperature [K]

E: Young's or elasticity modulus [Pa]

R: Drop radius of curvature [m]

M_{w}: Molecular weight [Da]

Contents

List of Abbreviations and Symbols 105
A. Introduction 107
B. Background, Materials and Methods 108
    1. Drop in Equilibrium 108
    2. Evaporating Drop 109
    3. Deformable Surfaces 110
    4. Evaporation of Solvent Droplets on Polymer Surfaces 111
C. Applications of Evaporating Solvent Drops on Deformable/Soluble Substrates 114
    1. Microlenses on Polymer Surfaces by ink jetting]Ink Jetting Solvent Drops 114
    2. Microstructures on Polymer Surfaces by mixtures of solvents]Mixtures of Solvents 115
    3. Microlithography and Other Applications 116
D. Decoupling the Processes 119
    1. Mass Transport Inside the Drop: the Coffee Stain Effect 120
    2. Dissolution and Swelling of the Substrate 121
    3. Etching and Collapse of the Substrate 122
    4. Elastic Deformation of the Substrate 122
E. Brief Summary 124
F. Acknowledgements 125
G. References 125

A. Introduction

The evaporation of drops has developed lately to an extremely active and prolific area of research, since it is relevant for a vast number of technological applications. Processes such as wetting, dewetting or drying occur in a many industrial processes, like coatings, crop spraying, printing, or biotechnological applications. Evaporation of microdrops is also of interest in the study and development of heat transfer devices for cooling microelectronic devices, or in combustion technologies. The use of droplets as microreactors in nanochemistry or as liquid media for DNA optical mapping as well demands a detailed knowledge of the evaporating mechanism of sessile drops. In fundamental research, the study of evaporation drops is used in order to understand the properties of interfaces between solids, liquids and gases.

The first theory of the evaporation of a free spherical liquid drop in an infinitely extending surrounding gas was proposed by Maxwell more than a century ago [1]. Assuming that evaporation is only limited by diffusion of the vapor molecules away from the drop, the rate of mass change could be well described [2, 3]. The same theoretical approach also applies to sessile drops. Since then, the evaporation of large drops from solid surfaces is by now widely studied and understood. Picknett and Bexon [4] proposed evaporation models for sessile drops observing that there are two principal modes of evaporation: one where the contact area of the drop remains constant while the contact angle decreases, and the other one where the contact area diminishes with a constant contact angle. They also observed a mixed mode close to the end of evaporation, where the mode changes from one to the other in the course of evaporation. Shanahan and Bourges [5] suggested that a sessile drop in open-air conditions evaporates in three stages: In the first one, the contact radius of the drop remains constant, while the height and the contact angle decreases. In the second one, provided that the surface is sufficiently smooth, the contact radius diminishes while the contact angle is approximately constant. The final stage shows the drop disappearing in an irregular fashion, which is difficult to track experimentally.

Several models have been proposed to the evaporation kinetics of sessile drops. Lebedev [6] and later Picknett and Bexon—like Maxwell, Morse, and Langmuir did for spherical drops—assumed that the evaporation of a sessile drop is diffusion driven only. They independently established that for a drop with the shape of a spherical cap, the vapour concentration field around the top half of an equiconvex lens is equivalent to the electrostatic potential field. Hu and Larson compared the two previous models and implemented them in a FEM simulation [7] for the case of constant contact radius evaporation. The agreement between the analytic models, the simulations, and experimental data was very good. As well, Meric and Erbil proposed a model considering a sessile droplet as having 'pseudo-spherical cap' geometry, introducing a flatness parameter of the droplet, with good agreement with experimental data on large drops [8].

In the last few years more and more groups started investigating also wetting and evaporation of drops on deformable [9–21] or on soluble substrates [22–39].

In this work we intend to report the state-of-the-art research on evaporating droplets from soluble polymer surfaces along with the physico-chemical processes involved. In this respect, we want to address but a few peculiar properties of soluble polymers [40], like that: (i) in the presence of a solvent they become soft to a degree depending on the relative vapour pressure of the solvent in the surrounding atmosphere; (ii) solvent diffuses into the polymer thereby swelling and softening it; (iii) solvent and polymer can mix at any ratio since there is not a clearly cut solubility for polymers.

B. Background, Materials and Methods

1. Drop in Equilibrium

For the sake of simplicity, let us first consider a drop deposited onto a non-deformable and non-soluble substrate (Fig. 1). In equilibrium, i.e., when the drop is not evaporating, Young's equation must hold. It establishes the relation among the three surface tensions acting at the rim of the drop (three phase contact line, TPCL).

$$\gamma_S - \gamma_{SL} = \gamma_L \cos \Theta, \tag{1}$$

γ_L is the surface tension at the interface liquid/gas, γ_S is the surface tension at the interface solid/gas, and γ_{SL} is the surface tension at the interface solid/liquid. Θ is called contact angle, or wetting angle, of the liquid on the solid. Equation (1) is strictly valid only if the drop is not evaporating and if gravity can be neglected.

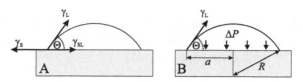

Figure 1. (A) Sessile drop in equilibrium on a solid surface, with contact angle Θ and surface tensions γ_L, γ_S, and γ_{SL}. (B) Action of liquid surface tension γ_L and Laplace pressure ΔP.

We can neglect gravity, and thus the effect of the hydrostatic pressure which would flatten the drop, when the drop is smaller than the capillary length

$$K = \sqrt{\frac{\gamma_L}{\rho g}}, \qquad (2)$$

where ρ is the density of the liquid and g the gravitational acceleration. For water, $\gamma_L = 0.072$ N/m, $\rho = 1$ g/cm^3 and $g = 9.8$ m/s^2. The shape of the drop is thus not influenced by gravity if the radius of curvature is well below $\sim$2 mm.

In addition to the surface tensions acting at the TPCL, another force plays a major role. Due to its curvature, the pressure inside the drop is higher than outside. The difference between in and out is called Laplace or capillary pressure [41]

$$\Delta P = \frac{2\gamma_L}{R} = \frac{2\gamma_L \sin \Theta}{a}, \qquad (3)$$

where R is the radius of curvature of the drop, which is related to the contact radius a and to the contact angle Θ (Fig. 1).

As an example, for a drop of water forming a contact angle of 60° with a surface, the pressure difference is $\Delta P = 1.2$ mbar when $a = 1$ mm, and $\Delta P = 1200$ mbar when $a = 1$ μm.

Thus, when a small, non evaporating microdrop is sitting on a surface and forms a finite contact angle with it, the macroscopic picture is:

 (i) the drop has a spherical shape,

 (ii) Young's equation accounts for the in plane (horizontal) balance of forces at the TPCL,

 (iii) the vertical component of the surface tension $\gamma_L \sin \Theta$ is pulling upwards at the TPCL and is counterbalanced by the Laplace Pressure ΔP, which is pushing uniformly downwards over the whole contact area πa^2.

2. Evaporating Drop

Until now we discussed drops in thermodynamic equilibrium. What changes when a drop evaporates, and why does a drop evaporate at all? A liquid with a planar surface evaporates only when its vapour pressure P_0 is higher than the pressure of its vapour in its surroundings. Thus, if the surroundings are saturated with its vapour the liquid does not evaporate. It is in equilibrium, because at any time the number of molecules evaporating from and condensing to the surface is similar. However, drops have a slightly higher vapour pressure in comparison to a planar surface due to their curvature. For this reason, they evaporate also in a saturated atmosphere. This is quantified by the Kelvin Equation [41]

$$P_V = P_0 e^{\lambda/R}, \tag{4}$$

where the vapour pressure of the liquid in the drop is P_V, and the parameter λ is a function of the temperature and the nature of the liquid. Thus the vapour pressure increases with decreasing drop size. As an example, a planar water surface has a vapour pressure $P_0 = 31.69$ mbar at NPT. If the surface is curved and the radius of curvature is $R = 1$ µm, the vapour pressure is $P_V = 31.72$ mbar, and if $R = 100$ nm, $P_V = 32.02$ mbar. The difference between the planar and the curved surfaces is small, but it is high enough for the drop to evaporate.

3. Deformable Surfaces

The profile of a deformable elastic surface close to the TPCL was first calculated by Lester [42]. He shows that the Neumann's force triangle is valid to account for the surface force distribution, and Young's equation is assumed valid only for small deformations. Later Rusanov [43] developed expressions for the complete profile of an elastic surface by the action of a sessile drop. He considered a stress vector **P** as a combination of the surface tension at the TPCL and the capillary pressure at the liquid/solid interface. With this, he calculated the vector $\mathbf{z}(x, y)$ of the vertical displacement at all points of the surface by using the theory of elasticity [44]. In a one-dimensional representation (see Fig. 2), Rusanov calculated the surface profile $z(r)$ for three surface sections: (i) the deformation underneath the drop ($r \leqslant a$); (ii) the deformation at the TPCL. The TPCL is assumed to be a thin line, but of finite thickness t ($a \leqslant r \leqslant a + t$); and (iii) the deformation of the surface not covered by the drop ($r \geqslant a + t$):

$$z(r) = \frac{4(1-v^2)}{\pi E} \left\{ \Delta P a E\left(\frac{r}{a}\right) + \frac{\gamma_L \sin \Theta}{t} \left[a E\left(\frac{r}{a}\right) - (a+t) E\left(\frac{r}{a+t}\right) \right] \right.$$
$$\left. + \frac{\pi(1-2v)}{4(1-v)} \gamma_L \Delta \cos \Theta \right\},$$
$$r \leqslant a, \tag{5a}$$

$$z(r) = \frac{4(1-v^2)}{\pi E} \left\{ \Delta P r G\left(\frac{a}{r}\right) + \frac{\gamma_L \sin \Theta}{t} \left[r G\left(\frac{a}{r}\right) - (a+t) E\left(\frac{r}{a+t}\right) \right] \right.$$
$$\left. + \frac{\pi(1-2v)}{4t(1-v)} \gamma_L \Delta \cos \Theta (a+t-r) \right\},$$
$$a \leqslant r \leqslant a+t, \tag{5b}$$

$$z(r) = \frac{4(1-v^2)}{\pi E} \left\{ \Delta P r G\left(\frac{a}{r}\right) + \frac{\gamma_L \sin \Theta}{t} r \left[G\left(\frac{a}{r}\right) - G\left(\frac{a+t}{r}\right) \right] \right\},$$
$$r \geqslant a+t, \tag{5c}$$

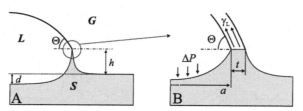

Figure 2. (A) Scheme of the deformation of an elastic surface by a sessile droplet. L, S and G stand for liquid, solid and gas. The dotted line displays the solid surface without deformation. Θ is the equilibrium contact angle of the droplet, h is the maximum height of the ridge and d is the maximum depth of the depression under the drop. (B) Magnified scheme of the contact point at the TPCL. The surface tension acts over a surface layer of thickness t.

where $G(k) \equiv E(k) - (1 - k^2)K(k)$, and $E(k)$ and $K(k)$ are the total normal elliptical Legendre integrals of the first and second kind respectively, $\Delta \cos \Theta$ is the difference between the macroscopic contact angle and the equilibrium contact angle given by the Young's equation, t is the thickness of the liquid–gas interfacial layer, v is the Poisson's ratio of the material, and the other parameters have been defined before. Later, White [45] extended this model introducing the influence of the surface stress transmission of a free liquid interface at the microscopic TPCL on the elastic surface deformation, i.e., he considered the action of the disjoining pressure. He concluded that the microscopic contact angle is zero and that the macroscopic one obeys Young's equation after introducing a correction factor that depends on the microscopic deformation height at the TPCL and the deformation as a result of the disjoining pressure.

4. Evaporation of Solvent Droplets on Polymer Surfaces

In recent times, several technological applications based on the deposition of solvent microdrops on soluble polymer substrates by inkjet devices have been presented, and the microstructures generated on the surface after solvent evaporation have been discussed. The literature is vast and growing, therefore here is listed a representative, but by no means complete, experimental [22–25, 27–35, 37–39, 46–50] and theoretical [26, 36, 51, 52] catalogue. The interest towards a more fundamental understanding of the processes involved is increasing. This will allow for an improved knowledge of the technology and a better control of the resulting microstructures.

As it has been pointed out, among others in [52], the evaporation of a solvent droplet from a soluble polymer surface is a very complex process, where several independent, but correlated, physico-chemical phenomena occur. A sketch of the situation, addressing some of the processes involved, is displayed in Fig. 3.

We can divide these processes into three main groups of action:

(i) First, when a solvent drop is deposited on a soluble surface, this wets the surface and takes the shape of a flattened lens due to its high affinity with the substrate material. The drop is pinned at the rim due to the roughness of the

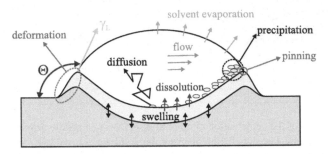

Figure 3. Physico-chemical processes involved in the evaporation of solvent drops from soluble polymer surfaces.

surface. The solvent diffuses into the material to a certain depth. At the same time the polymer swells due to the solvent diffusing into it, and dissolves at the interface with the solvent. After some time, the drop consists of a more or less viscous polymer/solvent solution, and the polymer underneath the drop is softened.

(ii) Second, the drop deforms the softened polymer surface. It pulls the surface upwards at the rim, due to the surface tension at the pinned TPCL. As well, due to the Laplace pressure, the drop pushes downwards the bottom of the drop.

(iii) Third, the 'coffee-stain effect'. It is known that the evaporation rate of a drop which is pinned at the TPCL is higher at the rim than in the middle. Therefore a radial flow occurs from the centre of the drop outwards. This flow carries dissolved polymer from the centre to the rim, and eventually deposits it there.

We want to point out that wetting by the solvent, diffusion of solvent and softening of the polymer, and solvation of the polymer are all processes characterized by quite different time constants.

When the solvent finally evaporates completely, a structure is generated at the area where the drop was deposited. Its characteristics are strongly dependent on the materials used and on the parameters set.

For instance, Li *et al.* [30] studied the structures produced by depositing toluene drops on polystyrene (PS) surfaces, and ethyl acetate drops on polyethylmethacrylate (PEMA) surfaces. The authors varied the molecular weight of the polymer and the number of deposited drops, and analysed the structures obtained. When ethyl acetate drops are deposited on PEMA, a crater-like shape with a high rim of deposited material is produced. The bottom of the structure becomes deeper and the rim higher when the number of drops is increased. The bottom of the crater is always flat. The reproducibility of this shape for all explored molecular masses of PEMA and number of drops shows that the coffee-stain effect is predominant in the system. On the other hand, the structures produced by depositing toluene drops on PS surfaces are much more diverse, especially in the case of PS molecular masses

Nr. of drops - toluene

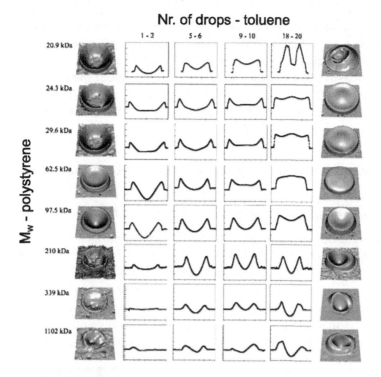

Figure 4. Surface plots (left and right) and cross-sections (middle) of structures left by toluene drops deposited by ink jetting on polystyrene with different molar masses, 1–2 h after drying. The height in the surface plots is automatically rescaled so that height is only in relative units. For the cross-sections the ticks are spaced 1 µm vertically and 50 µm horizontally. The four height profiles for each molar mass refer to depositions of 1–2, 5–6, 9–10, and 18–20 drops deposited at intervals of 0.5 s on the same spot. [Reprinted figure from: Li, G.; Graf, K.; Bonaccurso, E.; Golovko, D. S.; Best, A.; Butt, H. J., *Macromolecular Chemistry and Physics* **2007**, 208, 2134. Copyright (2007) by Wiley-VCH Verlag GmbH & Co. KGaA. Reproduced with permission.]

lower than 100 kDa (Fig. 4). When the number of drops deposited is small the structure is crater-like with a ring of material around. The central depression is spherical cap shaped. Similarly to the crater-like structures in PEMA, these structures show the predominance of the coffee-stain effect for small drop numbers. In contrast, when the number of deposited drops is increased, a dot-like elevation is formed in the centre of the structure. This situation is explained as due to an excess of dissolved polymer at the bottom of the drop, impeding the radial flow towards the outer rim. For high molar masses, the structures become asymmetric.

In summary, evaporation structures left by microscopic solvent drops on polymer surfaces are characteristic of the specific polymer/solvent combination and the molar mass of the polymer. If the solvent evaporates with a constant contact radius from the polymer surface, the shape of the evaporation structures is determined by the processes mentioned above. Depending on which of these processes is dominating, different structures can be generated. For the system polystyrene/toluene, if the dissolved molecules diffuse a larger distance than covered by evaporation driven flow, convex structures are formed. If the evaporation driven flow dominates, concave structures are formed. The amount of dissolved polymer influences this relation by increasing the viscosity and reducing the vapour pressure of the solution, thus reducing also the evaporation driven flow. In the following paragraph, various applications of ink jetting solvent microdrops on polymer surfaces will be presented along with the microstructures obtained.

C. Applications of Evaporating Solvent Drops on Deformable/Soluble Substrates

1. Microlenses on Polymer Surfaces by Ink Jetting Solvent Drops

By depositing a tiny drop of solvent onto a polymer surface (volume between 0.1 and 1 nL) by, e.g., ink jetting [53, 54] it is possible to fabricate a spherically shaped microcrater (Fig. 5A). Such concave structures can be used as moulds for another polymer to fabricate convex microlenses in a second processing step. As seen above, the solvent has three distinct effects on the polymer: (i) it dissolves part of it, so that the drop is made of a polymer solution rather than of pure solvent after some time; (ii) it diffuses into it and makes it swell; (iii) it softens the material, so that it

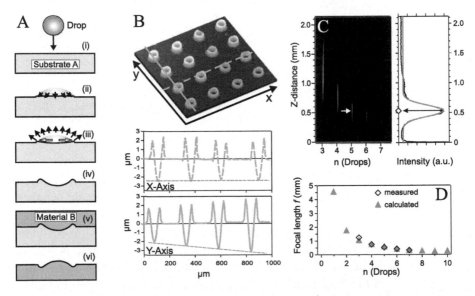

Figure 5. (A) Technology to fabricate concave and convex lenses by solvent microdrops. (B) Detail of a micrograph and X and Y profiles of an array of 10×10 concave microlenses made by toluene drops on polystyrene ($M_w = 200$ kDa). In each line a different number of drops was deposited. (C) Experimental determination of the focal length of a row of convex PDMS microlenses by LSCM and intensity profile of the light focussed by one lens (white arrow). (D) Calculated and measured focal lengths of an array of convex PDMS microlenses. [Reprinted figure from: Pericet-Camara, R.; Best, A.; Nett, S. K.; Gutmann, J. S.; Bonaccurso, E., *Optics Express* **2007**, 15, 9877. Copyright (2007) by the Optical Society of America. Reproduced with permission.]

becomes easier to be deformed by small surface forces. At the same time the drop evaporates. After evaporation of the solvent, a concave microcrater is left behind [25]. This crater is formed according to the physico-chemical processes described above.

Using an automatised XYZ-controlled inkjet head one can make two dimensional arrays of microcraters (Fig. 5B). Diameter and depth of the craters can be controlled by the number of microdrops, as demonstrated in the surface profiles in Fig. 5B: along the X-axis the number of drops was similar and thus also the depth of the craters (see horizontal dashed line connecting the bottoms); along the Y-axis the number of drops was continuously increased and thus also the depths went deeper (see inclined dashed line connecting the bottoms). This array was used as a master for moulding with the elastomer polydimethylsiloxane (PDMS) and making plano-convex microlenses. The focal length of such lenses was directly measured with a laser scanning confocal microscope (LSCM) (Fig. 5C). The lenses are irradiated with a parallel laser beam from the flat side, which is focussed by the lenses. The intensity of the light is maximum in the focus, as shown by the intensity profile of the transmitted light. The distance between the point of highest intensity and the flat surface yields the focal length. The focal length can also be calculated from geometrical parameters, like the refractive index and the radius of curvature of the lens. Both methods yield similar values (Fig. 5D). Microlens arrays fabricated with this technique possess extremely good optical qualities, small aberration, and a surface roughness in the nanometer scale [31].

2. Microstructures on Polymer Surfaces by Mixtures of Solvents

If instead of using pure solvents mixtures of solvents are used, the shape of the produced microstructures can be additionally controlled by the mixing ratio. This is also known for droplets of polymer solutions dissolved in two solvents evaporating from hard, inert surfaces [46, 55]. Not only diameter and depth can be controlled, but also the shape, e.g., the sphericity/asphericity of the microcraters. So did Karabasheva *et al.* [27] by using a mixture of acetophenone (AP) and ethyl acetate (EA) and ink jetting drops of the two pure solvents and of mixtures of them on PS surfaces with a molecular weight $M_w = 200$ kDa. The two solvents are fully miscible in any proportion. They have different surface tensions, which affect the contact angle of the deposited drops on the surface, and vapour pressures, which affect the evaporation rate. AP, e.g., evaporates much slower than EA. In Fig. 6 are shown micrographs of the structures resulting from the evaporation of the drops, together with representative profiles. Despite the differences in volatility, in both cases a concave shape remains (Figs. 6A, B), as was seen in the previous paragraph for toluene. The crater from the relatively fast evaporating EA shows a lower depth-to-width ratio than AP. This was explained by the fact that less polymer can be dissolved by the drop if it evaporates faster, and thus less polymer is accumulated at the rim. When the authors used a mixture of solvents the shape of the microstructures drastically changed. With a mixture of EA and AP (5:1 ratio by volume), PS

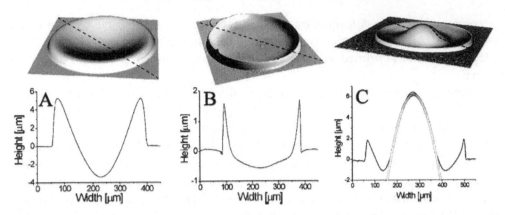

Figure 6. Microvessels on a polystyrene substrate ($M_W = 200$ kDa) with corresponding height profiles across the dashed lines. Structures are obtained with deposition of drops of (A) acetophenone, (B) ethylacetate, and (C) 5:1 mixture (Vol) of ethylacetate and acetophenone. [Reprinted figure from: Karabasheva, S.; Baluschev, S.; Graf, K., *Applied Physics Letters* **2006**, 89, 031110. Copyright (2006) by the American Institute of Physics. Reproduced with permission.]

is accumulated at the rim as well as in the centre of the structure (Fig. 6C). The effect occurred not only for microstructures, but also for structures with diameters ranging from tens of micrometers to several millimetres. The authors tentatively explain the concave/convex structures stating that the solvent is transported from the centre to the rim to replenish the loss due to evaporation. As AP evaporates much slower than EA, the concentration of AP at the pinned TPCL increases, leading to a concentration gradient of AP that is directed opposite to the concentration gradient of EA. Thus, differences in the evaporation rates might cause an additional flow from the rim to the centre. Additionally, a gradient of surface tension occurs, adding a surface flow of Marangoni type to the system. Both are leading to a redistribution of dissolved polymer into the centre of the drop. As a consequence, an increased EA content should enhance the accumulation in the centre. The author verified this by varying the mixing ratios of AP and EA: for mixing ratios from 2:1 to 6:1 the accumulation of material in the centre becomes more pronounced. For higher mixing ratios, the final structure becomes more complicated to interpret, due to the formation of a second crater in the middle.

3. Microlithography and Other Applications

3.1. Via-Hole Etching

Kawase *et al.* first employed inkjet etching of polymers with solvent drops to fabricate via holes through a 500 nm thick polyvinylphenol (PVP) film. They were looking for fast and low-cost alternatives to conventional microlithography for making a connection between two conducting polymer films, or between two electrodes more in general, across an insulating layer [22, 56]. They deposited small ethanol drops onto the insulator with an inkjet printing head. The drops dissolve the insulator locally and evaporate within a second. Here too, when a drop dried, the

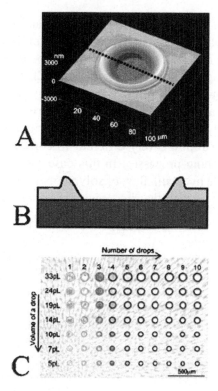

Figure 7. (A) AFM image of a crater-like hole formed by ink jetted ethanol drops. A series of drops (5 pL for each) is deposited onto a PVP film with a thickness of 500 nm. The depth of the hole corresponds to the thickness. A ridge is created at the rim of the hole and the height is greater than 1 μm. (B) Schematic profile of the crater along the dashed black line. (C) Micrograph of inkjet-printed via-holes with an increasing number of ethanol drops (horizontal) and with a series of different drop volumes (vertical). Different colours represent different thickness. The colours show that via holes through a 500 nm thick film are formed with deposition of 5 to 7 drops. [Reprinted figure from: Kawase, T.; Sirringhaus, H.; Friend, R. H.; Shimoda, T., *Advanced Materials* **2001**, 13, (21), 1601. Copyright (2001) by Wiley-VCH Verlag GmbH & Co. KGaA. Reproduced with permission.]

insulating material was preferentially re-deposited at the TPCL, resulting in a decrease of thickness in the central area and in the formation of a ridge at the former TPCL. When the deposition of a drop was repeated at the same position, the hole became deeper and the ridge higher, and after several repetition steps a via-hole was formed. The AFM image of such hole (Fig. 7A) confirms this. Holes could be formed with different numbers of solvent drops (horizontal direction) and with different drop volumes (vertical direction). The etching depth is indicated by the colour generated by light interference, and the via-hole is attained after 5–7 drop depositions. Further deposition of solvent drops after reaching the bottom of the film does not change the shape or diameter of the hole.

The process of hole formation can be explained by the action of a number of processes already mentioned above: (i) the solvent dissolves part of the polymer, so that the drop is made of a polymer solution rather than of pure solvent; (ii) it diffuses partially into the polymer making the thin polymer film swell slightly; (iii) the solvent also softens the material, making it prone to be deformed by small surface forces; (iv) the drop evaporates, and faster at the TPCL than at the centre, causing a net flow of solvent towards the rim of the drop; (iv) the fluid flow is faster than diffusion, so that a net amount of polymer is transferred to the rim by the end of the evaporation. The physical phenomena of hole formation via inkjet printing of solvents onto a polymer film are very different to conventional photolithography, soft-lithography and etching processes. In this case the etched material is not removed from the substrate by a bulk flow of solvent, but it is transferred locally from the central region of the drop to the rim. It is thus rather a re-arrangement than an etching.

3.2. Microstructure-Etching

De Gans *et al.* used a technology similar as in the paragraph above to etch microstructures in thin polymer films spincoated on hard substrates, like glass or silicon [47]. They used a XYZ-controlled inkjet device to generate and deposit microdrops of various solvents onto thin PS films (thickness of around 100 nm). They did not simply fabricate round holes, but also lines and even more complex patterns (Fig. 8). These geometries were a result of the type of etchant, of the type of polymer, and of the deposition pattern formed by the solvent droplets. The authors obtained the different geometries mainly by varying the spacing between the drops and by using different solvents. They assume that the size of a hole or a line

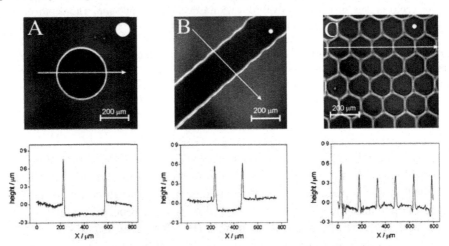

Figure 8. (A) Confocal scanning microscopy image of a hole etched into a PS film by a 100 μm drop of isopropyl acetate (shown to scale as the white circle in the upper right). (B) Groove etched in poly(benzyl methacrylate) film by a line of 30 μm *n*-butyl acetate droplets (circle, upper right) with a spacing of 120 μm. (C) Hexagonal holes etched into a PS film by a hexagonal array of 30 μm isopropyl acetate droplets. [Reprinted figure from: de Gans, B. J.; Hoeppener, S.; Schubert, U. S., *Advanced Materials* **2006**, 18, (7), 910. Copyright (2006) by Wiley-VCH Verlag GmbH & Co. KGaA. Reproduced with permission.]

depends on a combination of three factors: (i) the spreading of the solvent on the substrate; (ii) the evaporation rate of the solvent; (iii) the rate of dissolution of the polymer in the solvent. The size of holes formed in a PS film using a number of good solvents was investigated. Ethyl acetate, isopropyl acetate, *n*-butyl acetate, toluene, anisole, and acetophenone drops were deposited on thin films. The initial size of the drop and/or the surface tension was found to determine the diameter of the holes. On the other hand, no clear trend was observed for the vapour pressure: even if the vapour pressure of anisole ($P_0 = 4.7$ mbar) is ten times larger than the vapour pressure of acetophenone ($P_0 = 0.46$ mbar), it had no effect on the diameter of the hole. This was a surprising finding, since the vapour pressure was expected to influence the hole etching process in terms of the very different evaporation speeds of the solvents. The authors thus conclude that the rate of dissolution of PS in the solvents used always exceeds the rate of evaporation of the polymer/solvent combination. Thus the wetting of the polystyrene film by the solvent seems to be the only factor that determines the size of the microstructures.

3.3. Microlithography

In a way similar as above, more complex microstructures than spherical craters can be produced also on bulk polymer substrates. Again PS was used with toluene as the solvent. A two dimensional logo was chosen and the inkjet device was programmed to redraw it on the polymer surface either as single dots, or with continuous lines (Fig. 9). As in the previous paragraphs, the initial size of a droplet determines the minimum line width that can be achieved, and the number of droplets determines the depth of the structures.

D. Decoupling the Processes

The process of structure formation by inkjet etching of polymer surfaces is a complex one, as is demonstrated by the examples given in the previous paragraph. It is rather a combination of a series of processes, as described in Section B.3. The first step towards the comprehension is thus to analyse and understand the single processes, as we try to do in the following paragraphs.

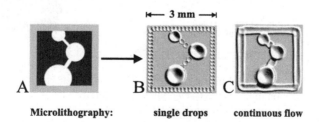

Figure 9. Reproduction of a logo (A) by inkjet etching lithography. Confocal scanning microscopy image of the structure etched into a polystyrene surface by single, consecutively deposited toluene microdrops (B) or by a continuous flow of toluene microdrops. (Picture by courtesy of Ramon Pericet–Camara and Sebastian Nett.)

1. Mass Transport Inside the Drop: the Coffee Stain Effect

It is common experience that when a spilled drop of coffee dries on a solid surface, it leaves a dense, ring-like deposit along the perimeter, or TPCL. The coffee, initially dispersed homogeneously in the entire drop, becomes concentrated into a small fraction of it. Such ring deposits are found wherever drops containing dispersed solids or polymers evaporate on a surface. They affect processes such as printing, washing and coating. By now the mechanism responsible for the coffee-stains on surfaces has been widely understood [48, 57–69]. The characteristic pattern of the deposition are due to a form of capillary flow in which pinning of the TPCL of the evaporating drop ensures that liquid evaporating from the rim is replenished by liquid from the interior. The resulting outward flow can carry virtually all the dispersed material to the rim. This mechanism, first described quantitatively by Deegan *et al.* [57], predicts a distinctive power-law growth of the ring mass with time, a law independent of the particular substrate, liquid or deposited solids. The model has since then been substantiated by various experimental works and further theoretical studies. The qualitative observations show that rings are formed for a wide variety of substrates, dispersed materials (solutes), and carrier liquids (solvents), as long as (i) the solvent drops forms a non-zero contact angle with the surface; (ii) the TPCL is pinned to its initial position for almost the entire evaporation time; and (iii) the solvent evaporates, i.e., its vapour pressure is higher than zero. Deegan *et al.* [57] also could rule-out other mechanisms which typically are responsible for solute transport in liquids, like surface tension gradients, solute diffusion, electrostatics and gravity, and found that they can be neglected in the ring formation process. The phenomenon is basically due to a geometrical constraint: the free surface, constrained by the pinned TPCL, pushes the fluid outwards to compensate for the losses due to evaporation, taking place all over the free surface of the droplet.

Figure 10 illustrates the factors leading to the outward flow in a small, thin, dilute, circular drop. The evaporative flux continuously reduces the height of the

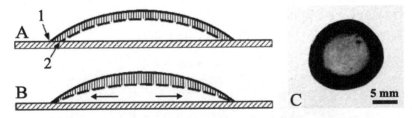

Figure 10. Schematic illustration of the origin of the advective current. (A) When the TPCL is not pinned, uniform evaporation removes the hashed layer, the interface moves from the solid line to the dashed line, and the TPCL moves from position 1 to 2. (B) If the TPCL is pinned the motion from 1 to 2 must be prevented by an outflow to replenish the liquid removed from the edge (arrows). The interface of the drop thus moves differently in order to keep the TPCL fixed. (C) Dried coffee ring on a table. [A and B: Reprinted figure from: Deegan, R. D.; Bakajin, O.; Dupont, T. F.; Huber, G.; Nagel, S. R.; Witten, T. A., *Physical Review E* **2000**, 62, (1), 756. Copyright (2000) by the American Physical Society. Reproduced with permission. C: a coffee stain on the authors' desk.]

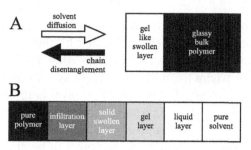

Figure 11. (A) Schematic illustration of the one-dimensional solvent diffusion, swelling, and dissolution process; (B) Schematic composition of the surface layer. [Adapted figure from: Miller-Chou, B. A.; Koenig, J. L., *Progress in Polymer Science* **2003**, 28, 1223. Copyright (2003) by Elsevier B.V.. Reproduced with permission.]

drop. If there were no flow, the evaporation would alter the height profile (Fig. 10A): at the perimeter, all the liquid would be removed and the drop would shrink. Since the radius of the drop cannot shrink, as its TPCL is held pinned, to prevent the shrinkage the liquid must flow outwards as in Fig. 10B. The height profile must maintain the spherical cap shape dictated by surface tension for small drops. Thus during a short time the hashed area must be removed. This area is different from that in Fig. 10A, so that the radial flow must make up for this difference. The flow drags the solute along with it, so that at the end of the evaporation nearly all of it is deposited at the former TPCL.

2. Dissolution and Swelling of the Substrate

The dissolution and the swelling of polymers in solvents is a key area in polymer and material science. A large number of phenomenological, experimental, and theoretical literature has been published on this topic over the years [40, 70–73]. The book, and the few papers and reviews cited here are only representative, but by no means exhaustive. The phenomenon is especially interesting because it is different from non-polymeric materials, which dissolve instantaneously and where the dissolution process is generally controlled by the external mass transfer resistance through a liquid layer adjacent to the solid/liquid interface. The dissolution of a polymer into a solvent, on the other hand, involves two transport processes, namely solvent diffusion and chain disentanglement (Fig. 11A). When an uncrosslinked, amorphous, glassy polymer is in contact with a solvent, the solvent will diffuse into it. Due to plasticization of the polymer by the solvent, a gel-like swollen layer is formed. It has two separate interfaces, one between the glassy polymer and the gel layer, and the other between the gel layer and the solvent. After a certain time the polymer dissolves. One of the earliest contributors to the study of polymer dissolution was Ueberreiter [74] who outlined the surface layer formation process. He described the structure of the surface layers of glassy polymers during dissolution from the pure polymer to the pure solvent as follows: the infiltration layer, the solid swollen layer, the gel layer, and the liquid layer (Fig. 11B). The infiltration layer is the first layer adjacent to the pure polymer. A polymer in the glassy state contains free volume in the form of a number of channels and holes of molecular dimensions, and the first penetrating solvent molecules fill these empty spaces and start the diffusion process without any necessity for creating new holes. The next layer is the solid swollen layer where the polymer/solvent system building up in this layer is still in the glassy state. Next, the solid swollen layer is followed by the gel layer, which contains swollen polymer material in a rubber-like state, and a liquid layer, which surrounds every solid in a liquid.

The dissolution/swelling process has been found to be affected by a large number of factors, like the molecular weight and the polydispersity of the polymer; the composition, conformation and the structure of the polymer; the type of solvent and additives to the polymer; the effects of the environmental parameters, like temperature, stirring of the solvent, etc.; the processing conditions of the polymer. To this end, several models have been proposed to explain the experimentally observed dissolution and swelling behaviour and have been reviewed extensively, among others by Narasimhan *et al.* [75, 76].

In conclusion, as for the coffee-stain effect, the mechanisms responsible for polymer dissolution and swelling in a solvent have been singled out and are understood to a large extent.

3. Etching and Collapse of the Substrate

A substrate could also be etched by a solvent, and collapse afterwards. This is not the case for synthetic polymers, but has been described for the special case of biopolymers, like proteins. Ionescu *et al.* deposited microdrops of an aqueous solution containing the proteolytic enzyme trypsin onto a thin layer of bovine serum albumin (BSA), and were able to generate small microcraters in the protein layer [77, 78]. An enzyme like trypsin is able to cleave proteins at specific sites, thus a major collapse of the three dimensional structure of the BSA layer is expected after deposition of the drop. The authors were able to tune the size and depth of the microcraters by the size of the microdrops and by repeated drop depositions. The action of the solvent is twofold: the water causes the BSA to swell underneath the drop, thus permitting the enzyme to diffuse to the cleavage sites and to cause the local collapse of the protein layer. Monte Carlo simulations confirmed the experimentally observed process (Fig. 12). This process of etching and considerable collapsing of the material is less significant for polymer–solvent systems, although the solvent could as well cause a local rearrangement or compaction of polymer chains underneath the drop, as has been observed by Bates *et al.* [49].

4. Elastic Deformation of the Substrate

The surface effects described in Section B.1 are mostly negligible for sufficiently hard materials, such as mineral surfaces. However, the deformation may be of micrometer size for surfaces with a Young's modulus below around 100 kPa. The ridge as result of the surface tension of the droplet pulling upwards at the contact line has been observed *in-situ* [13, 16, 79] and *ex-situ* [11, 14, 17, 18, 80, 81] in

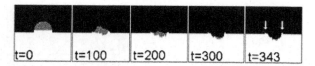

Figure 12. A series of images from a Monte Carlo simulation. The white portion represents the BSA substrate, dark grey represents the water drop, light grey dots within the drop represent trypsin molecules, and the black background represents air. The snapshots show the various stages of trypsin etching the BSA sample that has initially swollen. The times indicated are the iteration step of the simulation. [Reprinted figure from: Ionescu, R. E.; Marks, R. S.; Gheber, L. A., *Nano Letters* **2003**, 3, 1639. Copyright (2003) by the American Chemical Society. Reproduced with permission.]

previous works. Carre' *et al.* [79] demonstrated that the spreading dynamics of a liquid on a soft elastic surface is modified, due to viscoelastic dissipation induced by the presence of the 'wetting ridge'. They measured the surface profile close to the TPCL during the spreading of tricresylphosphate and formamide on a soft silicone elastomer and on natural rubber with a scanning white-light interferometer microscope. They concluded that the spreading time of a drop can vary as much as an order of magnitude for various conditions. Extrand and Kumagai [11] studied the contact angle of sessile drops of ethylene glycol and water on elastomer surfaces. They observed that the hysteresis between the receding and the advancing contact angle was enhanced for lower surface elastic moduli. By imaging the wetting ridge with optical microscopy after fast removal of the sessile drop, they established that the contact angle hysteresis was not negligible when the ridge height exceeded the local surface roughness. Pu et al. [14, 17, 18, 81] studied the wetting of acrylic thermoplastics surface by sessile drops or wetting fronts. Formation of wetting ridges at quasi-periodic distances after complete removal of the water from the surface was observed. The contact line causes a plastic deformation of the surface, as it is deduced from an enhanced rim height as compared to previous measurements. This deformation hinders the advancement of the wetting front, until the liquid mass is too large to be supported by the meniscus, and the drop breaks and flows to establish a new TPCL. The periodic wetting ridges are corroborated by using the Wilhelmy plate method to obtain force curves as a function of the plate displacement. The maximum peak periodicity at the force curves coincides with the spatial periodicity of the wetting ridges observed on the surfaces.

Pericet-Camara *et al.* [13, 16] measured the complete deformation of a soft polydimethylsiloxane (PDMS) substrate under a fluorophore-dyed ionic liquid drop. By using simultaneous fluorescence and reflection laser scanning confocal microscopy (LSCM) it was possible to obtain three-dimensional images of the fluorescing drop and the reflection from the solid–gas interface (Fig. 13). The reflection image confirmed the occurrence of a ridge at the TPCL. Moreover, a depression of the substrate underneath the drop was observed. This depression confirmed the effect of the capillary pressure of a sessile drop on an elastic surface. Due to the negligible vapour pressure of ionic liquids, the drop does not evaporate during the typical experimental time, i.e., it is in thermodynamic equilibrium. The complete

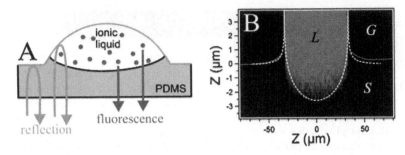

Figure 13. (A) Schematic of the working principle of the measurement: one wavelength (light grey) is reflected at the solid–gas or liquid–gas interface, the other wavelength (dark grey) excites the fluorophore. (B) Experimental image of a section of an ionic liquid droplet labelled with a fluorophore on a soft PDMS surface obtained with LSCM. The grey region displays the fluorescing droplet, the grey solid line is the reflection from the solid–gas interface, and the white dashed line is the deformation profile according to Rusanov's model. [Reprinted figure from: Pericet-Camara, R.; Best, A.; Butt, H. J.; Bonaccurso, E., *Langmuir* **2008**, 24, 10565. Copyright (2008) by the American Chemical Society. Reproduced with permission.]

strain profile fitted very well with the theoretical one plotted from Rusanov's model (Eqs. (6a–c)). By using PDMS substrates with various cross-linking densities it was possible to deduce that substrates with lower Young's moduli bear a more intense strain.

In summary, in this last paragraph the multifaceted evaporation process of microdrops from soluble and/or deformable surfaces has been tentatively described, with the aim of demonstrating that it is a much more complex phenomenon than the evaporation from undeformable surfaces. Some of the individual processes involved have been presented and briefly discussed. Some of them are understood to a fair extent, qualitatively as well as quantitatively. For some others experimental and modelling efforts have still to be made. The same holds for the understanding of the entire process.

E. Brief Summary

In this chapter we have reported on the state-of-the-art research on evaporating droplets on deformable and/or soluble polymer surfaces along with the physico-chemical processes involved. Since in recent times a number of inkjet technologies based on the deposition of solvent microdrops on soluble polymer substrates have been presented, an understanding of the phenomenon is required. This will allow for an improved knowledge of the technology and a better control of the resulting microstructures. As we pointed out the evaporation of a solvent droplet from a soluble and deformable polymer surface is a very complex process during which several independent physico-chemical phenomena occur. Three main actions occur at different time scales: (i) the solvent wets the polymer surface; (ii) it diffuses into the polymer and swells and softens it; and (iii) it dissolves the polymer. The drop can deform the softened polymer by the action of capillary forces. Or due to evaporation a capillary flow is generated inside the drop and transports material from the centre to the rim and eventually deposits it there. We pointed out that some of these processes are by now understood to a fair extent, while for some others experimental and modelling efforts are still required.

F. Acknowledgements

EB and RPC acknowledge the Max Planck Society (MPG) for financial support. RPC further thanks the Swiss National Science Foundation for funding (Contract Nr. PBGE2-112884).

G. References

1. Maxwell, J. C., Diffusion. Scientific Papers, II (1890) 625–647.
2. Morse, H. W., On evaporation from the surface of a solid sphere. Proc. Am. Acad. Arts Sci., 45 (1910) 361–367.
3. Langmuir, I., The evaporation of small spheres. Physical Review, 12(5) (1918) 368–370.
4. Picknett, R. G. and Bexon, R., The evaporation of sessile or pendant drops in still air. Journal of Colloid and Interface Science, 61(2) (1977) 336–350.
5. Shanahan, M. E. R. and Bourges, C., Effects of evaporation on contact angles on polymer surfaces. International Journal of Adhesion and Adhesives, 14(3) (1994) 201–205.
6. Lebedev, N. N., Special Functions and Their Application. Prentice-Hall, Englewood Cliffs, New Jersey, 1965.
7. Hu, H. and Larson, R. G., Evaporation of a sessile droplet on a substrate. Journal of Physical Chemistry B, 106(6) (2002) 1334–1344.
8. Meric, R. A. and Erbil, H. Y., Evaporation of sessile drops on solid surfaces: pseudospherical cap geometry. Langmuir, 14(7) (1998) 1915–1920.
9. Carre, A. and Shanahan, M. E. R., Influence of the "wetting ridge" in dry patch formation. Langmuir, 11 (1995) 3572–3575.
10. Carre, A., Gastel, J. C. and Shanahan, M. E. R., Viscoelastic effects in the spreading of liquids. Nature, 379(6564) (1996) 432–434.
11. Extrand, C. W. and Kumagai, Y., Contact angles and hysteresis on soft surfaces. Journal of Colloid and Interface Science, 184 (1996) 191–200.
12. Pu, G., Guo, J. H., Gwin, L. E. and Severtson, S. J., Mechanical pinning of liquids through inelastic wetting ridge formation on thermally stripped acrylic polymers. Langmuir, 23(24) (2007) 12142–12146.
13. Pericet-Camara, R., Best, A., Butt, H. J. and Bonaccurso, E., Effect of capillary pressure and surface tension on the deformation of elastic surfaces by sessile liquid microdrops: An experimental investigation. Langmuir, 24(19) (2008) 10565–10568.
14. Pu, G. and Severtson, S. J., Characterization of dynamic stick-and-break wetting behavior for various liquids on the surface of a highly viscoelastic polymer. Langmuir, 24(9) (2008) 4685–4692.
15. Yu, Y. S., Yang, Z. Y. and Zhao, Y. P., Role of vertical component of surface tension of the droplet on the elastic deformation of PDMS membrane. J. Adhes. Sci. Technol., 22(7) (2008) 687–698.
16. Pericet-Camara, R., Auernhammer, G. K., Koynov, K., Lorenzoni, S., Raiteri, R. and Bonaccurso, E., Solid-supported thin elastomer films deformed by microdrops. Soft Matter, 5(19) (2009) 3611–3617.

17. Pu, G. and Severtson, S. J., Variety of wetting line propagations demonstrated on viscoelastic substrate. Applied Physics Letters, 94(13) (2009).

18. Pu, G., Ai, J. and Severtson, S. J., Drop behavior on a thermally-stripped acrylic polymer: influence of surface tension induced wetting ridge formation on retention and running. Langmuir, 26(15) (2010) 12696–12702.

19. Sokuler, M., Auernhammer, G. K., Roth, M., Liu, C. J., Bonaccurso, E. and Butt, H. J., The softer the better: fast condensation on soft surfaces. Langmuir, 26(3) (2010) 1544–1547.

20. Jerison, E. R., Xu, Y., Wilen, L. A. and Dufresne, E. R., The deformation of an elastic substrate by a three-phase contact line. Physical Review Letters, 2011, in print.

21. Chen, L. Q., Auernhammer, G. K. and Bonaccurso, E., Short time wetting dynamics on soft surfaces. Soft Matter, 2011, in press.

22. Kawase, T., Sirringhaus, H., Friend, R. H. and Shimoda, T., Inkjet printed via-hole interconnections and resistors for all-polymer transistor circuits. Advanced Materials, 13(21) (2001) 1601–1605.

23. Saiz, E. and Tomsia, A. P., Atomic dynamics and Marangoni films during liquid-metal spreading. Nature Materials, 3(12) (2004) 903–909.

24. de Gans, B.-J., Duineveld, P. C. and Schubert, U. S., Inkjet printing of polymers: State of the art and future developments. Advanced Materials, 16(3) (2004) 203–213.

25. Bonaccurso, E., Butt, H. J., Hankeln, B., Niesenhaus, B. and Graf, K., Fabrication of microvessels and microlenses from polymers by solvent droplets. Applied Physics Letters, 86 (2005) 124101.

26. Cordeiro, R. M. and Pakula, T., Behavior of evaporating droplets at nonsoluble and soluble surfaces: Modeling with molecular resolution. Journal of Physical Chemistry B, 109(9) (2005) 4152–4161.

27. Karabasheva, S., Baluschev, S. and Graf, K., Microstructures on soluble polymer surfaces via drop deposition of solvent mixtures. Applied Physics Letters, 89 (2006) 031110.

28. Li, G. F., Hohn, N. and Graf, K., Microtopologies in polymer surfaces by solvent drops in contact and noncontact mode. Applied Physics Letters, 89(24) (2006).

29. Li, G. F., Butt, H. J. and Graf, K., Microstructures by solvent drop evaporation on polymer surfaces: Dependence on molar mass. Langmuir, 22(26) (2006) 11395–11399.

30. Li, G., Graf, K., Bonaccurso, E., Golovko, D. S., Best, A. and Butt, H. J., Evaporation structures of solvent drops evaporating from polymer surfaces: influence of molar mass. Macromolecular Chemistry and Physics, 208 (2007) 2134–2144.

31. Pericet-Camara, R., Best, A., Nett, S. K., Gutmann, J. S. and Bonaccurso, E., Arrays of microlenses with variable focal lengths fabricated by restructuring polymer surfaces with an ink-jet device. Optics Express, 15(15) (2007) 9877–9882.

32. Pericet-Camara, R., Bonaccurso, E. and Graf, K., Microstructuring of polystyrene surfaces with nonsolvent sessile droplets. Chemphyschem, 9(12) (2008) 1738–1746.

33. Monteux, C., Tay, A., Narita, T., De Wilde, Y. and Lequeux, F., The role of hydration in the wetting of a soluble polymer. Soft Matter, 5(19) (2009) 3713–3717.

34. Li, G. F. and Graf, K., Microstructures formation by deposition of toluene drops on polystyrene surface. Physical Chemistry Chemical Physics, 11(33) (2009) 7137–7144.

35. Saiz, E., Benhassine, M., De Coninck, J. and Tomsia, A. P., Early stages of dissolutive spreading. Scripta Materialia 62(12) (2010) 934–938.

36. Tay, A., Monteux, C., Bendejacq, D. and Lequeux, F., How a coating is hydrated ahead of the advancing contact line of a volatile solvent droplet. European Physical Journal E, 33(3) (2010) 203–210.

37. Zhang, X. H., Wei, X. X. and Ducker, W., Formation of nanodents by deposition of nanodroplets at the polymer–liquid interface. Langmuir, 26(7) (2010) 4776–4781.
38. Tay, A., Lequeux, F., Bendejacq, D. and Monteux, C., Wetting properties of charged and uncharged polymeric coatings-effect of the osmotic pressure at the contact line. Soft Matter, 7(10) (2011) 4715–4722.
39. Muralidhar, P., Bonaccurso, E., Auernhammer, G. K. and Butt, H.-J., Fast dynamic wetting of polymer surfaces by miscible and immiscible liquids. Colloid and Polymer Science, 289 (2011) 1609–1615.
40. Miller-Chou, B. A. and Koenig, J. L., A review of polymer dissolution. Progress in Polymer Science, 28(8) (2003) 1223–1270.
41. Atkins, P. and de Paula, J., Physical Chemistry, 7th ed. Oxford University Press, New York, 2002.
42. Lester, G. R., Contact angles of liquids at deformable solid surfaces. Journal of Colloid Science, 16(4) (1961) 315-&.
43. Rusanov, A. I., Theory of wetting of elastically deformed bodies, 1. Deformation with a finite contact-angle. Colloid Journal Of The Ussr, 37(4) (1975) 614–622.
44. Landau, L. D. and Lifshitz, E. M., Theory of Elasticity, 3rd ed., Vol. 7. Butterworth-Heinemann, 2002, p 187.
45. White, L. R., The contact angle on an elastic substrate. 1. The role of disjoining pressure in the surface mechanics. Journal of Colloid and Interface Science, 258 (2003) 82–96.
46. Tekin, E., de Gans, B. J. and Schubert, U. S., Ink-jet printing of polymers—from single dots to thin film libraries. Journal of Materials Chemistry, 14(17) (2004) 2627–2632.
47. de Gans, B. J., Hoeppener, S. and Schubert, U. S., Polymer-relief microstructures by inkjet etching. Advanced Materials, 18(7) (2006) 910–914.
48. Soltman, D. and Subramian, V., Inkjet-printed line morphologies and temperature control of the coffee ring effect. Langmuir, 2008 in print.
49. Bates, C. M., Stevens, F., Langford, S. C. and Dickinson, J. T., Nanoscale craters in poly(methyl methacrylate) formed by exposure to condensing solvent vapor. Journal of Materials Research, 22(12) (2007) 3360–3370.
50. Khan, F., Zhang, R., Unciti-Broceta, A., Diaz-Mochon, J. J. and Bradley, M., Flexible fabrication of microarrays of microwells. Advanced Materials, 19(21) (2007) 3524.
51. Stupperich-Sequeira, C., Graf, K. and Wiechert, W., Modelling and simulation of micro-well formation. Mathematical and Computer Modelling of Dynamical Systems, 12(4) (2006) 263–276.
52. Haschke, T., Wiechert, W., Graf, K., Bonaccurso, E., Li, G. and Suttmeier, F. T., Evaporation of solvent microdrops on polymer substrates: From well controlled experiments to mathematical models and back. Nanoscale And Microscale Thermophysical Engineering, 11(1-2) (2007) 31–41.
53. Lee, F. C., Mills, R. N. and Talke, F. E., The application of drop-on-demand ink jet technology to color printing. IBM Journal of Research and Development, 28(3) (1984) 307–313.
54. Lee, E. R., Microdrop Generation, Vol. 5. CRC Press, Taylor and Francis, 2003.
55. de Gans, B.-J. and Schubert, U. S., Inkjet printing of well-defined polymer dots and arrays. Langmuir, 20(18) (2004) 7789–7793.
56. Kawase, T., Shimoda, T., Newsome, C., Sirringhaus, H. and Friend, R. H., Inkjet printing of polymer thin film transistors. Thin Solid Films, 438 (2003) 279–287.
57. Deegan, R. D., Bakajin, O., Dupont, T. F., Huber, G., Nagel, S. R. and Witten, T. A., Capillary flow as the cause of ring stains from dried liquid drops. Nature, 389 (1997) 827–829.
58. Deegan, R. D., Bakajin, O., Dupont, T. F., Huber, G., Nagel, S. R. and Witten, T. A., Contact line deposits in an evaporating drop. Physical Review E, 62(1) (2000) 756–765.
59. Deegan, R. D., Pattern formation in drying drops. Physical Review E, 61(1) (2000) 475–485.

60. Popov, Y. O. and Witten, T. A., Characteristic angles in the wetting of an angular region: Deposit growth. Physical Review E, 68(3) (2003).
61. Popov, Y. O., Evaporative deposition patterns: Spatial dimensions of the deposit. Physical Review E, 71(3) (2005).
62. Bigioni, T. P., Lin, X. M., Nguyen, T. T., Corwin, E. I., Witten, T. A. and Jaeger, H. M., Kinetically driven self assembly of highly ordered nanoparticle monolayers. Nature Materials, 5(4) (2006) 265–270.
63. Hu, H. and Larson, R. G., Marangoni effect reverses coffee-ring depositions. Journal of Physical Chemistry B, 110(14) (2006) 7090–7094.
64. Park, J. and Moon, J., Control of colloidal particle deposit patterns within picoliter droplets ejected by ink-jet printing. Langmuir, 22(8) (2006) 3506–3513.
65. Bormashenko, E., Bormashenko, Y., Pogreb, R., Stanevsky, O. and Whyman, G., Droplet behavior on flat and textured surfaces: Co-occurrence of Deegan outward flow with Marangoni solute instability. Journal of Colloid and Interface Science, 306(1) (2007) 128–132.
66. Kinge, S., Crego-Calama, M. and Reinhoudt, D. N., Self-assembling nanoparticles at surfaces and interfaces. Chemphyschem, 9(1) (2008) 20–42.
67. Bonn, D., Eggers, J., Indekeu, J., Meunier, J. and Rolley, E., Wetting and spreading. Reviews of Modern Physics, 81(2) (2009) 739–805.
68. Craster, R. V. and Matar, O. K., Dynamics and stability of thin liquid films. Reviews of Modern Physics, 81(3) (2009) 1131–1198.
69. Bhardwaj, R., Fang, X. H., Somasundaran, P. and Attinger, D., Self-assembly of colloidal particles from evaporating droplets: role of DLVO interactions and proposition of a phase diagram. Langmuir, 26(11) (2010) 7833–7842.
70. Flory, P. J., Principles of Polymer Chemistry. Cornell University Press, London, 1953.
71. Ilyina, E. and Sillescu, H., Toluene self-diffusion in solutions of linear and cross-linked polystyrene. Polymer, 36(1) (1995) 137–141.
72. Masaro, L. and Zhu, X. X., Physical models of diffusion for polymer solutions, gels and solids. Progress in Polymer Science, 24(5) (1999) 731–775.
73. McDonald, P. J., Godward, J., Sackin, R. and Sear, R. P., Surface flux limited diffusion of solvent into polymer. Macromolecules, 34(4) (2001) 1048–1057.
74. Ueberreiter, K., The Solution Process. Academic Press, New York, 1968, pp. 219–257.
75. Narasimhan, B. and Peppas, N. A., The physics of polymer dissolution: Modeling approaches and experimental behavior, in: Polymer Analysis—Polymer Physics, Vol. 128, 1997, pp. 157–207.
76. Narasimhan, B., Mathematical models describing polymer dissolution: consequences for drug delivery. Advanced Drug Delivery Reviews, 48(2–3) (2001) 195–210.
77. Ionescu, R. E., Marks, R. S. and Gheber, L. A., Nanolithography using protease etching of protein surfaces. Nano Letters, 3(12) (2003) 1639–1642.
78. Ionescu, R. E., Marks, R. S. and Gheber, L. A., Manufacturing of nanochannels with controlled dimensions using protease nanolithography. Nano Letters, 5(5) (2005) 821–827.
79. Carré, A., Gastel, J.-C. and Shanahan, M. E. R., Viscoelastic effects in the spreading of liquids. Nature, 379 (1996) 432–434.
80. Saiz, E., Tomsia, A. P. and Cannon, R. M., Ridging effects on wetting and spreading of liquids on solids. Acta Materialia, 46(7) (1998) 2349–2361.
81. Pu, G., Guo, J., Gwin, L. E. and Severtson, S. J., Mechanical pinning of liquids through inelastic wetting ridge formation on thermally stripped acrylic polymers. Langmuir, 23(24) (2007) 12142–12146.

Modeling Approaches and Challenges of Evaporating Sessile Droplets

M. Antoni [a] **and K. Sefiane** [b]

[a] Aix - Marseille Université - Université Paul Cézanne, UMR - CNRS 6263 ISM^2
Centre St. Jérôme - BP 461 - Marseille 13397 Cedex 20, France
[b] School of Engineering, University of Edinburgh, Kings Buildings Edinburgh,
EH9 3JL, United Kingdom

Contents

A. Introduction 129
B. Challenges and Limitations in 135
C. FEM Approach and Illustration of Evaporating Droplets 139
D. Theoretical Model and Parameters 140
E. Simulation of the Hydrodynamics of Heated Water Droplets 146
F. Conclusion 154
G. References 155

A. Introduction

Droplet evaporation has historically been of interest in many different branches of science and engineering and describes the process whereby a single fluid droplet transforms into a vapour phase, by the absorption of the required latent heat from a heated element or the ambient. The heat transfer rate that can be achieved by drop-wise evaporative cooling is much greater than that of air cooling techniques, so is of fundamental importance in many areas involving heat transfers. The recent expansion in the micro-electronics industry and Microsystems in general has for example reignited interest in drop-wise evaporation as a potential cooling mechanism for microchips and microreactors. The number of transistors per integrated circuit board has increased exponentially since its invention and thus the cooling requirements have increased correspondingly to the point where traditional air cooled systems cannot meet anymore the required thermal load. More generally fuel droplets play significant roles in the spray and combustion processes. The performance and operation of gas turbines, industrial furnaces, diesel engines and rocket engines all depend on understanding droplet evaporation and combustion, and the ability to

Drops and Bubbles in Contact with Solid Surfaces
© Koninklijke Brill NV, Leiden, 2012

model and predict the evaporation of burning droplet behaviour. Worth mentioning also the more traditional sciences, like biology where droplet evaporation has many practical applications. The rate of droplet evaporation is for example critical to successfully develop herbicides and fertilisers in order to predict accurate dosing as well as determining how a droplet behaves upon contacting with a plant surface. More recently drop wise evaporation has been utilised as a means of "stretching" DNA molecules, used in DNA and gene analysis.

Because of the growing applications, many attempts have been made to successfully model the rate of droplet evaporation. But the complicated interaction of heat and mass transfer, have so far produced results with limited use. This is for example the case of the physical phenomenon of phase change that has proved to be historically one of the most fascinating topics in heat transfer. Experimental work has intensified over the past fifty years as more sophisticated experimental techniques, such as high speed imaging and infrared tools, allow better understanding of the evaporation regimes and their underlying physics. Experimental studies revealed that the evaporation of a single droplet is a complicated process that results from the interplay between fluid mechanics, heat transfer and surface chemistry. In this context, the detailed understanding of the physical and chemical mechanisms involved in the evaporation of heated droplets on a substrate is still in many aspects an open problem [1–11] in particular for the precise description of the heat/mass transfers that occur at liquid/vapour and liquid/solid interfaces.

The purpose of this chapter is to provide a review highlighting some challenges and answers in the kinetics of evaporating droplets in the case of pure droplets of millimeter size deposited on a heating substrate. This study is further motivated by the large variety of industrial applications involving droplet evaporation. Such systems are indeed important in biochemistry [12], new materials development [13, 14], paint industry and printing [15] as well as in innovative nuclear decontamination processes with low effluent volumes. All these systems have in common a bulk hydrodynamics coupled with interfacial properties. This coupling is actually one of the difficulties in the description of the mechanisms occurring in the evaporation of fluids and motivated in the past decades extensive fundamental research in molecular chemistry, fluid mechanics and thermodynamics. Molecules in fluids are subjected to electrostatic and Van der Waals forces. These interactions are superimposed to the ones governing the evolution of atoms within molecules and described by the time evolution of intra-molecular degrees of freedom. All these contributions lead to complex chaotic dynamics that appears at the macroscopic scale as Brownian motion. At liquid/gas interfaces, intermolecular forces are anisotropic in the normal direction to the interface and the molecules having a sufficient kinetic energy will overcome Van der Waals attractive forces and escape in the gas phase surrounding the droplets. The corresponding energy loss of the liquid leads, at the macroscopic scale, to evaporative cooling that might generate temperature discontinuities at the interface [16]. Finally, it is the balance between mass loss, due to evaporation, and mass increase, due to the condensation of molecules back on the interface, that determines the droplets mass evolution and therefore the change in time of their volume [17–19].

Evaporation and vaporization phenomena correspond both to a transition of a substance from its liquid phase to its gaseous phase. But the difference between these two transformations is their characteristic time. Evaporation is a slow process when compared to vaporization. Quasi-equilibrium hypotheses can hence be invoked. This will allow important simplifications of the models used in particular for the description of unsteadiness and to overcome difficulties in the implementation of time evolution. Due to the complexity of evaporative processes, a broad range of spatial and temporal scales are involved. This problem has led to the development of specific numerical approaches to simulate either hydrodynamic phenomena or the dynamics at the molecular scales. Despite well adapted modeling [20] and numerical methods, the available computing power is still too limited to include simultaneously in the simulations both the full detail of microscopic phenomena and hydrodynamics occurring at macroscopic scales. This is why we will propose in this chapter to illustrate our purpose with a simplified model dealing with droplet evaporation on the basis of a quasi-steady hydrodynamics. Molecular phenomena like liquid density changes due the anisotropy at liquid/air interfaces is not described explicitly as in molecular dynamic simulations although they contribute indirectly to the model through liquid/gas interfacial properties.

The evaporation of a volatile liquid in contact with unsaturated air is inevitable because of the diffusivity of the molecules of this latter in the surrounding atmosphere [21–23]. Generally it is a relatively slow process, as long as ambient and substrate temperature, remain sufficiently smaller than the saturation one. Evaporation can even stop when the atmosphere becomes saturated with the liquid vapor. The study of evaporation phenomena hence requires not only a precise description of the liquid but also needs to account for its vapor concentration when studying mass transfers. These latter are governed by molecular mechanisms that have a characteristic timescale of the order of 10^{-10} s [24], which is the typical timescale for molecular evaporation mechanisms. Another typical time is determined by the diffusion of molecules in the gas phase and given by the ratio R^2/D, where D is the diffusion coefficient of the vapor–liquid molecules in the air (typically $D = 0.2443$ cm^2 s^{-1} for water vapor in air) and R is the contact line radius that corresponds to the coexistence line between the three phases: liquid, solid and gas. R is about 1 mm for the droplets that will be considered in this chapter. These two estimates give $R^2/D > 10^{-5}$ s and show that the time characteristics of molecular mechanisms and diffusion phenomena differ by five orders of magnitude. The slowest mechanism will control evaporation and here it is diffusion in the gaseous phase. These estimates explain why millimeter droplets evaporate only within few minutes for moderate substrate temperature [25] and why they rapidly reach a quasi-steady state as evaporation takes place [26].

Besides the previous time scale estimations and quasi-steady properties, heated droplets also raise the specific problem of their coupling with the heating substrate on which they are deposited [27–29] and of the role of their wetting properties in the overall evaporation process [30]. A water droplet of few micro-liters on a hydrophobic substrate will adopt an almost spherical geometry, whereas in the case of wetting substrates (such as copper or aluminum) the same droplet will take a spherical cap shape. From the point of view of thermodynamics, such a system must exhibit a unique equilibrium contact angle θ given by Young's equation (in the limit of validity of this latter). However for evaporating droplets in a nonsaturated atmosphere, the mass balance is in favor of the gaseous phase where the chemical potential of liquid is smaller. The system is therefore out of equilibrium and the contact angle deviates from the equilibrium one. It becomes an out of equilibrium quantity and will exhibit a time evolution that will depend on the droplet characteristics.

The temperature T_s of the substrate plays key role in this process. If it is larger than the saturation temperature, the droplets evolve too quickly for a simplified model to be applicable for its description. For smaller temperatures however, the water comes in close contact with the substrate and, if it is made of copper or aluminum, the droplet will wet it. In this chapter, the surrounding air has atmospheric pressure. Under these conditions, the evaporation dynamics shows several well identified evolution regimes that do not depend solely on the hydrophobic or hydrophilic nature of the substrate. Experimental conditions and in particular the way the droplet is heated are also important inputs here. Extensive experiments have been devoted to the description of droplets evaporation dynamics [31] and their analysis shows three main evaporation regimes [32]. In the first one, evaporation leads to a progressive reduction of the height (h) of the droplet and of its contact angle. The droplet keeps a spherical cap shape and its contact line is almost a circle. This regime is called pinned droplet regime since R remains constant. For water on aluminum or copper substrates, this pinned regime is also the longest since it corresponds to the evaporation of more than 80% of the initial droplet volume for moderate values of T_s. This first regime ends when the θ reaches a critical receding value that depends on temperature, on the nature of the substrate and of course on the liquid. For water droplets on copper or aluminum substrates, pinned regimes can be observed for contact angles as small as $\theta \approx 10$ degrees. The second evaporation regime is a receding contact angle regime. The contact line is no more pinned and both h and R decrease. The droplet progressively shrinks but with values of θ than can be constant or increase if there is recoil. Pinned and recoil regimes and are illustrated in Fig. 1. In the last evaporation phase, h, θ and R finally all simultaneously vanish. Among these three regimes, the first one is of fundamental interest not only due the important volume of evaporated liquid but also because of the possibility to use simplified hydrodynamic models for the droplet simulation [25, 26]. It allows indeed a step by step description of the hydrodynamic occurring within the droplet including the properties at the liquid/gas interface governing the heat and mass transfers.

Figure 1. Sketch representing a pinned evaporative regime (left) and a depinning one (right).

Theoretical models currently available in the literature treat evaporating pinned droplets having either spherical or ellipsoidal geometries [33]. These models give accurate predictions of evaporative rates in isothermal conditions (i.e., conditions where the droplet, the air and substrate are all at the same temperature). Droplet evaporation is then essentially controlled by the diffusion of molecules from the liquid phase into the gaseous one. Although giving good predictions, isothermal models do not always give a satisfactory agreement with the experiments [27, 29, 34] in particular when trying to adapt them for droplets on moderately heated substrates. From this perspective, full understanding of droplet evaporation phenomena is still incomplete. The hydrodynamics that might develop in heated substrate conditions is still, for example, in its exploratory phase. If heat dissipation by convective cells is well known for large volumes, only a very limited number of studies have been devoted to the hydrodynamics of small droplets [22–28, 38]. For such systems, the main difficulty is due to the fact that evaporation processes strongly depend on interfacial phenomena such as Marangoni driven convection that becomes fundamental when energy supply from wetting substrates have to be accounted for. The interfacial temperature difference between the contact line of a deposited droplet and its apex is generally less than a fraction of a degree Celcius, but such a small temperature difference is sufficient to lead to the development of thermocapillary instabilities [25] having their origin in the surface tension differences between 'warm' and 'cold' areas of the interface. These instabilities generate a Marangoni flow on the surface of the droplet that creates not only a shear on the adjacent liquid but also modifies heat transfers as well as liquid circulation. All the above discussions show that transport mechanisms of heat in small droplets are more complex than what is usually expected for millimeter size objects. Droplets involve not only diffusive based heat transfers but also convective phenomena. In droplet evaporation, we are hence primarily concerned with the Marangoni effect caused by temperature gradients within the droplet. They can be the result of direct heat transfer processes or of mass transfer processes involving enthalpy changes or of any other type of natural or forced flow. There have been several attempts to produce numerical simulations of droplet evaporation including or not the role of Marangoni effects. Most notable are the models proposed by Liao [35], Di Marzo *et al.* [36], and Lorenz and Mickic [17], which assume conduction into the droplet from the heated surface is the dominant mode of heat transfer. Each model produces results with varying degrees of success, but as far back as 1958, Zuiderweg and Harmens [37] reported the influence that internal fluid motion, and in particular the fluid circulation caused by surface tension gradients, had on the rate of mass transfer in unsupported distillation columns. Besides Marangoni driven convection, Rayleigh or gravity controlled convection can also occur as a result of the same temperature and concentration gradients which give rise to Marangoni convection. Temperature and concentration gradients can indeed cause density stratification within the fluid which can generate a fluid circulation balancing Marangoni convection. Recent works suggests that in situations in which the characteristic length is greater than the capillary length, Rayleigh convection can be the dominant mechanism. In contrast, Marangoni convection is dominant in situations in which the characteristic length is smaller than the capillary one.

Ruiz and Black modelled Marangoni convection in a hemispherical droplet placed in non-flow surroundings [38]. Their model, allowed the prediction of the evaporation rate in small diameter hemispherical droplets with pinned contact line. As a droplet evaporates, the contact line between solid and liquid travels over the heated surface making the contact line vary with time. Ruiz and Black also point to the fact that when the contact angle is less than 90° the evaporation kinetics can be split into two main regimes. For the most of the droplet lifetime, the contact line remains pinned and the contact angle decreases. In the second period the contact line recedes but only once the contact angle has become small. As the contact angle decreases the droplet approaches a thin film configuration and three dimensional motion of the droplet begins to occur. Craft and Black [34], have shown that the time taken for a droplet to de-pin is influenced by the temperature of the substrate. These authors also highlight the fact that after the drop begins to de-pin there is a step change in the rate of evaporation, which is not accurately modelled in the model of Ruiz and Black [38]. The authors themselves are quick to highlight the inadequacies in some of the assumptions used in their model. In particular, the assumption that the mass transfer coefficient from liquid to vapour can be estimated using the Reynolds analogy proposed by Incropera *et al.* [18]. This analogy assumes the rate of mass transfer from liquid to vapour can be estimated from the heat transfer characteristics. Using analogies in this manner is somewhat inaccurate as they are generally based on intuition rather than rigorous analysis. Anyhow, this model produced results with reasonable agreement with experimental observations for small diameter evaporating droplets and clearly indicates that the physics of mass transfer into the vapour phase can be better described using the diffusion equation as an alternative of the Reynolds analogy currently employed. Hu and Larson [39], have shown that as a droplet evaporates, the vapour concentration distribution satisfies the Laplace equation with time varying droplet surface. Hu and Larson's numerical simulations agree well with experimental data, implying that the rate of evaporation (i.e., mass transfer from the liquid into vapour) can be accurately predicted using a simple diffusion equation.

Despite still on-going important theoretical modelling and experimental efforts, it is clear that the best way to gain insight in the evaporation kinetics and hydrodynamics of volatile droplets is through accurate and robust numerical techniques. Developing such techniques is however far from being fully achieved and many attempts are still under way. This chapter critically appraises the merits of models developed to predict the evaporation of small sessile droplets with pinned contact line placed upon a heated surface. Furthermore, the results of a Finite Element Method (FEM) approach applied to specific evaporation conditions are presented and discussed in Paragraph D. This example will illustrate how far numerical CFD simulations can address droplet evaporation mechanisms and will simultaneously emphasize the inherent limitations of FEM.

B. Challenges and Limitations in CFD Simulations

This Paragraph discusses some still open questions in the numerical simulations of evaporating droplets deposited on substrates. An evaporating droplet is a typical example of unsteady multiphase system with free interfaces for which the liquid phase can be assumed as viscous and incompressible. Boussinesq approximation can be used here since temperature gradients in liquids are small due to their large heat conductivity (when compared to the one of gases). The explicit time dependence of the equations have, however, to be accounted for in the numerical schemes as well as heat and mass transfer mechanisms generated in particular by the phase change. Moreover, subsequent droplet volume reduction requires adapted front localization techniques to track the moving water/air free interface that, from the numerical point of view, corresponds to discontinuities where derivatives have to be handled and computed carefully. This is one of the first important challenges in the simulation of unsteady, multiphase systems with an additional complexity in evaporating heated droplets since liquid/air, liquid/solid and solid/air interfaces are all simultaneously involved. In free interface systems, it is of paramount importance to capture interfaces numerically, to keep their sharpness finite and, while processing that way, satisfy overall mass and volume conservation laws. As mentioned above, in the case where droplets are wetting the substrate, pinned evaporating regime is first observed and the only free moving boundary is the liquid–gas interface. Such moving interfaces are the most studied cases in two phase systems where no solid phase has to be accounted for, like sprays, atomization problems or emulsions (in this last case gas being replaced by the continuous liquid phase). A second important challenge in the simulation of droplets on substrates arises when contact angles become small enough for the fast contact line de-pinning regime to occur. All the aforementioned interfaces, and not only liquid–air, are then rapidly time-evolving. The interplay between the interactions involved in this evaporative regime still raises a number of open numerical questions that range from mass conservation constraints to a reliable description of all moving free interfaces. Also still challenging is the accurate description of interfacial properties, such as surface tension or adhesion forces of the liquid on the substrate [41, 42].

A complete description of droplet evaporation dynamics involves almost all possible spatial and temporal scales from macroscopic down to microscopic: macroscopic, for hydrodynamics (convection, heat transfer) and microscopic, for molecular contributions (liquid–solid adhesion forces, surfactant layers kinetics, etc.). This has led to the development of specific numerical methods for intensive CPU calculations to simulate either the hydrodynamic phenomena [43] with computational fluid dynamics (CFD) techniques or the mechanisms at the molecular scale with molecular dynamics (MD) [43]. In most of the investigations of evaporation processes, quasi-steady hypotheses are assumed. They have the important advantage to allow the dissociation of hydrodynamic time scales from the molecular ones. This approximation is usually well justified here since molecular processes are fast when compared to hydrodynamic ones and hence do not need to be described explicitly as done in mesoscopic approaches [44] or in MD simulations [45]. In continuum models, all molecular properties are averaged out and contribute only through equations of state. However, although justified in most of the studies, such a description is no more adapted when investigating contact line dynamics. Depending on the wetting properties of the droplet liquid on its substrate, we have already mentioned that de-pinning phenomena can be observed for sufficiently small contact angles. They are at the origin of complex contact line motions for which adhesion forces have to be accounted for. These forces act at the molecular level and should be investigated, strictly speaking, with a self-consistent molecular force field describing the interactions between the molecules of the liquid and the substrate (atoms or molecules) at the atomic scale. Despite intense algorithmic research to speed up MD simulations in particular with the use of averaged or effective self-consistent force fields, the available computing power is still too limited to investigate molecular systems with more than few tens of thousands of (small) molecules for more than few milliseconds [45]. This is still too limited to be efficiently coupled with CFD simulations in the specific case of evaporating systems. Indeed, the objective of any numerical algorithms is to make possible undertaking simulations which are as accurate as possible with a minimal CPU time. This goal is usually well achieved for CFD and MD when considered separately. But when studying evaporating systems, the question is how to associate both simulation techniques when such large scale separations have to be described. This still remains challenging in the simulation of evaporating systems, although models are available in the literature [46] and similar multi-scale problems (but at the nuclear and atomic scales) have been resolved for more than two decades in numerical chemistry [47].

The pinned droplet evaporative regime can be described with CFD using the Navier–Stokes and heat transfer equations coupled with continuity and specific boundary conditions equations. The key is then the development of adapted solvers for their integration in order to obtain well converged and stable hydrodynamic solutions. Important efforts are still devoted to solvers optimization but, among the actual challenges in computer science, optimal methods for the continuum formulation of sharp free moving interfaces are still one of the main concerns. The first difficulty in this context is to track interfaces with fast and robust algorithms in a way to limit numerical diffusion and to avoid their artificial smearing. Several methods, having each specific advantages and limitations, are available in the literature.

Boundary integrals allow for example a high order of accuracy in the description of interfaces for potential flows (irrotational or Stokes flows) but are not easily applicable to viscous flows governed by Navier–Stokes equations such as the one describing droplet evaporation [48, 49]. On the other hand, Lagrangian approaches allow an explicit description of the interface location and rely on the resolution of the interface equations on a small number of grid nodes. However, they impose non trivial handling of these latter for mass conservation purposes, when highly distorted or complex geometries have to be treated. Approaches coupling finite volume methods with a structured grid together with an immersed set of connected marker points for explicit interface tracking have also been developed [50]. Always for front tracking, Volume of Fluid (VOF) methods are mass conservative and allow the description of systems with non trivial geometries without complex grid nodes restructuring [51]. They belong to the family of Eulerian techniques and treat the simulation domain as a whole (i.e., a single fluid), which makes them easier to formulate for full 3D geometries. But the price of their simplicity is the loss of interface sharpness and the necessity to introduce specific methods for its preservation. In the case of multiphase systems like evaporating droplets, this approach uses a function acting like a fluid tracer and taking a specific value for each fluid. This function is advected as time evolves and used to compute the interface curvature which is incorporated into the Navier–Stokes equation where it contributes as a source term. Each of the previous numerical approaches is more or less adapted to specific problems but, among all, VOF methods have the notable advantage of robustness even for basic structured discretization meshes when using implicit finite volume schemes. When used with high order integration schemes, VOF methods have the ability to handle most of the changes in the shape of the interfaces [52]. They hence appear to be adapted to catch important topological transitions in the interface geometry such as the appearance of residual droplets in experiments that sometimes show up in the ultimate evaporation regime. But, the use of implicit integration schemes leads to constraints in time stepping that have to be treated carefully. The requirement to fulfill the CFL condition for numerical convergence can indeed lead to prohibitive CPU times due to too small time steps. Satisfactory compromises between time and space discretization steps can usually be found to overcome this difficulty.

In the context of free interfaces capturing, the generalization of VOF approaches to systems with more than two phases and moving boundaries like in evaporating droplets is still to be developed. Adapted expressions for the sources to be introduced in the momentum equation are yet expected to provide stable simulations and new insights not only for receding contact line regimes but also for the pinned one. Indeed, besides liquid–gas interfaces, solid–liquid ones can also be time evolving in this regime due to chemical reactions such as corrosion or electrolytic deposition. The actual contribution of these effects can be regarded as marginal in pinned regimes, but are key in contact line de-pinning processes. They indeed modify the substrate roughness and consequently contribute to enhance or weaken the intensity of contact line adhesion forces.

Besides interfaces tracking, a second important problem in evaporating droplets is the numerical description of the liquid–gas surface tension. This task can be quite naturally achieved in Lagrangian methods (for example FEM) whereas for the VOF techniques, it first requires to map interfacial forces to volume ones. As already discussed, Lagrangian methods are better adapted here, but only as long as droplets keep a simple geometry for all the duration of their evaporation. As a result, cusps or topological changes like the formation of smaller droplets after splitting of a large one cannot be simply assessed. In VOF approaches, this becomes possible naturally since all the phases are treated on the same basis, but the computation of the surface tension contributions requires more CPU efforts. In fact, the pressure jump across the interface is proportional to its curvature and needs the computation of a large number of spatial derivatives of the colour function [53] imposing the use of higher order schemes. This task can be accurately achieved with weighted, essentially non oscillatory schemes [54] that are well adapted to hyperbolic problems like colour function advection and sharpness preservation of interfaces with only little additional CPU time. But, still due to numerical finiteness, unwanted numerical artifacts consisting in spurious velocities (the so called parasitic currents) show up in the vicinity of the interface. Fortunately, these latter become non negligible only for slowly evolving interfaces (slowly with respect to thermocapillary convection time scales). This means that contact line regions of pinned droplet regimes will be strongly affected by such numerical artifacts. The magnitude of the parasitic currents can be reduced by both the inertial and viscous terms in the Navier–Stokes equations but not with increased mesh refinement or decreased computational time step [55]. Efforts are still devoted to interface sharpness conservation with improved interface reconstruction algorithms (PLIC-VOF) [56], second gradient techniques [57] or hybrid methods [57–59]. Despite much work, modeling interfacial tension and capillary phenomena is still a maturing domain in computer science. Many new improvements in the numerical algorithms are still to come and will provide in the next future new techniques to overcome parasitic currents and hence allow VOF description of both fast and slow evolving free interfaces.

Progress is also expected when incorporating soluble or insoluble surface active agents (surfactants) in the theoretical models. While very important literature has focused on the numerical investigation of clean interfaces, only a very limited number of studies attempted to include the effect of surfactants. For small concentrations, surfactants are all collected on the interfaces and form molecular layers at the origin of capillary instabilities due to the decreasing surface tensions. They generate similar phenomena to the ones resulting from temperature changes and lead to non uniform Marangoni stresses that significantly modify not only droplets hydrodynamics but also evaporative rates [60]. Like for the previously discussed interfacial tracking or control of parasitic currents, computation of interfacial flows with surfactants still raises challenging questions on both algorithmic and theoretical modeling levels. The first condition here is clearly the preservation of interface sharpness. Unless this condition is fulfilled, finite size effects will enter into play and numerical diffusion takes over. This will lead up to numerical outputs where insoluble surfactants can, for example, drop off the interface. Surfactant concentration follows a convection-diffusion equation that describes its time evolution on the interface as well as its solubilisation. The problem of surfactant incorporation in numerical models is therefore mapped in a convective-diffusive equation coupled with the Navier–Stokes equation in a highly non-linear way. Computational studies involving surfactants have first focused on Stokes flows using hybrid methods with unstructured grids [61] or boundary integral techniques [62] and it is only less than one decade ago that interest has really grown for the use Navier–Stokes equation and VOF methods [63–65]. Among the still ongoing algorithmic improvements, one important question here is the actual possibility for CFD techniques to describe surfactant layers in evaporating droplets. The reducing volume of the liquid phase indeed implies evolving adsorbed and solubilized surfactant concentrations. Assuming local equilibrium hypothesis to hold, the adsorbed surfactant concentration value can be obtained in the simplest case from the Langmuir isotherm. But in practice, surfactant layers contain a tangential stress generated by the molecular interactions that can make them unstable if an additional external stress brings them to exceed the equilibrium spreading pressure. Beyond this pressure, surfactant layers become too stiff to remain on the interface and water–air surfactant monolayers are for example known to collapse through a fracturing process [66]. Such collapses are expected to happen in evaporating systems where thermocapillary convection and mass reduction contribute both to locally enhance interfacial stress. In the presence of surfactants, interfaces can thus behave as inelastic systems including rupture thresholds. This clearly means that it is not always sufficient to assume only convective and diffusive contributions in surfactant layers modeling and simulations. The implementation in CFD methods of additional force contributions like the ones used for the description of self organized criticality could be one possible way for modeling the collapse of surfactant layers.

C. FEM Approach and Illustration of Evaporating Droplets

Amongst the different numerical techniques presented above, the finite element method (FEM) with remeshing of the simulation domain is the one that allows the simplest tracking of free interfaces. But it can be CPU demanding in particular for too complex interface topologies because of the prohibitive number of mesh cells necessary to describe such configurations. This technique is hence adapted to simulations involving interfaces with simple geometries where the number of mesh cells remains reasonably small. One first condition for FEM to be efficient is therefore to focus on interfaces with simple topologies. Another important condition is to avoid too frequent re-meshing of the simulation domain. This might be necessary to properly track the interface and capture its evolution with time. FEM approaches are indeed suitable when time scales can be separated in a way to finally focus on a system presenting a time evolution dominated by a single and well defined phenomenon. Both previous conditions are to be satisfied when considering the evaporation of small pinned droplets on moderately heated substrates. In the case of water droplets of few micro-liters deposited on copper rough substrates, evaporation shows a pinned regime where droplets keep simple spherical cap geometries and where the relevant time scale is given by the time evolution of θ (or equivalently h). The latter turns out to be relatively slow and lasts from few tens of seconds up to several tens of minutes for moderate substrate superheats. All other time scales, usually which are much faster like the ones associated with contact line fluctuations and molecular diffusion, can be averaged and further used to justify quasi-steadiness hypotheses used for model simplifications.

D. Theoretical Model and Parameters

In this Paragraph we present the approximations and the basic equations that are implemented numerically to illustrate this chapter with simulations of water droplet evaporation on heated copper substrates. The volume of the droplets that we will consider in the following is of a few micro-liters. This makes negligible the hydrostatic pressure contribution when compared to surface tension one and explains why such small droplets keep a spherical cap geometry in their first evaporative regime [8, 33, 38]. This system reduces to a problem with an axial symmetry that presents the important advantage to allow a full description of the droplets by limiting its study to only one of its section (i.e., to a single plane containing the symmetry axis). This will simplify the equations of the theoretical model, ease their numerical handling and significantly speed up simulations that will be restricted to a two dimensional space instead of a full three dimensional one. The wetting properties of water droplets on substrates such as aluminum or copper show the existence of pinned evaporation regimes where the contact angle θ decreases with time while contact radius R remains constant. As discussed previously, the parameters θ and R will be two of the basic variables used for droplet description and are illustrated in Fig. 2. A third parameter denoted by L_s is also necessary here and presented in Fig. 2 and Fig. 3. It corresponds to the radius of the heated zone of the substrate (substrate heater) that we assume to be a disk located in such a way to preserve the axial symmetry of the droplet/substrate system as well as the planar shape of the substrate (see Fig. 3). When $L_s/R < 1$ the substrate heater is smaller than R as illustrated in Fig. 2(a) and it is only the central part of the droplet basis that is heated up. Conversely, when $L_s/R > 1$ (Fig. 2(b)), the substrate heater is greater than the contact radius and it is not only the complete droplet/substrate interface that is heated but also the air nearby the contact line. L_s is a fundamental parameter in evaporation studies in particular for the description of evaporative mass flows [34, 48] and further opens interesting perspectives

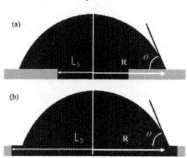

Figure 2. Droplet on a heated substrate for two different sizes of the heater (in dark gray). In (a) the heater is smaller than the contact radius ($L_s < R$) while in (b) it is larger ($L_s > R$). The contact angle θ, contact radius R are also represented as well as the symmetry axis (white vertical line).

Figure 3. Substrate characteristics. The heated zone (substrate heater) is at temperature T_s and is represented in black.

in the understanding of the hydrodynamics inside the droplets as will be illustrated below.

Due to the spherical cap geometry of the droplets and their pinned dynamics, mechanisms including the contact line dynamics and Van der Waals forces acting between water and the substrate do not have to be explicitly described here. Depinning or receding contact lines situations will also run out of the scope of the present chapter [25, 30]. As we aim to describe here the droplet hydrodynamics, it results here that Navier–Stokes and heat equations for incompressible viscous fluid will be sufficient for our purpose [23, 66]. Boussinesq approximation is another important simplification that will be used. It was discussed above that its use is justified by the temperature differences that are too small to generate significant changes in the thermodynamic properties (density, heat conductivity, dynamic viscosity) of water. As only pinned droplets will be considered R is fixed. Finally, water evaporation is assumed to be a slow process when compared to the equilibration time at the water/air interface. One direct consequence of these approximations is that partial derivatives with respect to time can be neglected in both Navier–Stokes and heat equations. In this latter, viscous dissipative contributions will also be neglected. In this context water circulation and heat transfer inside the droplet can be described by the two stationary equations [67]:

$$(v \bullet \mathrm{grad})v = \nu \Delta v, \tag{1}$$

$$(v \bullet \mathrm{grad})T = k \Delta T, \tag{2}$$

where ν (resp., k) is the kinematic viscosity (resp., thermal diffusivity) of liquid water.

For the gas phase, the same assumptions as for water will be used. It is treated as an ideal gas mixture with a vapor saturation that is known for all temperatures [68]. As liquid water and air have viscosities that differ by orders of magnitude, the shear stress generated by Marangoni circulation on the gas phase can be neglected. Air is therefore assumed to remain still above the droplets for the complete duration of evaporation. It is hence not necessary to use the full Navier–Stokes equation for its description since the convective term does not come into play. This approximation might be restrictive considering the possible development of convective motion in air due to the temperature differences between air adjacent to the substrate and in the far field or to relative density differences between air and the vapor water. This latter is lighter than air and can convect away from the droplet to the far field because of buoyancy forces. Dragging forces due to this flow can then generate circulating motion in air. This effect will be neglected in the following. With all the previous approximations, description of air can finally be limited to a simple Laplace equation [26]:

$$\Delta T = 0. \tag{3}$$

Equations (1)–(3) have to be further completed with the appropriate boundary conditions. In the far field of the simulation domain, air is assumed to have everywhere the same temperature denoted by T_∞. The substrate is assumed to be constituted by two embedded ideal heat conductors that are supposed to be thermally isolated from one another (see Fig. 3). The inner part is at temperature T_s whereas the temperature of the outer one is fixed either by the one of the water droplet (when $R > L_s$) or air (when $R < L_s$).

Continuous temperature conditions are imposed at water/air boundaries for all times but at the initial condition (this will be discussed below). Temperature derivatives on the symmetry axis of the system are kept to zero while one of all the other boundaries (droplet/air interface and droplet/non heated substrate region) is determined from the solutions of the Eqs (1), (2) and (3). Concerning velocity boundary conditions, normal (respectively radial) components are set to zero in the substrate vicinity (respectively on the symmetry axis) whereas tangential constraints are set up at the water/air interface. Finally radial derivatives of the velocity field are fixed to zero on the symmetry axis.

For the FEM approach efficiency and to avoid numerical precision problems, we use a dimensionless expression of the previous equations. We introduce to this end the usual Reynolds (Re) and Prandtl (Pr) numbers, dimensionless radius r^* and heigh z^* defined respectively by $r^* = r/R$ and $z^* = z/R$ as well as dimensionless velocity (v^*) and temperature (T^*) defined by:

$$v^* = \frac{v}{U}, \tag{4}$$

$$T^* = \frac{T - T_\infty}{T_s - T_\infty}, \tag{5}$$

where U is the characteristic velocity of the system. Equations (1)–(3) then write:

$$(v^* \bullet \text{grad})v^* = \frac{1}{\text{Re}}\Delta v^*, \tag{6}$$

$$(v^* \bullet \text{grad})T^* = \frac{1}{\text{Re}\,\text{Pr}}\Delta T^*, \tag{7}$$

$$\Delta T^* = 0. \tag{8}$$

Due to the axi-symmetry of the system, these equations are expressed in dimensionless cylindrical coordinates (r^*, z^*). The detailed construction of equations (6), (7) and (8) is described in Ref. [26]. They constitute the basic equations that will be simulated in the following Paragraph. It is important to note here that due to the approximations that are used for their construction and the fact that the substrate is assumed to be ideal, these equations describe the evaporating droplet problem on the basis of three quasi-steady subsystems: the droplet, the substrate and the surrounding air. The coupling between these subsystems is determined by the properties at each of the interfaces and in particular at the water/air one where evaporation takes place. This separation into subsystems is justified as long as the substrate can be assumed to be an ideal heat conductor and evaporation remains slow when compared to the setting of local equilibrium in both air and water. As the typical time in this problem is the time for the complete evaporation of the droplet, this hypothesis holds as long as T_s remains small compared to the water vaporization temperature. This is the reason why T_s is always set to values smaller than 70°C in all the numerical results presented hereafter. Another important point to be discussed is the use of Eq. (8) for temperature in air. This equation implies that hydrodynamics is neglected but the use of this equation also ignores the fact that air is a gaseous mixture in which vapor water is expected to diffuse from regions where the relative humidity is high (i.e., near the droplet interface) towards regions where it is smaller (i.e., droplet far field). The characteristic time of this diffusive phenomenon is given by water concentration gradients in air. The larger they are, the faster will be evaporation when the droplet is isothermal [11, 19, 39, 69]. In non isothermal situations, temperature gradients will contribute as an addition and reinforce concentration gradients due to changes in the value of the local saturated vapor concentration. The vapor water diffusion in air is hence driven by temperature gradients in the neighborhood of the droplet interface where they will be enhanced due to the small thermal conductivity of air.

In all the above equations time does not appear any more due to the quasi-steady hypotheses that have been adopted. However, evaporative phenomena are clearly time evolving and explicit time dependence has to be reintroduced in the model. This is done using the same approach as the one for isothermal systems when neglecting evaporative cooling [8, 70]. In nonsaturated air and in isothermal situations, the driving mechanism for evaporation is the nonhomogeneity of relative humidity in air. Air is saturated with water vapor at the droplet interface and in the far field relative humidity takes a value denoted by H. In this context it is actually possible to treat the mass transfer for an evaporating droplet on a rigorous mathematical basis. This problem can indeed be mapped into the calculation of the electrostatic potential generated by a conductive surface obtained by the intersection of two eccentric spheres forming a biconvex lens. The analogy with droplet evaporation provides here the following equation for the mass loss [39, 70, 71]:

$$\frac{dm}{dt} = -DR(c(T_s) - Hc(T_\infty))\phi(\theta), \tag{9}$$

where m is the droplet's mass, $c(T)$ is the vapor water concentration at temperature T, D water vapour diffusion coefficient in air and $\phi(\theta)$ is a dimensionless function of the contact angle reflecting geometrical effects. All other notations have been defined. The main advantage of this equation is to bring back the time dependence in the model and, in doing so, helps address the problem of water circulation in deposited and heated droplets. This approach is justified by the fact that temperature differences inside the droplet and along its interface are all small for the initial conditions that will be considered in the following. It is hence reasonable to assume that Eq. (9) also holds here, even if it has been essentially developed in the literature for isothermal situations. The small temperature gradients that the system will develop will only slightly perturb the vapor water concentration in the neighborhood of the interface. But overall thermocapillary phenomena will ensue. Surface tension temperature dependence in the temperature range of the present section ($30°C \leqslant T_s \leqslant 70°C$) is described by equation:

$$\sigma = \sigma_0 + \beta \times (T - T_0), \tag{10}$$

where $\sigma_0 = 72 \times 10^{-3}$ N/m and $\beta \approx -1.6 \times 10^{-4}$ N/(m K) are respectively the surface tension at temperature $T_0 = 298$ K and the temperature coefficient of surface tension that is supposed to be temperature independent. Temperature gradients along the interface are key here and this is where special numerical efforts have to be devoted in particular for an accurate computation of v^* and the precise integration of Eqs (6) and (7) in the computational domain describing the water droplet (see Fig. 4). Droplet evaporation is always accompanied by a temperature decrease due to evaporative cooling [7, 31]. This effect has been accounted for in the literature and is of first importance in isothermal systems where there is no heat input to balance the corresponding energy loss. It might also become important in the neighborhood of the contact line due to large mass flux. Considering the small size of the droplets and the relative large difference $T_s - T_\infty$ (it is always larger than $10°C$), evaporative cooling is assumed to be compensated rapidly and at all times by the important substrate incoming heat flow.

As discussed before, the numerical method that is used to simulate Eqs (6)–(8) is based on stationary hypotheses and hence treats evaporating droplets as coupled thereafter subsystems. No free interfaces have to be accounted for in this approach. The copper/water and copper/air interfaces are fixed in time whereas the water/air evolves according to the mass evolution given in Eq. (9). For each time step, specific finite elements calculation stages are performed for their description. The first

(a)   (b)

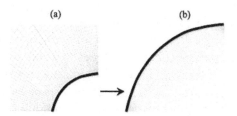

Figure 4. Presentation of the unstructured mesh grid that is used for the integration of the model. (a) Mesh grid in the neighborhood of the droplet. (b) Shows a magnification of the mesh used to describe the droplet. Black line is the water/air interface.

step consists in the resolution of Eq. (8) that gives the temperature everywhere in the vapor phase and therefore also nearby the water/air interface. The computed temperature at the interface is then used as a boundary condition input for Eq. (7) that is then integrated together with Eqs (6), (7) and (10). From this second step, the function $\phi(\theta)$ is obtained. In the last step, the time evolution of the droplet mass is computed from Eq. (9) with a basic leap-frog scheme. As droplets are assumed to have a spherical cap geometry, the mass loss corresponds to a decrease of θ that can be readily computed from the droplet's mass and that is finally used to evaluate the new location of the water/air interface for the next time step round.

Hydrodynamic simulations are based on the discretization of space by a mesh grid that has to be adapted to the problem under investigation. For the simulation of water/air interfacial properties, a special care has to be devoted to the mesh grid structure nearby this interface and at both the droplet apex and contact line. Singularities in the hydrodynamics show up in these two locations and unstructured meshes have to be used to properly catch all the features occurring in these interfacial regions. Figure 4 illustrates the mesh structure that will be used in the following as well as its refinement in the neighborhood of the interfaces. Inside the droplet and in the far field a coarse meshing is imposed and will be sufficient for our purpose. Figure 4 displays only part of the simulation domain and in all the runs, far field boundary is actually rejected at a distance of $160R$ that was shown to be large enough for the finite size of the simulation domain to become negligible. As substrate is assumed to be ideal here, there is no need to describe it explicitly since its influence appears only through the water/substrate boundary conditions described by T_s and L_s [48]. Several works avoid ideal substrate approximations and deal with substrates with finite heat conductivity. It was shown that temperature gradients can occur at the liquid/substrate interface as well as inside the substrates and that they can contribute to changes in the hydrodynamics of the droplets [34].

As evaporation proceeds, the droplet volume decreases and a remeshing is used for each time step to keep track of the water/air interface. This procedure is tedious and CPU time consuming but it ensures the accurate description of the interface location as well as properties especially for the implementation of Eq. (10). As this latter only comes into play at the level of the water/air interface, is it discretized only with the mesh cells located at this boundary (see black line in Fig. 4).

The model and the treatment of evaporation under consideration here are adapted to only small pinned droplets keeping a spherical cap shape and evaporating slowly. This excludes all the movements of the interfaces that could lead to complex geometries and renders the front tracking FEM approach inadequate. This is in particular true for contact line dynamics where molecular forces have to be accounted for when describing receding contact angle regimes. Although they are known to play a key role in evaporation of liquids on substrates, they only come into play either in the very early or very last evolution stage of droplets evaporation and hence run out of the scope of the illustration of FEM's we aim to give here. The quasi-steady approximations used to obtain Eqs (6), (7) and (8) and the FEM on which numerics will be based on are hence applicable only to deposited droplets on substrates with pinned contact lines that is to say to moderate substrate temperatures. This is why T_s will be restricted to the range 30°C–70°C in the following and with a far field temperature that will always be set to 20°C. Although restrictive, we will illustrate in the next Paragraph that these conditions and the FEM approach allows a precise study of the circulation of water in the droplets and the description of the properties of thermocapillary effects. These conditions have also the advantage to avoid the implementation of unsteady multiphase algorithms for which interface tracking and numerical diffusion control require much more efforts [73] than for FEM techniques in particular when phase changes have to be accounted for.

E. Simulation of the Hydrodynamics of Heated Water Droplets

When performing numerical simulations, whatever the numerical technique is, one first problem to tackle is the initial conditions of the system. For evaporating droplets investigations, some variables are fixed for the complete duration of the simulations (T_s, T_∞, H, L_s and R) whereas others evolve in time. Among these, some are initially free (θ, T, v) and others (m) are deduced from these free variables. In the following $T_\infty = 20°C$, $H = 20\%$ and $R = 1$ mm for all the simulations. Initial values of θ, T and v have also to be set for the simulations to start with. They will be the same for all the runs and are given in the table below:

Pressure will be set to the atmospheric one. With these conditions the only free parameters will be associated with the substrate properties T_s and L_s. Finally as

Table 1.
Initial values of the free variables θ, T, v

Variable	Initial value
θ	$= 80°$
T	$= T_s$ everywhere inside the droplet
	$= T_\infty$ everywhere outside the droplet
v	$= 0$ m/s

there is a one to one correspondence between time and contact angle we will only consider in the following contact angles. The smaller they will be the closer the droplet will be to complete evaporation.

This choice of initial conditions implies that the droplet is initially at the temperature of the substrate (T_s) and that air is at the temperature of the far field (T_∞). The temperature across the water/air interface is hence initially discontinuous but will be smoothed out right after the first time step of the simulations. Simulations are coherent with the temperature continuity condition imposed in the model. Practically, such an initial condition corresponds to an experimental procedure where water droplets would itself be initially at temperature T_s before being deposited on the substrate that would be rapidly heated up to T_s right before this deposition. This initial condition has been used for the validation of the model of Paragraph D. Literature proposes numerous experimental data of evaporating droplets describing more or less precisely working conditions [10, 30, 32, 34]. The large number of parameters required for a complete experimental description of such a system and especially the technical difficulty to properly control all of them is an important limiting factor regarding to simulation validation. This difficulty is probably one of the major obstacles to carry out reliable comparisons between experimental results and simulations where all the parameters have to be known. Controlled experimental conditions coherent with the model presented here are available with $T_s = 60°C$, $T_\infty = 26°C$, $H = 40\%$, $R = 0.515$ mm for micro-liter volume droplets [34]. FEM numerical simulations carried out for these systems show a good agreement for the time evolution of the droplets volume provided the radius of the substrate heater L_s is set to $L_s \approx 1.5R$ [25, 48]. This indicates that, although very simplified, the model of Paragraph D is sufficient to provide relevant insights for pinned evaporating water droplets.

Figure 5 displays the dimensionless temperature T^* in air when $L_s = 1$ mm and $T_s = 60°C$ for different contact angles (i.e., different successive times). This representation is restricted to the left half section of the droplet (the complete configuration can be readily obtained by axi-symmetry). At a given contact angle (i.e., a given time) a fast decrease of T^* due to the poor heat conductivity of air is observed when moving from the droplet interface to the far field in the normal direction to the interface [26, 76]. The corresponding gradient is almost constant everywhere when $\theta = 80°$ and increases at the contact line level as contact angle decreases. This increasing trend is controlled by the value of L_s and occurs only for $L_s/R \leqslant 1$ while when $L_s/R > 1$, air nearby the contact line is heated by the substrate up to temperature T_s, and temperature gradients become smaller. It results here that when $L_s/R \leqslant 1$ (resp., $L_s/R > 1$) mass flux is increased (respectively decreased) in the vicinity of the contact line.

The rising trend of the mass flux nearby the contact line can generate important contributions to the overall mass loss of the droplets. In the case $L_s/R = 1$ illustrated in Fig. 5, it is initially the same as everywhere else on the interface as suggested by the uniform normal gradients of T^* of Fig. 5(a) and as discussed. But

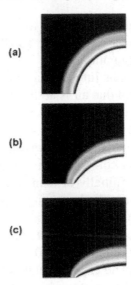

Figure 5. Dimensionless temperature T^* in air for $T_s = 60°C$ and $\theta = 80°$ (a), $\theta = 50°$ (b), $\theta = 20°$ (c). The water/air interface is at a temperature close to $60°C$ whereas the far field is at $T_\infty = 20°C$, $R = L_s = 1$ mm. This representation is limited to the half left section of the system for clarity. The water droplet shows up in the images as a white spherical cap in the lower right corner of each image.

this situation changes as time elapses and Fig. 5(b) and (c) indicate that contact line temperature gradients become larger as θ decreases. Evaporation is therefore governed by two competitive effects: a first one associated with the progressive area reduction of the water/air interface that tends to reduce the available surface for the mass flow to occur, and a second one due to the diverging mass flux close to the contact line. When $\theta > 25°$ this balance is in favor of the area reduction and mass flow decreases whereas when $\theta < 25°$ it is increased due to the strong increase in mass flux nearby the contact line. The model presented in Paragraph D and the FEM simulations hence suggest that evaporation kinetics of a heated droplet is determined by the balance between the progressive inhibition of mass flow due to smaller water/air interface available for the evaporation to take place across and an enhancement of this latter due to the rise in mass flux at the contact line.

Quantitative values of T^* are not plotted in Fig. 5 but can be obtained from Fig. 6 where radial dependence of T^* is displayed for $T_s = 30°C$ to $70°C$ with $L_s/R = 1$. Each curve of (a) (respectively (b)) corresponds to the value of T^* along the water/air interface as a function of substrate temperature when the contact angle $\theta \approx 70°$ (resp., $\theta \approx 30°$) is reached. Both figures indicate very small temperature differences between the droplet apex and the contact line. When $T_s = 70°C$ for example, the temperature at the droplet apex is $T^* = 0.994$ ($T \approx 69.7°C$) and it increases to $T^* = 0.996$ ($T \approx 69.8°C$) when $\theta = 30°$ due to the increasing substrate proximity. Temperature differences along the interface are only few fractions of degrees. These relatively small values can be explained by the initial conditions that

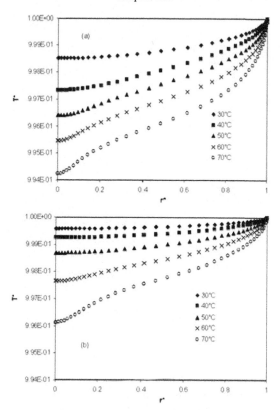

Figure 6. Snapshot of the dimensionless temperature T^* as a function of dimensionless radius r^* for different values of the substrate temperature T_s, with $L_s = R = 1$ mm. (a) (respectively (b)) shows T^* along the interface when $\theta \approx 70°$ (respectively $\theta \approx 30°$). The value of T_s is given in legend. $r^* = 0$ is the droplet apex and $r^* = 1$ is the contact line.

are considered here. The droplets are indeed initially exactly at the same temperature as that of the substrate and the only way to create temperature gradients is heat transfer from the droplet to cooler surrounding air. As the droplet apex is farthest from the substrate (and thereafter from the heat source) it is clear that the cooling effect of air, although very limited, will be most efficient in this region. Conversely, in the contact line region, such a cooling is no more possible due to the dominating influence of the substrate.

Although relatively small, temperature differences along the droplet interface illustrated in Fig. 6 will locally modify the interfacial tension according to Eq. (10). The resulting Marangoni stress will generate a shear in the adjacent water and finally induce thermocapillary motion [26, 34, 48, 70, 75]. The water streamlines are shown in Fig. 7 together with the corresponding isotherms. As water is initially at temperature T_s, temperature gradients inside and at the droplet interface are small (Fig. 7 right section) but sufficient to trigger thermocapillary convection (Fig. 7 left section). This figure also shows the existence of a convective cell that flattens

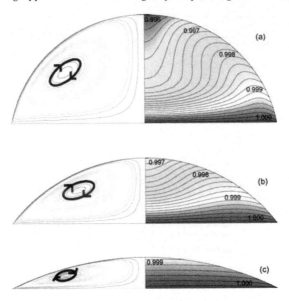

Figure 7. Streamlines (left section) and isotherms (right section) in a droplet for $T_s = 60°C$ and several values of θ. $\theta = 80°$ (a), $\theta = 50°$ (b), $\theta = 20°$ (c). Black arrows show the circulation of the velocity field. Some values of T^* are also indicated. $L_s = R = 1$ mm.

as droplet height decreases. Regarding Marangoni velocity, its largest magnitude is several tens of mm/s and indicates that water circulation in the convective cells can be relatively fast and still accurately described by FEM' approach. This circulation will enhance the heat transfer from the substrate to the droplet interface. But, although important for thermocapillary phenomena, it has been shown that it does not influence significantly the overall evaporative mass flow and hence, the evaporation kinetics of the droplets [6].

One more point that can be discussed in the frame of this chapter is the precise influence of the substrate geometry in the droplet hydrodynamics. In the example proposed here, the substrate is ideal and presents the peculiar geometry illustrated in Fig. 3. Such geometry is technically challenging due to the absence of heat transfer between the substrate heater and the rest of it. From the numerical point of view, it has however the important interest to provide the possibility to emphasize the precise impact of such large temperature differences on the droplet hydrodynamics. In the above discussion, the condition $L_s = R$ was imposed and the substrate heater was exactly of the same size as that of the droplet contact radius R. The ratio $\Delta = L_s/R$ that was set to 1 with Δ the relevant parameter here since it is the one that allows the control of the substrate heat flow inside the droplet. When $\Delta < 1$ the substrate heater is smaller than the contact line radius and all the heat it provides flows into the droplets and contributes to the evaporation kinetics. When $\Delta > 1$, the substrate heater is larger and only part of the heat flow is used for water heating and evaporation. Δ thereafter controls the adiabatic character of the heat exchanges in the substrate/droplet system.

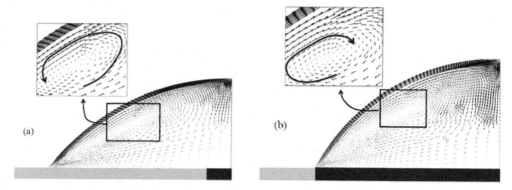

Figure 8. Water circulation inside the droplet for $T_s = 50°C$ when $\theta = 60°$. (a) $\Delta = 0.1$ and (b) $\Delta = 1.0$. Insets illustrate the details of the velocity field in the vicinity of the water/air interface and bold arrows show the direction of water circulation. The values of the velocity can be obtained using the full visible arrows in the bottom right corner of the insets and are respectively 3.0 mm/s (a) and 1.5 mm/s (b). The dark grey area of the substrate is the heater.

The influence of the value of Δ on the hydrodynamical evolution of the droplet is illustrated in Fig. 8 for $\Delta = 0.1$ and $\Delta = 1.0$. This last case is again presented here to ease comparison and was shown to exhibit a clockwise rotating convective cell (cf. Fig. 7). A similar convective cell shows up when $\Delta = 0.1$ but with the important difference that it now rotates counter clockwise due to reversed thermocapillary effects. The apex of the droplet is indeed now warmer than its contact line that turns out to be too far from the substrate heater to rapidly reach temperature T_s. Surface tension is therfore larger in its vicinity leading to a thermocapillary circulation that now flows from the droplet apex to the contact line. This temperature and flow inversion clearly indicates that the substrate heater size plays key role in heated droplet hydrodynamics. A detailed study of the evaporative mass flux and droplet volume evolution for such substrates has been discussed in [26, 34, 48]. It has been shown for example that the overall droplet evaporation time is almost independent of Δ as long as $\Delta \leqslant 1$. The velocity scales in Fig. 8 can be estimated from the velocity vector in the lower right corner of the insets. They differ by a factor of two with faster velocities when $\Delta = 0.1$ due to larger interfacial temperature gradients. Indeed, when contact angle $\theta = 60°$ is reached, the absolute value of the temperature difference between the droplet apex and contact line is close to 0.1°C when $\Delta = 1.0$ and reaches 0.4°C when $\Delta = 0.1$. The driving force for thermocapillary convection is therefore stronger in this last case. This is actually not always true since Marangoni flow can be strongly modified for small values of Δ as will be discussed next in this chapter.

Figure 9 displays the Marangoni velocity profile v_{mar} along the water/air interface for two different contact angles with the initial condition of table 1 and for $\Delta = 0.1$. It indicates important modifications as contact angle decreases. Initially, when $\theta = 80°$, it shows a single maximal value close to $r^* = 0.5$ that is a typical situation realized initially for almost all values of T_s and L_s. In the case of large values

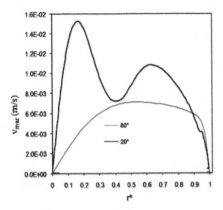

Figure 9. Marangoni velocity profile v_{mar} at the surface of the droplet as a function of r^* for two different contact angles (see legend) with $T_s = 50°C$ and $L_s/R = 0.1$.

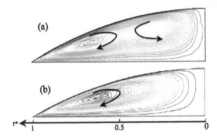

Figure 10. Streamlines for $\Delta = 0.1$ (a) and $\Delta = 1.0$ (b) when $\theta = 20°$ with $T_s = 50°C$. Black arrows indicate the orientation of water circulation.

of Δ, only one such maximum shows up for the entire evaporation time and, in the last stage of the evaporation, it drifts towards the apex of the droplet. This drift is explained by the large deformation of the convective cells presented in Fig. 7 and Fig. 8 as the droplet volume decreases. When the droplet height h becomes too small, the available space for convective cells shrinks and the corresponding Reynolds number follows the same trend. Consequently, for not too fast evaporating droplets, the hydrodynamics evolves from convective to diffusive. This transition is usually not observable experimentally due to receding contact line regimes that finally prevails in the very last phase of droplet evaporation. Back to $\Delta = 0.1$, Fig. 9 indicates a relatively complex Marangoni circulation with the appearance of two maxima for decreasing contact angles. Although no stagnation point ($v_{mar} = 0$ m/s) can be observed here, these two maxima are reminiscent of bifurcations in the hydrodynamics of the droplets similar to the one illustrated in Fig. 10(a) that is a snapshot of the water streamlines when $\theta = 20°$ and $\Delta = 0.1$. The case $\Delta = 1.0$ is also displayed (Fig. 10(b)). The comparison between these two figures indicate qualitatively similar water clockwise circulation in the droplet part where $r^* > 0.5$. But for $r^* < 0.5$, water circulation is completely different with a counter clockwise rotating convective cell when $\Delta = 0.1$ due to the larger temperature at the droplet apex that was not present when $\Delta = 1.0$.

The two convective cells displayed in Fig. 10(a) result from a bifurcation in the droplet hydrodynamics from a single cell configuration, common to all droplets for large contact angles, to a two cell configuration for small contact angles. The critical contact angle is $\theta \approx 35°$. Numerical simulations suggest that subsequent bifurcations show up as contact angles become smaller with an increasing number of convective cells. The existence of this bifurcation scenario results from the conjugated effect of thermocapillary convection and an increasingly confined geometry due to the vanishing droplet volume with time. But it raises important questions in the interpretation of the simulation results in particular due to the difficulty here to properly mesh such flat droplets and to ensure numerical convergence. For small values of θ water circulation appears indeed to be unstable even with very refined unstructured mesh grids. Efforts still have to be devoted to analyse this problem and to determine whether hydrodynamics remains laminar or if turbulent regimes are finally supported by the droplet at small contact angles.

This illustration of the evaporating history of water droplets on copper substrates shows that FEM based approaches and simplified models can provide insights into the phenomenology of such a system. The specificity of FEM being the condition for the droplets to be pinned and exhibiting a spherical cap geometry, allowed important simplifications and fast simulations. Still, this example presents a number of important limitations. Among them, quasi-steadiness of the hydrodynamics and heat/mass transfer, axi-symetry, absence of evaporative cooling or pinned contact lines, etc. All these approximations have been discussed and were shown to be justified for specific experimental conditions. Important efforts therefore still have to be dedicated to improve both the modelling and the numerical techniques in order to provide a more accurate and general description of evaporating droplets. First of all the full 3D geometry has to be implemented since droplet evaporation is known to involve non axi-symetric effects even in pinned regimes. This is for example the case of thermal waves recently observed [77]. The description of thermal waves will require important efforts in both modelling and numerics to significantly speed up 3D computations. The second important limiting factor of the present illustration is the omission of evaporative cooling that has clearly to be accounted for. As for front tracking, it leads to potential problems in the present FEM illustration n particular for the local mass fluxes simulation with robust and stable algorithms. Last but not least is the description of the contact line dynamics that was assumed to be time independent while receding contact line regimes are known to show up [78]. One efficient way to overcome these limitations is the use of Eulerian single fluid techniques. But, as for FEM, these approaches have their own restrictions discussed in Paragraph B. Spurious currents induced by interfacial tension algorithms in such approaches is the most important issue. They generate parasitic sources and a numerical noise that can create, by their cumulative effect, a thickening of the interfaces and, overall, to the destabilization of the simulations. Spurious currents are actually intimately related to the multiphase character of evaporating systems and to the fundamental constraint to numerically catch free evolving interfaces.

The issue of free interfaces tracking motivated more than two decades of research activity and still remains one of the most challenging field in CFD. In this context, it appears that the coupling of Eulerian approaches together with Level Set techniques are the most promising for robust and CPU effective multiphase simulations. For evaporating droplets on heated substrates, such an approach would present the great interest to allow the treatment of the solid substrate, the evaporating liquid and the surrounding gas on the same basis without the necessity to invoke quasi-steadiness hypotheses or spherical cap geometries.

F. Conclusion

The evaporation and wetting of sessile droplets is a growing area of interest to scientists and engineering. This increasing interest is driven by many recent applications making use of this fundamental phenomenon. With these exciting opportunities there are some serious challenges concerning the understanding of underlying physical mechanisms. In studying this phenomenon two main approaches have been adopted over the last decade, experimental techniques and numerical modeling approaches. A variety of experimental techniques have been utilized to reveal the trends and behavior associated with phase changes and wetting of sessile droplets, whether on heated substrates or evaporating under ambient conditions. To measure evaporation rates and droplet profile evolution, some of the experimental techniques used include optical measurements, microbalances and Atomic Force Microscopes cantilevers. To measure interfacial and bulk temperatures, microthermocouples and infra red thermography have been implemented. The hydrodynamics within evaporating droplets has also been characterized using Particle Image Velocimetry (PIV) techniques. The use of these experimental techniques by numerous research groups over the last decade has led to the establishment of some accepted trends and behaviors regarding the evaporation rates, wetting tendencies and hydrodynamics within volatile droplets. Following these established physical trends, the challenge has been to develop adequate theoretical models and numerical schemes to predict and confirm experimental observations. Many numerical approaches have been developed and tested by confronting the predicted results to the experimental data with some mixed success.

In this chapter we have reviewed some of the modeling approaches developed to describe droplet evaporation. We pointed to the challenges posed when it gets to developing accurate numerical approaches. Coping with free moving interfaces and being able to integrate this dynamics numerically in an accurate manner is one of these challenges. Contact lines and physical length scales in this region where the three phase meet is another important challenge to properly model the problem. The physical phenomena near the contact line span over length scales ranging from continuum to molecular scales. This poses a very serious challenge to any numerical approach to model the evaporation of heated droplets. Some hybrid approaches combining CFD and molecular dynamics could be tested but it is still too early to conclude about the success of such approaches. Three dimensional description of the phenomenon is also essential to capture some of the experimental observations. This is still not achieved in the current numerical models due to too demanding CPU efforts.

At this stage, our conclusion is that there is still much to be done in order to develop satisfactory numerical models capable of accurately describing the phenomenon of wetting and evaporation of sessile droplets. This chapter points to some of the improvements to be made and some of the tasks to be undertaken in order to make a step further towards this goal.

G. References

1. Sefiane, K., Wilson, S. K., David, S., Duffy, B. R. and Dunn, G., Physics of Fluids, 21(6) (2009) 062101.
2. Buffone, C., Sefiane, K. and Easson, W., Physical Review E, 71 (2005) 056302.
3. Sefiane, K., David, S. and Shanahan, M. E. R., Journal of Physical Chemistry B, 112(36) (2008) 11317.
4. Sefiane, K. and Bennacer, R., Journal of Fluid Mechanics, 667 (2011) 260.
5. Girard, F., Antoni, M. and Sefiane, K., Langmuir, 26(7) (2010) 4576.
6. Hu, H. and Larson, R. G., J. Phys. Chem. B, 110 (2006) 7090.
7. Popov, Y. O., Phys. Rev. E, 71 (2005) 036313.
8. Birdi, K. S., Vu, D. T. and Winter, A., J. Phys. Chem., 93 (1989) 3702.
9. Sefiane, K. and Petrol, J., Sci. Eng., 51 (2006) 238.
10. Rowan, S. M. et al., J. Phys. Chem. B, 104 (2000) 8217.
11. Popov, Y. O. and Witten, T. A., Phys. Rev. E, 68 (2003) 036306.
12. Jing, J. P. et al., Proc. Natl. Acad. Sci., 95 (1998) 8046.
13. Gonuguntla, M. and Sharma, A., Langmuir, 20 (2004) 3456.
14. Norris, D. J., Arlinghaus, E. G., Meng, L. L., Heiny, R. and Scriven, L. E., Adv. Mater., 16 (2004) 1393.
15. Kawase, T., Sirringhaus, H., Friend, R. H. and Shimoda, T., Adv. Mater., 13 (2001) 1601.
16. Fang, G. and Ward, C. A., Phys. Rev. E, 59(1) (1999) 417.
17. Lorenz, J. and Mickic, B., ASME J. Heat Transfer, 93 (1970) 46.
18. Incropera, F. P. and DeWitt, D. P., Fundamentals of Heat and Mass Transfer. John Wiley and Sons, New York, 1990.
19. Shanahan, M. E. R., Langmuir, 18 (2002) 7763.
20. Girard, F., Antoni, M., Faure, S. and Steinchen, A., Microgravity Sci. Tec., 18(3/4) (2006) 42.
21. Cachile, M., Benichou, O. and Cazabat, A. M., Langmuir, 18 (2002) 7985.
22. Sefiane, K., Tadrist, L. and Douglas, M., Int. J. Heat Mass Transfer, 46 (2003) 4527.
23. Erbil, H. Y. and Meric, R. A., J. Phys. Chem. B, 101 (1997) 6867.
24. Cachile, M., Benichou, O., Poulard, C. and Cazabat, A. M., Langmuir, 18 (2002) 8070.
25. Girard, F., Antoni, M. and Sefiane, K., Langmuir, 24(17) (2008) 9207.
26. Girard, F., Antoni, M., Faure, S. and Steinchen, A., Langmuir, 22(26) (2006) 11085.
27. Dunn, G. J., Wilson, S. K., Duffy, B. R., David, S. and Sefiane, K., Colloids and Surfaces A, 323(1–3) (2008) 50.

28. Dunn, G. J., Wilson, S. K., Duffy, B. R., David, S. and Sefiane, K., Journal of Fluid of Mechanics, 623 (2009) 329.
29. Sefiane, K., Wilson, S. K., David, S., Duffy, B. R. and Dunn, G., Physics of Fluids, 21(6) (2009) 062101.
30. Bourges-Monnier, C. and Shanahan, M. E. R., Langmuir, 11 (1995) 2820.
31. Ristenpart, W. D. et al., Phys. Rev. Lett., 99 (2007) 234502.
32. Rowan, S. M., McHale, G., Newton, M. I. and Toorneman, M., J. Phys. Chem. B, 101 (1997) 1265.
33. Erbil, H. Y. and Dogan, M., Langmuir, 16 (2000) 9267.
34. Crafton, E. F. and Black, W. Z., Int. J. of Heat Mass Transfer, 47 (2004) 1187.
35. Liao, Y. Dropwise evaporative cooling of solid surfaces. Ph.D. thesis, University of Maryland, College Park, MD, 1 (992).
36. Di Marzo, M., Tartarini, P. and Liao, Y., Int. J. Heat Mass Transfer, 36 (1993) 4133.
37. Zuiderweg, F. J. and Harmens, A., Chem Eng Science, 9(2/3) (1958) 89.
38. Ruiz, O. E. and Black, W. Z., Int. J. Heat Transfer, 124 (2002) 854.
39. Hu, H. and Larson, R. G., J. Phys. Chem., 106 (2002) 1334.
40. Shanahan, M. E. R. and de Gennes, P. G., Comptes Rendus Acad. Sci. 302(II) (1986) 517.
41. Carré, A. and Shanahan, M. E. R., Langmuir, 11 (1995) 24.
42. Scardovelli, R. and Zaleski, S., Annual Review of Fluid Mechanics, 31 (1999) 567.
43. Schlick, T., Molecular Modeling and Simulation: An Interdisciplinary Guide. Interdisciplinary Applied Mathematics, Mathematical Biology, 21. Springer-Verlag, 2002.
44. Rednikov, A. and Colinet, P., Langmuir, 27(5) (2011) 1758.
45. Cuny, V., Antoni, M., Arbelot, M. and Liggieri, L., Colloids and Surfaces A, 323 (2008) 180.
46. Car, R. and Parrinello, M., Phys. Rev. Lett., 55 (1985) 2471.
47. Hou, T. Y., Lowengrub, J. S. and Shelley, M. J., Journal of Computational Physics, 169 (2001) 302.
48. Girard, F. and Antoni, M., Langmuir, 24 (2008) 11342.
49. Tryggvason, G. et al., Journal of Computational Physics, 169 (2001) 708.
50. Hirt, C. W. and Nichols, B. D., Journal of Computational Physics, 39 (1981) 201.
51. Sweby, P. K., J. Siam Numer. Analysis, 21 (1984) 995.
52. Brackbill, J. U., Kothe, D. B. and Zemach, C., Journal of Computational Physics, 100 (1992) 405.
53. Balsara, D. S. and Shu, C. W., Monotonicity, Journal of Computational Physics, 160 (2000) 405.
54. Harvie, D. J. E., Davidson, M. R. and Rudman, M., Applied Mathematical Modeling, 30 (2006) 1056.
55. Meier, M., Yadigaroglu, G. and Smith, B. L., European Journal of Mechanics—B, 21 (2002) 61.
56. Jamet, D., Torresand, D., Brackbill, J. U., Journal of Computational Physics, 182 (2002) 262.
57. Ménard, T., Tanguy, S. and Berlemont, A., International Journal of Multiphase Flow, 33 (2007) 510.
58. Tong, A. Y. and Wang, Z., Journal of Computational Physics, 221 (2007) 506.
59. Tsuji, M., Nakahara, H., Moroi, Y. and Shibata, O., Journal of Colloid and Interface Science, 318 (2008) 322.
60. Yon, S. and Pozrikidis, C., Computer & Fluids, 27 (1998) 879.
61. Eggleton, C. D., Tsai, T. M. and Stebe, K. J., Phys. Rev. Lett. 87 (2001) 048302.
62. Renardy, Y., Renardy, M. and Cristini, V., European Journal of Mechanics—B/Fluids, 21 (2002) 49.
63. James, A. J. and Lowengrub, J., Journal of Computational Physics, 201 (2004) 685.
64. Alke, A. and Bothe, D., Fluid Dynamics & Materials Processing, 5 (2009) 345.

65. Smith, R. D. and Berg, J. C., Journal of Colloid and Interface Science, 74 (1980) 273.

66. Chandrasekhar, S., Hydrodynamic and Hydromagnetic Stability. Dover, 1981.

67. Landau, L. and Lifchitz, E., Cours de physique théorique, mécanique des fluides. MIR, 1989.

68. Perry, R. H., Green, D. W. and Maloney, J. O., Perry's Chemical Engineers' Handbook. M.-H.P. Publishing, 1997.

69. Rymkiewicz, J. and Zbigniew, Z., Int. Comm. Heat Mass Transfer, 20(5) (1993) 687.

70. Schwarz, J. et al., J. Aerosol Sci., 30 (1999) 399.

71. Hu, H. and Larson, R. G., Langmuir, 21 (2005) 3972.

72. Lebedev, N. N., Special Functions and Their Applications. Prentice-Hall, 1965.

73. Semenov, S., Starov, V. M., Rubio, R. G. and Velarde, M. G., Colloids and Surfaces A, 372 (2010) 127.

74. Lekhlifi, A., Antoni, M. and Ouazzani, J., Colloids and surfaces A, 365 (2010) 70.

75. Hu, H. and Larson, R. G., Langmuir, 21 (2005) 3963.

76. Sefiane, K. and Cameron, J., Progress in Computational Fluid Dynamics, 6(6) (2006) 363.

77. Sefiane, K., Moffat, J. R., Matar, O. K. and Craster, R., Applied Physics Letters, 93(7) (2008) 074103.

78. Girard, F., Antoni, M. and Sefiane, K., Langmuir, 27(11) (2011) 6744.

Nanobubbles at Hydrophobic Surfaces

Vincent S. J. Craig [a], Xuehua Zhang [b] and Jun Hu [c,d]

[a] Department of Applied Mathematics, Research School of Physical Sciences and Engineering, Australian National University
[b] Department of Chemical and Biomolecular Engineering, University of Melbourne, VIC 3010 Australia
[c] Shanghai Institute of Applied Physics, Chinese Academy of Sciences, Shanghai China
[d] Bioo-X Life Science Research Center, Life Science and Technology College, Shanghai Jiao Tong University, 200030, Shanghai China

Abstract
The very existence of nanobubbles at hydrophobic surfaces is controversial. On theoretical grounds such bubbles should be short-lived, dissolving under their own internal pressure. Further, a number of highly surface sensitive experiments have reported results inconsistent with the presence of nanobubbles, yet other researchers observe nanobubbles, most notably using Atomic Force Microscopy. Here I attempt to reconcile these results, describe the unusual properties of nanobubbles and the compelling evidence for their existence as well as describe some new surprising nanobubble forms. Finally, applications for nanobubbles are explored.

Contents

A. Introduction 159
B. Early Evidence for Nanobubbles 160
C. Confirmation of the Existence of Nanobubbles 161
D. Nanobubble Characteristics 163
E. How to Produce Nanobubbles 167
F. Why Were Thought not to Exist 168
G. The Stability of Nanobubbles 170
H. Uses for Nanobubbles 171
I. Conclusion 171
J. References 172

A. Introduction

Nanobubbles are tiny gas pockets that are known to be present under some conditions on hydrophobic surfaces. Their very existence is somewhat surprising as it is generally thought that small bubbles are unstable. Small bubbles are characterized by a high internal gas pressure. For a bubble of given radius this can be calculated using the Laplace equation.

$$\Delta P = \gamma \left(\frac{1}{r_1} + \frac{1}{r_2} \right), \tag{1}$$

where ΔP is the pressure change across the interface, γ is the surface tension and r_1 and r_2 are the principal radii of curvature. Consequently for a spherical bubble of radius 100 nm in water the internal pressure is ~14 atmospheres. An internal pressure of such magnitude has severe consequences for the existence of such a bubble as the solubility of the gas in the aqueous phase is significantly enhanced. An increase in solubility results in loss of gas to the surrounding medium, a consequent decrease in radius and a further increase in pressure and an even greater driving force for dissolution. Thus lifetimes are expected to be short in the high pressure world of nano-sized bubbles. For a bubble of radius 100 nm it is calculated that the bubble lifetime is only 100 μs, [1] therefore it is expected that if nanobubbles were produced they would rapidly dissolve under their own internal pressure. However the substrate has an important role to play in the stability of nanobubbles. The hydrophobic surface influences the shape of nanobubbles such that their radius of curvature is much greater. A nanobubble 50 nm in height may have a radius of curvature of 500 nm. The theoretical lifetime associated with a bubble of this size is still very short at ~1 ms but these calculations assume that there is no supersaturation of gas in the aqueous phase and no material at the interface opposing gas transport or reducing the surface tension.

B. Early Evidence for Nanobubbles

The first report of the existence of nanobubbles in the literature appeared in 1994. Parker, Claesson and Attard [2] were using a highly sensitive force measurement device to measure the attractive force between two hydrophobic surfaces in water. They found that the force exhibited clear steps and they interpreted the attraction and the presence of these steps as arising from the presence of nanobubbles on the surfaces. As each nanobubble bridged the gap between the surfaces it would give rise to a stepwise increase in the attractive force. From the experimental data they could give an estimate of the height of the nanobubbles and found them to be < 100 nm. At the time this report was highly controversial as there were theoretical objections to the existence of stable nanobubbles as we have seen above. Significantly this study tied the existence of nanobubbles to the mysterious long-range hydrophobic attraction which is measurable between hydrophobic surfaces in aqueous solutions [3].

A much earlier study by Kitchener and Blake [4] into the interaction of macroscopic bubbles with glass surfaces could also be cited as possible evidence for the existence of nanobubbles. They found that when a macroscopic bubble was pushed towards a native glass (hydrophilic) surface the repulsive electrostatic forces acted to prevent drainage of the liquid film between the bubble and the surface, as predicted by the DLVO theory of colloidal interaction. However, if the silica surface was made hydrophobic by silanation, the liquid film became unstable at thicknesses of the order of 100 nm. This is notable as the silanation process was found to have almost no effect on the surface charge of the silica surface and therefore one expects that the long-range forces would be unchanged. At the time the authors were concerned that their data may have been affected by the existence of small particles on the surfaces, though it now seems likely that the presence of nanobubbles may have influenced their results.

A significant step in the development of research into nanobubbles came with the arrival of the new millennium when images of nanobubbles obtained using the Atomic Force Microscope (AFM) were published by two independent groups working in China [5] and Japan [6]. It is noteworthy that both of these reports demonstrated that the presence, or absence, of nanobubbles is strongly dependent upon the history of the sample, though the wider community still often fails to grasp the significance of this. In the work by Ishida *et al.* [6] they were able to demonstrate that the range of the hydrophobic attractive force was extended when nanobubbles were present on the surface. Nanobubbles were found to have a diameter of ~650 nm and a height of ~40 nm. Importantly they reported that nanobubbles were found when a hydrophobic surface was immersed in water, whereas if the surface was initially hydrophilic and rendered hydrophobic by chemical reaction without exposure to air, no nanobubbles were found. Further evidence that the features in the images were actually gas pockets was provided by employing degassed water, in which no nanobubbles were found. The work by the Shanghai group [5], focused on the production and imaging of nanobubbles and they presented some high quality detailed images of nanobubbles on both mica surfaces and Highly Ordered Pyrolytic Graphite (HOPG) surfaces (see Fig. 1). One presumes that the mica surface bore a small amount of contamination that rendered it hydrophobic, at least in parts. Indeed the authors report that the production of nanobubbles on mica was intermittent. Importantly in this work the authors describe the first method of controllably producing nanobubbles. This is achieved by a solvent exchange method which is discussed in more detail below. Both of these reports state that the nanobubbles observed were stable for hours.

C. Confirmation of the Existence of Nanobubbles

Surface force measurements have inferred the presence of nanobubbles on hydrophobic surfaces. Steps are reported in plots of force *versus* separation that are not described by any conventional surface force. Whilst an estimate of the height of nanobubbles can be obtained from such measurements very little other information is revealed, therefore other techniques are required in order to demonstrate the existence of nanobubbles and to investigate their properties.

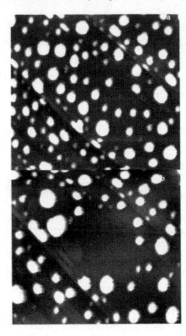

Figure 1. AFM tapping mode images of nanobubbles on HOPG in water. Image size 10 μm × 10 μm. In the lower image it can be seen that the imaging process has led to the removal of bubbles in a 4 μm × 4 μm area. Image taken from reference [5].

The most direct evidence for the existence of nanobubbles is provided by Atomic Force Microscope images [5–9]. These images are obtained in 'tapping mode' in aqueous solution whereby the cantilever and tip are oscillated at a frequency generally in excess of ~10 kHz and the amplitude of the oscillation is used to control the feedback of the instrument. That is, the surface separation is adjusted to keep the oscillation amplitude at a predetermined level and the adjustment in separation required is mapped in two dimensions to produce a height image. This technique is commonly employed to image soft surfaces, such as biological samples, in fluids. Tapping mode imaging of nanobubbles is challenging and the images are influenced by the imaging conditions [9], nonetheless careful analysis has revealed a great deal about the morphology of nanobubbles. The other common mode of imaging using the AFM is contact mode, which is known to be more suitable for hard surfaces. In general nanobubbles are too soft to be revealed by contact mode imaging.

Additional evidence for nanobubbles has been provided by rapid cryofixation and freeze fracture [10], which has been used to study the interface between water and a silicon substrate in both the native state and following treatment with hexamethyldisilazane vapor, which renders the surface hydrophobic. By freezing the sample at a very rapid rate, the ice is trapped in an amorphous state and any structures present are preserved without modification. It is then necessary to fracture the sample and image the interface. This is done using techniques that are well established for the imaging of biological samples. This study found that no structures were visible when a hydrophilic surface was employed or if the hydrophobic surface was used in conjunction with degassed water. In contrast, voids were found on hydrophobic substrates in contact with gassed water that are commensurate in size with the images of nanobubbles obtained by AFM studies. These findings parallel the imaging studies of Ishida *et al.* [6]. Whilst this study is important in that evidence for nanobubbles was found using an alternative technique, it was unable to provide additional morphological information.

Both AFM images and the cryofixation and freeze fracture techniques provide evidence for the existence of structures at the interface but they do not provide chemical information on the make-up of these structures. It was possible, for example, that these structures are formed by an organic contaminant rather than being gas-filled bubbles—though a range of evidence strongly suggested that they were indeed bubbles such as observation of coalescence [11] and production *in situ via* photocatalysis [12]. In an elegant experiment, Zhang, Khan and Ducker [13] have demonstrated that nanobubbles do indeed consist of gas. They used CO_2 saturated water to produce nanobubbles by the solvent exchange method [5]. CO_2 has an infrared spectrum that varies considerably between the gaseous and aqueous states due to the rotational fine structure that is visible in the gaseous state and therefore it can be used to directly reveal the phase state of the material in the nanobubbles. Furthermore, from the rotational fine structure the pressure of the CO_2 gas within the nanobubbles could be approximately determined and was found to be consistent with that expected from the nanobubbles.

D. Nanobubble Characteristics

The size, shape, contact angle, surface tension and internal pressure within a nanobubble are of interest not only in describing nanobubbles but also in addressing the issue of nanobubble stability. When looking at AFM images it should be remembered that most are presented with a considerable vertical exaggeration. The scale in the z dimension can be 100 times smaller than the scale in the x and y dimensions. This can easily give rise to an incorrect perception of the shape of nanobubbles. A sectional profile through a nanobubble can be very useful in gaining an understanding of their true shape. However, due to imaging artifacts caused by the deformability of nanobubbles a section taken through a standard tapping mode image can be misleading. In water the tip of the cantilever penetrates the nanobubble and leads to an uncertainty in the height. The degree of penetration is dependent upon the imaging conditions, in particular the amount of energy that the tip loses when contacting the surface (amplitude setpoint). Figure 2 illustrates the influence of setpoint on the image in surfactant solution. As the setpoint amplitude is reduced (a to g) the nanobubbles appear smaller due to the tip deforming the interface to a greater extent. Images h–j, show that the nanobubble recovers when the setpoint amplitude is increased.

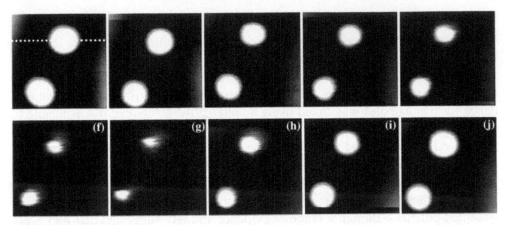

Figure 2. Images of the same two nanobubbles under different amplitude ratios in surfactant solution. The set-point amplitude is reduced on moving from image a to g and then increased again from h to j. The nanobubbles appear smaller as the setpoint amplitude is reduced as the nanobubble-tip force is increased and the tip deforms the nanobubble. Image taken from reference [9].

Zhang *et al.* [9] have shown that a truer measure of the shape of a nanobubble is gained by using the information in deflection *versus* piezo displacement data. This method which was first developed for measuring nano-droplets [14] allows the height of the nanobubble at a particular point to be determined accurately and can be obtained at each point of an image using a technique widely known as force-volume imaging.

Researchers have consistently reported that the nanobubble shape is accurately described as a spherical cap (except for the recent report on 'nanopancakes' which is discussed below). This infers that the curvature of the interface is everywhere the same and that the pressure drop across the interface is constant. If surface forces had a significant impact on the stability of nanobubbles one would expect that this would result in a change in curvature of the interface with separation from the surface [15]. The constant interfacial curvature is evidence that long-range surface forces are not responsible for the stability of nanobubbles as the magnitude of such forces reduce rapidly with distance. However, it is typically difficult to determine the shape of the interface within the last 20 nm of the surface due to the finite size of the cantilever tip, so the influence of surface forces within the last 20 nm cannot be gauged at this stage.

When the sectional profile of a nanobubble is accurately determined the contact angle can also be determined with reasonable accuracy. Surprisingly, the nanoscopic contact angle differs considerably from the macroscopic contact angle. In many cases the contact angle has been reported to be in the range of 10° to 40° [6, 7], where the angle has been measured through the gas phase. However the correct definition of the contact angle at the three phase line is to measure the angle through the more dense (i.e., liquid) phase, thus the true value is given by $180° - x°$, where $x°$ is the angle measured through the gas phase and the con-

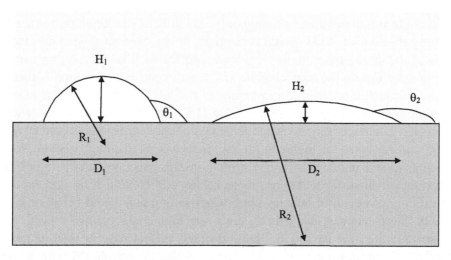

Figure 3. Schematic diagram indicating the influence of the contact angle on shape for two spheri-cal cap bubbles of equal volume (7.25×10^5 nm^3) with the normal macroscopic contact angle (on the left, $\theta_1 = 110°$) and the observed nanoscopic contact angle (R. H. S., $\theta_2 = 170°$). The radius of curvature is much smaller for the macroscopic contact angle, ($R_1 = 88$ nm $< R_2 = 1000$ nm) as is the bubble diameter ($D_1 = 83$ nm $< D_2 = 174$ nm). Whilst the maximum height is larger ($H_1 = 58$ nm $> H_2 = 15.2$ nm) for the bubble with the macroscopic contact angle. The increase in the radius of curvature associated with the nanoscopic contact angle results in a much reduced internal pressure and is responsible for the stability of nanobubbles.

tact angles are more correctly reported as being in the range of 140°–170° [6, 9]. These exceptionally high contact angles are much larger than the corresponding macroscopic contact angles which are in the range of 90°–120°. This has several significant implications. The first is on the shape of a nanobubble. This is illustrated in Fig. 3 where two bubbles of the same volume are depicted one with the macro-scopic contact angle and the other with the nanoscopic contact angle. The shape is influenced dramatically. When the contact angle is 170° the bubble has a maxi-mum height of only 15 nm and the diameter of the base is 347 nm, whereas for a contact angle of 110° the bubble has a maximum height of 58 nm and the diameter of the base is only 165 nm. The curvature of the interface is greatly reduced by an increase in nanobubble contact angle, in the example in Fig. 3 the radius of curva-ture is 1000 nm for a contact angle of 170° (see the bubble on the right), whereas the same volume with a contact angle of 110° has a radius of curvature of 88 nm (left bubble). Thus the Laplace pressure is also greatly reduced, for the high contact angle case it is ~1.4 atmospheres compared to ~16.4 atmospheres for the lower contact angle. In this case the greater contact angle reduces the Laplace pressure by a factor of 11.7. A reduction in interfacial curvature and a consequent significant reduction in the internal pressure within a bubble aids greatly in our understanding of the stability of nanobubbles, as this means that the driving force for dissolution is greatly reduced and with a small degree of supersaturation we can expect that the bubbles are kinetically stable for hours and even days. Whilst the surprisingly high contact angle aids in our understanding of the stability of nanobubbles it is not clear as to why the contact angle should be so high. It has been suggested that the line tension at the three-phase line is significant and for such small volumes can have a significant effect on the contact angle [7]. An alternative explanation is that very small scale roughness can lead to pinning of the interface and result in a contact angle that is far from the equilibrium value [16]—though one report of nanobubbles on HOPG surfaces reports nanobubbles with more conventional contact angles of 119° and attributes the high contact angles measured to contamination. We feel this explanation is unlikely as many different groups have measured similarly high contact angles on surfaces that are more robust that HOPG. Nam and Jua study the periodic nucleation of bubbles upon superheating and found behavior consistent with the presence of nanobubbles with very high contact angles ($\sim$170°) [17]. It is worth noting that nano-scale liquid droplets do not have contact angles that differ from the macroscopic value. This suggests that the nanobubble contact angle anomaly is associated with the low density of molecules in the gaseous phase and their confinement in the z direction. The height of a nanobubble is of the order of the mean free path, so in the z direction a molecule can travel across a nanobubble without colliding with any others [18].

The size range of stable surface nanobubbles remains unclear. Nanobubbles have not been observed with heights greater than 100 nm and base diameters of greater than 1000 nm [19]. It currently cannot be determined what the lower limit is to the size of nanobubbles due to difficulties associated with imaging soft materials using the AFM. Very small bubbles may exist but if they are penetrated by the tip such that the tip strikes the substrate they may be very difficult to discern.

The internal pressure in a nanobubble is not only controlled by the curvature of the interface but also the surface energy of that interface. One can imagine that the high energy air–water interface could readily adsorb contaminants from solution, particularly as it ages. This will result in a lower surface tension and a reduced Laplace pressure. Zhang *et al.* [9] have been able to estimate the surface tension of a nanobubble in a non-ionic surfactant solution by evaluating the deformation of the nanobubble interface under the pressure of the AFM tip. A surface tension of 43 mN/m was obtained which is commensurate with literature values for macroscopic interfaces with the same concentration of this surfactant. Consequently the Laplace pressure would be reduced to $\sim$60% of the Laplace pressure present for a clean water interface and the same bubble geometry. Unfortunately this method cannot be applied to nanobubbles in the absence of surfactant as the tip of the cantilever penetrates the nanobubble. The difference in the interaction between tip and nanobubbles in water and surfactant solutions suggests that nanobubbles in water have an interface that is not significantly contaminated [9], however a small amount of insoluble contaminant at the nanobubble surface may strongly influence the stability of nanobubbles [20]. The development of a general method to measure the surface tension of nanobubbles would be of great utility.

A remarkably different novel morphology for a nano-scale gaseous state at a hydrophobic surface has also been reported [16]. This differs markedly from the spherical caps that are generally reported and consists of a gaseous layer <5 nm thick that extends in the lateral direction for microns with constant height. This shape has led to these gaseous states being dubbed nanopancakes. Imaged on HOPG, the shape of these nanopancakes is clearly determined by the atomic layer steps in the underlying suHas thbstrate. The Van der Waals force across such nanopancakes should be attractive and contribute to the instability of such structures. It remains a mystery as to how this morphology may arise, but it is known that it can be present along with the formation of nanobubbles and it has only been observed on crystalline surfaces, such as HOPG, $MoS2$ and talc [21]. Recent studies show that micropancakes can be made up of discrete layers and that the micropancakes grow slowly on the timescale of hours [22]. Even more mysteriously, composite structures are seen where a nanobubble is present and free to move around on top of a nanopancake. This strongly suggests that a thin film remains between the two structures. Perhaps very small graphitic particles that arise from the substrate are operating to stabilize such films by the same mechanism as emulsions are stabilized by partially hydrophobic particles. This discovery poses several significant challenges to our understanding of stable gas states at hydrophobic surfaces and clearly warrants further investigation.

Another interesting feature of nanobubbles is their role as possible nuclei for macroscopic bubbles. Interestingly they have been found that they do not act as nucleation sites when shock waves are used to induce cavitation [23]. However they do nucleate vapor bubbles upon superheating. Their possible role as nuclei for macroscopic bubbles has important implications and warrants further investigation [17].

E. How to Produce Nanobubbles

One of the earliest publications reporting images of nanobubbles also included a clear description of a method by which nanobubbles could be produced on a hydrophobic surface [5] and several subsequent papers have adopted this methodology. Despite this many investigations of nanobubbles have been conducted without any effort to control the conditions that may influence nanobuble production, though degassing has commonly been employed to prevent the formation of nanobubbles or remove nanobubbles already present [6, 24]. This is the cause of much of the confusion in the literature, indeed many studies reporting null findings for nanobubbles did not employ conditions under which one might expect to find them.

The solvent exchange technique [5] for producing nanobubbles operates by inducing a supersaturation of gas at the interface. Common atmospheric gases such as nitrogen and oxygen are present in water in millimolar quantities at atmospheric pressure and room temperature. In many other solvents even greater quantities of gas are dissolved. A solvent miscible with water and with a greater solubility of atmospheric gases is chosen. Ethanol is most commonly used as it is available in very high purity. The substrate is immersed in ethanol and then the ethanol is displaced by water. Ethanol preferentially wets hydrophobic substrates and therefore the ethanol at the surface is not easily displaced, but as it is miscible with water it will mix with the aqueous phase. In doing so, the gases that were dissolved in the ethanol will be 'precipitated' at the interface, producing nanobubbles as the solubility of gas in the aqueous phase is exceeded. It is unclear at this stage if the solvent exchange method also leads to supersaturation of the bulk aqueous phase or even if it produces (if only momentarily) nanobubbles in the bulk aqueous phase. The generality of this technique has been demonstrated by using the same approach to precipitate liquid organic materials at hydrophobic interfaces [25].

Electrochemical methods have also been used for the generation of nanobubbles [26]. Using HOPG as a conducting hydrophobic substrate, a battery is used to complete an electrochemical cell such that hydrogen gas is formed at the surface of the HOPG by electrolysis of water. After only a small amount of current is allowed to flow the local concentration of gas is such that the solubility is exceeded and nanobubbles are produced. Interestingly nanobubbles are more easily produced at a neagative electrode. That is, the production of hydrogen nanobubbles is more effective than the production of oxygen nanobubbles [27]. This method was not initially employed as extensively as the solvent exchange method due to the necessity of having a conducting hydrophobic substrate. However, it is now attracting more interest due to the improved control it offers over the conditions of nanobubble production. A little used method for the production of nanobubbles makes use of the photocatalytic properties of titania. Under exposure to UV light in water/methanol solution hydrogen nanobubbles are produced [12].

Whilst these methods are well established and both are dependent upon the nucleation of a gas phase nothing is currently known about the nucleation process. Surprisingly it appears that nanobubbles do not act as nucleation sites for the production of macroscopic bubbles [23].

F. Why Nanobubbles Were Thought not to Exist

The existence of nanobubbles remains a controversial topic despite mounting evidence for their existence. The controversy was clearly demonstrated when in 2007 the prestigious journal *Nature* ran an article in the research highlights section which headlined 'Physics: No nanobubbles' [28]. Here we will address the theoretical and experimental work that has been cited as evidence that nanobubbles do not exist. We address these arguments and demonstrate that none of this work is able to rule out the existence of nanobubbles.

1. Bubble Lifetime

As described in the introduction, the existence of nanobubbles is thought to be precluded by their rapid dissolution under a high internal Laplace pressure. However, we have seen that the morphology of nanobubbles is such that the interfacial curvature only gives rise to a moderate internal pressure and nanobubbles can be maintained under these conditions for long periods with a moderate supersaturation of gas.

2. Surface Investigations

A number of surface sensitive experiments have been conducted on hydrophobic surfaces in water that should reveal the presence of nanobubbles. These can be categorized as ellipsometric studies and scattering studies.

2.1. Optical Methods

Nanobubbles present at a hydrophobic interface should provide excellent contrast in ellipsometric studies, as the refractive index of the gas phase will be near unity and as such it will be very different to both the aqueous phase ($n = 1.333$) and solid substrate ($n \approx 1.35$). However, at least two investigations that have sought to find vapor or air layers at the interface between hydrophobic surfaces and water have reported that no such layers are found [29, 30]. The sensitivity limits of these studies demand that a reasonable surface coverage of nanobubbles exist if they are to be revealed. It has been estimated that in each $10 \, \mu m \times 10 \, \mu m$ area more than 1000 nanobubbles would need to be present in order for them to be detected ellipsometrically [29]. In neither case were procedures such as solvent exchange purposely adopted to produce nanobubbles, so it is unsurprising that nanobubbles were not found in these studies, particularly as this was not their aim. A repetition of these investigations using the solvent exchange technique or electrochemical methods to produce nanobubbles is likely. Further evidence for the existence of interfacial air bubbles produced by solvent exchange was obtained from surface plasmon resonance (SPR) measurements [13]. After the formation of nanobubbles, the resonant angle shifts to a lower angle, demonstrating the displacement of water (refractive index ~ 1.33) by a material of lower refractive index.

2.2. Scattering at the Interface

There is much interest in the structure of water adjacent to hydrophobic interfaces and as such numerous scattering studies aimed at the interface have been conducted. Given the large mass density and electron density difference between a gas phase and either the hydrophobic substrate or water one might expect that nanobubbles would be readily revealed by such investigations, yet several scattering studies report no evidence for nanobubbles. Neutron reflectivity studies have reported results that have been interpreted as consistent with the precursor of nanobubbles [31, 32] but later investigations found no evidence for the existence of nanobubbles [33]. These latter experiments are supported by X-ray scattering studies [34, 35] that reveal evidence for a lower density of water adjacent to a hydrophobic surface but no evidence of nanobubbles. None of these investigations employed conditions that would be expected to produce nanobubbles, so it is unsurprising that evidence of nanobubbles was not found. A recent investigation employing solvent exchange to produce nanobubbles has revealed the presence of surface nanobubbles using X-ray scattering [36].

Some of the presented evidence for nanobubbles has been questioned. Tyrell and Attard [37, 38] reported AFM images depicting very high density coverage of features at a hydrophobic surface. These were interpreted to be nanobubbles. However the features presented in the images are consistent with a polymeric layer formed by reaction of the dichlorodimethylsilane (used to make the surface hydrophobic) with water vapor [39]. It is worth noting that the images of nanobubbles presented in this work have a morphology that differs from the many other reports of nanobubbles in that the contact line is highly convoluted and it is unfortunate that many cite this work when referring to images of nanobubbles when the prior investigations by both Chinese [5] and Japanese [6] groups report images that are true representations of nanobubbles.

During the imaging of nanobubbles using AFM it is necessary that the tip come into contact with the nanobubble. This brings the tip in close proximity to the substrate and it has been suggested that this process actually nucleates nanobubbles [29, 33]. That is, the belief is that nanobubbles are not present prior to the commencement of the imaging process but rather arise as an artifact of the imaging process. This idea has arisen from the force measurement community where the existence of nanobubbles has been associated with the measurement of the hydrophobic attraction between two hydrophobic surfaces. It is known that upon separating two hydrophobic surfaces in water a vapor phase can be produced [40]. However, AFM tips are hydrophilic and as such there is no reason to believe that proximity to another surface can nucleate bubbles. Further, one would expect that if the tip were nucleating bubbles that the distribution and size of bubbles would increase with continued imaging, whereas in most cases the image remains very stable for many hours [5, 6, 9] and in other cases the tip is seen to sweep the surface free of nanobubbles (see Fig. 1, lower frame) [5]. More evidence that the tip is not responsible for the nucleation of nanobubbles comes from investigations using the solvent exchange technique to produce nanobubbles. This technique causes supersaturation of gas in the aqueous phase and on occasion this causes the nucleation of micron sized bubbles, in areas adjacent to the presence of nanobubbles, which can be seen with standard optical microscopy [9].

G. The Stability of Nanobubbles

The stability of nanobubbles remains an open question. It has been variously argued that the stability arises from surface roughness [41], line tension [42], a dynamic equilibrium between gaseous depletion layer and the nanobubble [43] and the presence of insoluble material at the surface of the nanobubble that arrests dissolution [20]. The latter is an appealing explanation as one can imagine the presence of such material is unavoidable however it has been argued that this effect alone is insufficient to explain the extended stability of nanobubbles observed [44]. More challenging is substantial evidence that nanobubbles of radius 50 nm exist in solution for up to two weeks [45]. Whilst these solutions are significantly supersaturated they are not sufficiently so to explain such longevity and their presence in large concentrations and the concomitantly large surface area reduces the possibility of stabilization by impurities.

H. Uses for Nanobubbles

Nanobubbles are inherently interesting and are also proving to be useful. As we have seen we can already control the conditions under which they are produced and a recent report has described a means by which the size and location of nanobubbles can be controlled [46], so it appears that we will be able to manipulate nanobubbles to our advantage. It has been shown that nanobubbles can be used to prevent surface fouling by proteins [47] and are also effective at removing proteins from an already fouled surface [47, 48]. By combining these technologies it should be possible to controllably pattern proteins on a surface, where the nanobubbles are used as masks. The presence of a gas phase at the interface is crucial in superhydrophobicity and can lead to hydrodynamic boundary slip [49]. Thus the manipulation of nanobubbles could be used to actively change the contact angle of a surface or the flow properties of a fluid. Such manipulations are likely to be employed in microfluidic devices either as switching or mixing technologies. The nanobubbles for these applications could be produced on demand using electrolysis.

Recently nanobubbles have found application in medicine for drug delivery and the treatment of cancers often in concert with ultrasound [50]. The ultrasound is used to create nanobubbles or act on pre-existing nanobubbles to promote cellular uptake of drugs. Nanobubbles have been produced from short laser pulses impinging on metal nanoparticles resulting in the photothermal production of nanobubbles [51] with the aim of targeting cancer cells [52–54]. Nanobubbles have even been implicated in claims of homeopathic action taking place at extreme dilutions [55].

Further, there are circumstances where the presence of nanobubbles is unwanted. Their presence on sensors can cause significant measurement errors. Examples include magnetoelastic gravimetric sensors and normal operation of the quartz crystal microbalance [56, 57]. Nanobubbles are also thought to be the source of defects in electroplating and electropolymerised films [58, 59].

I. Conclusion

The investigation of nanobubbles is still in its infancy and many challenges remain. An understanding of the very high contact angle produced by nanobubbles is important as it is critical to their stability. The newly reported nanopancakes pose several challenges. Why are they stable? What is the internal pressure? How can we reconcile the rather flat interface with the high interfacial curvature at the edge of the pancake? How can a nanopancake and a nanobubble interact but remain as separate entities and why are they only observed on layered crystalline surfaces? One possibility is that nanobubbles delaminate the surface of HOPG to produce graphene layers which then coat the air–water interface and stabilize a range of exotic structure. Nanobubbles have been implicated in the formation of carbon nanostructures from HOPG [60]. Other questions arise such as how hydrophobic does a substrate need to be to support nanobubbles, recalling that the macroscopic and nanoscopic contact angles differ. This also has implications for super-hydrophobic surfaces where small gas pockets are trapped at surfaces. If these gas pockets are made very small what contact angle will they adopt and how will this influence the superhydrophobicity? Finally, enormous external pressures are used to 'crush' Harvey nuclei, how would nanobubbles survive such a high external pressure? What is clear is that understanding nanobubbles is going to demand input from scientists with a wide range of expertise. It is notable that to date very few nanobubble researchers have a background in the physics and chemistry of bubbles and therefore we hope that the presence of this report in this book will engage some of you in the challenge of nanobubble research.

J. References

1. Ljunggren, S. and Eriksson, J. C., Colloids and Surfaces a-Physicochemical and Engineering Aspects, 130 (1997) 151–155.
2. Parker, J. L., Claesson, P. M. and Attard, P., J. Phys. Chem., 98(34) (1994) 8468–8480.
3. Israelachvili, J. N. and Pashley, R. M., J. Colloid Interface Sci., 98(2) (1984) 500–514.
4. Blake, T. D. and Kitchener J., J. Chem. Soc. Faraday Trans. I, 68(8) (1972) 1435–1442.
5. Lou, S. T., Ouyang, Z. Q., Zhang, Y., Li, X. J., Hu, J., Li, M. Q. and Yang, F. J., Journal of Vacuum Science & Technology B, 18(5) (2000) 2573–2575.
6. Ishida, N., Inoue, T., Miyahara, M. and Higashitani, K., Langmuir, 16(16) (2000) 6377–6380.
7. Yang, J. W., Duan, J. M., Fornasiero, D. and Ralston, J., Journal of Physical Chemistry B, 107(25) (2003) 6139–6147.
8. Simonsen, A. C., Hansen, P. L. and Klosgen, B., Journal of Colloid and Interface Science, 273(1) (2004) 291–299.
9. Zhang, X. H., Maeda, N. and Craig, V. S. J., Langmuir, 22(11) (2006) 5025–5035.
10. Switkes, M. and Ruberti, J. W., Applied Physics Letters, 84(23) (2004) 4759–4761.
11. Bhushan, B., Wang, Y. and Maali, A., Journal of Physics-Condensed Matter, 20(48) (2008) 5004.
12. Shen, G. X., Zhang, X. H., Ming, Y., Zhang, L. J., Zhang, Y. and Hu, J., Journal of Physical Chemistry C, 112(11) (2008) 4029–4032.
13. Zhang, X. H., Khan, A. and Ducker, W. A., Physical Review Letters, 98(13) (2007) 136101.
14. Connell, S. D. A., Allen, S., Roberts, C. J., Davies, J., Davies, M. C., Tendler, S. J. B. and Williams, P. M., Langmuir, 18(5) (2002) 1719–1728.
15. Wayner, P. C., J. Colloid Interface Sci., 88(1) (1982) 294–295.
16. Zhang, X. H., Zhang, X. D., Sun, J. L., Zhang, Z. X., Li, G., Fang, H. P., Xiao, X. D., Zeng, X. C. and Hu, J., Langmuir, 23(4) (2007) 1778–1783.
17. Nam, Y. and Jua, Y. S., Applied Physics Letters, 93(10) (2008) 103115.
18. Ducker, W., Personal Communication, 2009.
19. Zhang, L. J., Zhang, X. H., Zhang, Y., Hu, J. and Fang, H. P., Soft Matter, 6(18) (2010) 4515–4519.

20. Ducker, W. A., Langmuir, 25(16) (2009) 8907–8910.

21. Zhang, X. H. and Maeda, N., J. Phys. Chem. C., 115 (2011) 736–743.

22. Seddon, J. R. T., Bliznyuk, O., Kooij, E. S., Poelsema, B., Zandvliet, H. J. W. and Lohse, D., Langmuir, 26(12) (2010) 9640–9644.

23. Borkent, B. M., Dammer, S. M., Schonherr, H., Vancso, G. J. and Lohse, D., Physical Review Letters, 98(20) (2007) 204502–1 to 204502–4.

24. Zhang, X. H., Li, G., Maeda, N. and Hu, J., Langmuir, 22(22) (2006) 9238–9243.

25. Zhang, X. H. and Ducker, W., Langmuir, 23(25) (2007) 12478–12480.

26. Zhang, L. J., Zhang, Y., Zhang, X. H., Li, Z. X., Shen, G. X., Ye, M., Fan, C. H., Fang, H. P. and Hu, J., Langmuir, 22(19) (2006) 8109–8113.

27. Yang, S. J., Tsai, P. C., Kooij, E. S., Prosperetti, A., Zandvliet, H. J. W. and Lohse, D., Langmuir, 25(3) (2009) 1466–1474.

28. VV.AA., Nature, 445(7124) (2007) 129–129.

29. Mao, M., Zhang, J. H., Yoon, R. H. and Ducker, W. A., Langmuir, 20(5) (2004) 1843–1849.

30. Takata, Y., Cho, J. H. J., Law, B. M. and Aratono, M., Langmuir, 22(4) (2006) 1715–1721.

31. Schwendel, D., Hayashi, T., Dahint, R., Pertsin, A., Grunze, M., Steitz, R. and Schreiber, F., Langmuir, 19(6) (2003) 2284–2293.

32. Steitz, R., Gutberlet, T., Hauss, T., Klosgen, B., Krastev, R., Schemmel, S., Simonsen, A. C. and Findenegg, G. H., Langmuir, 19(6) (2003) 2409–2418.

33. Doshi, D. A., Watkins, E. B., Israelachvili, J. N. and Majewski, J., Proceedings of the National Academy of Sciences of the United States of America, 102(27) (2005) 9458–9462.

34. Mezger, M., Reichert, H., Schoder, S., Okasinski, J., Schroder, H., Dosch, H., Palms, D., Ralston, J. and Honkimaki, V., Proceedings of the National Academy of Sciences of the United States of America, 103(49) (2006) 18401–18404.

35. Poynor, A., Hong, L., Robinson, I. K., Granick, S., Zhang, Z. and Fenter, P. A., Physical Review Letters, 97(26) (2006) 1–4.

36. Palmer, L. A., Cookson, D. and Lamb, R. N., Langmuir, 27(1) (2011) 144–147.

37. Tyrrell, J. W. G. and Attard, P., Physical Review Letters, 87(17) (2001) 176104.

38. Tyrrell, J. W. G. and Attard, P., Langmuir, 18(1) (2002) 160–167.

39. Evans, D. R., Craig, V. S. J. and Senden, T. J., Physica a-Statistical Mechanics and Its Applications, 339(1–2) (2004) 101–105.

40. Christenson, H. K. and Claesson, P. M., Science, 239(4838) (1988) 390–392.

41. Colaco, R., Serro, A. P. and Saramago, B., Surface Science, 603(18) (2009) 2870–2873.

42. Bormashenko, E., Chemical Physics Letters, 456(4–6) (2008) 186–188.

43. Brenner, M. P. and Lohse, D., Physical Review Letters, 101(21) (2008).

44. Das, S., Snoeijer, J. H. and Lohse, D., Physical Review E, 82(5) (2010) 056310.

45. Ohgaki, K., Khanh, N. Q., Joden, Y., Tsuji, A. and Nakagawa, T., Chemical Engineering Science, 65(3) (2010) 1296–1300.

46. Agrawal, A., Park, J., Ryu, D. Y., Hammond, P. T., Russell, T. P. and McKinley, G. H., Nano Letters, 5(9) (2005) 1751–1756.

47. Wu, Z. H., Chen, H. B., Dong, Y. M., Mao, H. L., Sun, J. L., Chen, S. F., Craig, V. S. J. and Hu, J., Journal of Colloid and Interface Science, 328(1) (2008) 10–14.

48. Liu, G. M. and Craig, V. S. J., Acs Applied Materials & Interfaces, 1(2) (2009) 481–487.

49. Vinogradova, O. I., Langmuir, 11(6) (1995) 2213–2220.

50. Ikeda-Dantsuji, Y., Feril, L. B., Tachibana, K., Ogawa, K., Endo, H., Harada, Y., Suzuki, R. and Maruyama, K., Ultrasonics Sonochemistry, 18(1) (2011) 425–430.

51. Lukianova-Hleb, E., Hu, Y., Latterini, L., Tarpani, L., Lee, S., Drezek, R. A., Hafner, J. H. and Lapotko, D. O., Acs Nano, 4(4) (2010) 2109–2123.
52. Lukianova-Hleb, E. Y., Hanna, E. Y., Hafner, J. H. and Lapotko, D. O., Nanotechnology, 21(8) (2010) 85102.
53. Wagner, D. S., Delk, N. A., Lukianova-Hleb, E. Y., Hafner, J. H., Farach-Carson, M. C. and Lapotko, D. O., Biomaterials, 31(29) (2010) 7567–7574.
54. Wen, D. S., International Journal of Hyperthermia, 25(7) (2009) 533–541.
55. Fisher, P., Human & Experimental Toxicology, 29(7) (2010) 555–560.
56. Feng, X. J., Roy, S. C. and Grimes, C. A., Langmuir, 24(8) (2008) 3918–3921.
57. Voinova, M. V., Jonson, M. and Kasemo, B., Biosensors & Bioelectronics, 17(10) (2002) 835–841.
58. Gao, Y. Y., Zhao, L., Bai, H., Chen, Q. and Shi, G. Q., Journal of Electroanalytical Chemistry, 597(1) (2006) 13–18.
59. Gupta, S., Applied Physics Letters, 88(6) (2006) 863108.
60. Janda, P., Frank, O., Bastl, Z., Klementova, M., Tarabkova, H. and Kavan, L., Nanotechnology, 21(9) (2010) 095707.

Model and Experimental Study of Surfactant Solutions and Pure Liquids Contact Angles on Complex Surfaces

A. J. B. Milne [a], **Karina Grundke** [b], **Mirko Nitschke** [b], **Ralf Frenzel** [b] and
Alidad Amirfazli [a]

[a] Department of Mechanical Engineering, University of Alberta Edmonton, AB, T6G 2G8, Canada.
E-mail: a.amirfazli@ualberta.ca

[b] Leibniz Institute of Polymer Research Dresden, D-01005, Dresden, Germany

Contents

A. Introduction 175
    1. Wetting of Pure Liquids 176
B. Literature on Surfactant Solution Wetting 179
C. Contact Angle Model Derivation 181
    1. Model Results and Theoretical Behavior 184
D. Experimental Procedure—Wetting of Smooth and Superhydrophobic Surfaces by Pure Liquids
  and Surfactant Solutions 189
E. Experimental Results and Discussion 192
    1. Smooth Surfaces—Surfactant Solutions 192
    2. Surfaces—Surfactant Solutions and Pure Liquids 193
    3. aluminum surfaces]Aluminum Surfaces—Surfactant Solutions and Pure Liquids 196
    4. AKD surfaces]AKD Surfaces–Surfactant Solutions and Pure Liquids 198
    5. Investigating the Role of intrinsic contact angle]Intrinsic Contact Angle 200
F. Closing Remarks 203
G. Acknowledgements 206
H. References 207

A. Introduction

In this chapter we study wetting (contact angles) on complex surfaces. We define a complex surface as one that is heterogeneous and/or rough (in contrast to a 'simple' surface which is smooth, chemically homogeneous, hard, non-reactive, etc.). Certain rough hydrophobic surfaces are superhydrophobic surfaces (SHS), displaying extremely high contact angles, repellency and mobility of liquids on the surface. The main focus of this chapter will be on SHS, but heterogeneous smooth sur-

Drops and Bubbles in Contact with Solid Surfaces

faces will also be considered. The questions of chemically heterogeneous rough surface, and rough hydrophilic surfaces, will be left aside for now. The wetting of complex surfaces will be probed with surfactant solutions as models of the impure liquids likely to be used in industrial applications (such as, e.g., antifouling and self cleaning properties [1–6], oil spill cleanup [7], drop and liquid actuation in microfluidics [8–13], decreasing fluid friction on immersed bodies and in channels [14–21], decreased icing/snow accumulation on structures [22–25], and the use of SHS as switches and sensors [26, 27]. Surfactant solution wetting will be compared with wetting by aqueous and non-aqueous pure liquids to examine the similarities and differences between the two. First the theories describing wetting by pure liquids are presented, below.

1. Wetting of Pure Liquids

When a drop of pure liquid rests on a smooth, flat and homogeneous surface, Young's equation relates the drop's intrinsic contact angle (θ_y) to the interfacial tensions (γ_{lv}, γ_{sl} and γ_{sv}, where l, v and s represent liquid, vapor and solid phases, respectively). These interfacial tensions depend upon solid and liquid chemistry and purity. Young's equation is:

$$\cos \theta_y = \frac{\gamma_{sv} - \gamma_{sl}}{\gamma_{lv}}. \tag{1}$$

The highest (thermodynamically relevant) contact angle reported [22, 28] for pure water on a smooth surface is $\sim 120°$. On complex surfaces, higher contact angles are possible, but their prediction is more difficult; traditionally, the two equations of Wenzel [29] and Cassie [30] have been used.

If the liquid completely wets the surface by contacting the entire solid interface beneath the drop, filling the pores/crevices, Wenzel's equation describes the Wenzel contact angle (θ_w) as:

$$\cos \theta_w = r \left(\frac{\gamma_{sv} - \gamma_{sl}}{\gamma_{lv}} \right) = r \cos \theta, \tag{2}$$

where the effect of topography is modeled by r, the roughness factor, which is the ratio of actual surface area to projected surface area. By Eq. (2), it is seen that for hydrophobic surfaces ($\theta_y > 90°$), r increases the apparent contact angle by increasing the contact area/energy of the drop on the surface, whereas for hydrophilic surfaces ($\theta_y < 90°$) roughness decreases the apparent contact angle.

The Cassie equation (3) is used to calculate the contact angle of a pure liquid on a heterogeneous, flat, rigid and chemically inert surface, made up of m different materials:

$$\cos \theta_c = \sum_{\forall m} f_m \cos \theta_y \bigg|_m , \tag{3a}$$

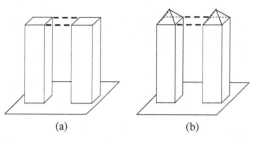

(a)　　　　　　　　(b)

Figure 1. Schematic close up showing surfaces decorated by (a) flat topped square pillars for which $f_1 = f$ and $f_2 = (1 - f)$, (b) the general case of an arbitrary surface with rough pillar tops, for which $f_1 \neq f$ and $f_2 \neq (1 - f)$.

$$\cos\theta_c = f_1 \cos\theta_y + f_2 \cos 180° \left(\underbrace{= f(\cos\theta_y + 1) - 1}_{\text{See text}} \right), \qquad (3b)$$

where θ_c is the Cassie contact angle. The weighting factors, f_m, are the *total* areas of each of the m materials under the drop, normalized by the total projected area. As such, $\sum_{\forall m} f_m \geqslant 1$. If a surface is rough and air is trapped under the drop, the trapped air can be treated as a surface with $\theta_y = 180°$, as illustrated in Eq. (3b). This is often employed to explain the superhydrophobicity of certain rough surfaces. Naive applications of Eq. (3b) express f_1 as f and f_2 as $(1 - f)$, with f defined as the *projected* solid–liquid area under the drop normalized by the total projected area. Obviously then, $f + (1 - f) = 1$. This simplification neglects the roughness of the solid–liquid interface and the curvature of the liquid–vapor interface under the drop, and is only valid for the case of flat topped pillars as seen in Fig. 1a. A more general case of rough pillar tops is shown in Fig. 1b, for which $f_1 \neq f$ and $f_2 \neq (1 - f)$. To consider the difference in another way, consider what Cassie and Baxter wrote [30]: "When [f_2 is zero] equation (3b)* reduces to Wenzel's equation for the apparent contact angle of a rough surface with the roughness factor f_1". Further, if $f_2 = 0$ and $f_1 = 1$, the Cassie equation reduces to the Young equation. This idea will be considered later in the analysis of superhydrophobic surface wetting by surfactant solutions, and in discussed in reference [31].

A given surface topography will have f_1, f_2 and r values. The question of which mode a drop will take is an interesting one. Following a similar strategy to Bico *et al.* [32], one can combine Eqs (2) and (3), to yield:

$$\cos\theta_{cr} = \left(\frac{f_2}{f_1 - r} \right). \qquad (4)$$

Equation (4) suggests that the wetting mode will be Cassie or Wenzel depending whether Young's (intrinsic) contact angle (controlled by liquid chemistry for a given

---

* In reference [30], Eq. (3) of this chapter (denoted as Eq. (5) in reference [30]) is what has become known as the Cassie equation. Further, Cassie and Baxter only considered a two material surface in their model.

surface chemistry) is greater ot less than the critical contact angle (θ_{cr}, constant for a given surface topography), respectively. Alternatively one can say that Eq. (4) suggests that the more stable wetting mode is the one that gives the lower contact angle. However, a more thorough examination of favorable modes requires thermodynamic analysis [33]. Experimental results for pure liquids or mixtures of pure liquids confirm that there is a generally smooth decrease in the intrinsic contact angle as liquid–vapor surface tension decreases [34]. On SHS, the decreasing intrinsic contact angle eventually leads to transition from Cassie to Wenzel wetting [7, 26, 35], often with abrupt decrease in contact angle [35, 36] due to the high roughness of the SHS amplifying the change in intrinsic contact angle. However, 'metastable' Cassie wetting modes have been suggested for surfaces which should exhibit the Wenzel mode with low contact angles [37–39], but instead display high contact angles and the Cassie mode. These researchers [37–39] suggest energy barriers that must be overcome for this transition to take place. This has again been seen in literature where intrinsic contact angles as low as 46.6° (reference [40]) and 70° (reference [34]) have yielded advancing contact angles of 105.3° (reference [40]) and 140° (reference [41]), respectively, when they should instead result in much lower contact angles due to the more favorable (according to Eq. (4)) Wenzel mode wetting (Eq. (2)).

Both Wenzel's and Cassie's equations describe how the intrinsic equilibrium value of contact angle is modified by topography/heterogeneities. However, advancing a drop across a surface increases the contact angle to the advancing contact angles, resulting in a higher energy metastable state compared to the Wenzel or Cassie angle. Likewise, receding a drop across a surface decreases the contact angle, resulting in another higher energy metastable state, resulting in the receding contact angle (Chapter 3 of reference [42]). The difference between the advancing and receding contact angle is called contact angle hysteresis (CAH). Prediction of the advancing and receding contact angles *a priori* is difficult, and has been the study of much work and debate, e.g., references [33] and [43–45]. However, Wenzel wetting mode is expected to have a higher CAH due to the higher work of adhesion necessary to remove the drop from the increased solid–liquid interface. In contrast, Cassie wetting mode is expected to have a lower CAH because of its reduced solid–liquid interface compared to the Wenzel mode. So, to design a SHS, researchers try to roughen a low energy surface in such a way that the Cassie wetting regime is favored. See references [33, 43] and [45] for more in-depth discussion.

All of the theories presented above were derived considering pure liquids on clean surfaces. Various models for surfactant solution wetting on simple (i.e., smooth and homogeneous) surfaces have been available for many years (e.g., references [46–49, 51]). Only recently however, were the first models for describing contact angles on rough/heterogeneous surfaces developed [27, 52]. In Section C, we show our derivations [27, 52] of modified Cassie and Wenzel models for the contact angle of surfactant solutions on rough and/or heterogeneous surfaces. Experimental protocols are briefly described in Section D. The models are then applied to understand and explain experimental results for wetting of 5 different SHS by various surfactant solutions in Section E, examining how topography, surface chemistry, and liquid type/purity affect wetting. But first, some of the literature available on wetting by surfactant solutions is briefly summarized in Section B.

B. Literature on Surfactant Solution Wetting

The wetting of simple surfaces by surfactant solutions is relatively well understood. On smooth hydrophilic surfaces (those with water contact angles less than 90°), surfactant solutions tend to spread, and can also demonstrate either a fingering instability leading to non-symmetric drop shapes [47, 53–65], or the autophobic effect where adsorption of surfactants create a more hydrophobic surface, leading to the spontaneous recede of a spreading drop [64–69]. The effect seen (fingering or autophobing) depends on the affinity of the surfactant for the surface. On smooth hydrophobic surfaces (those with water contact angles greater than 90°), the addition of surfactants tends to decrease contact angle [48–51, 70, 71]. Some researchers [48–51] have proposed an autophilic effect to occur on hydrophobic surfaces (with surfactants adsorbing ahead of the contact line on areas never touched by solution). However, solid recent evidence disproves this effect for the surfactants considered here [70] so the autophilic effect can be neglected. The decreased intrinsic contact angle of surfactant solutions on smooth hydrophobic surfaces (which all researchers agree occurs) could trigger a transition to the Wenzel mode of wetting in a SHS. Regarding receding contact angle, Varanasi and Garoff suggested [72] that re-organization of surfactants around the contact line during the recede can lead to pinning of the contact line and a decrease in the receding contact angle.

Very little consideration has been given to the use of surfactant solutions on SHS. Shirtcliffe *et al.* [26] studied superhydrophobic porous sol–gels for their use as switches (switching from superhydrophobic to hydrophilic behavior based on temperature, liquid type, or impurity concentration). Their study examined the wetting of only one surfactant solution on a single SHS. They reported a decrease in advancing contact angle from 140° with water to about 120° for concentrations of Sodium Dodecyl Sulphate (SDS) above Critical Micelle Concentration (CMC), and a decrease in receding contact angle of 140° for CMC of SDS, but did not present corresponding results with non-aqueous pure liquids for comparison. They did report that sufficiently low surface tension mixtures of ethanol in water penetrated the porous SHS. They suggested limited contact of the surfactant solution with the surface resulted in the high advancing angles, though they did not suggest why the solution did not penetrate the surface. They also suggest that surfactant films could be bridging the pores/crevices of the SHS and decreasing the receding contact angle due to pinning of the contact line.

Chang *et al.* [73] again studied a porous sol–gel, this time with a variety of surfactant solutions. They found that common single tailed surfactants could not decrease advancing contact angle below ~130–140° (i.e., could not cause a transition from Cassie to Wenzel mode wetting). Stronger wetting, double/branched tailed surfactants could decrease advancing contact angle to zero however, indicating a Cassie to Wenzel transition. They did not report much information on receding contact angle, and it is unclear whether their technique of decreasing drop volume by 20% would in fact access the receding mode of the drop. They explain their results by suggesting that the common surfactants cannot decrease the surface tension sufficiently to achieve the Wenzel mode (i.e., liquid penetrating the pores/crevices of the SHS), while the double/branched tailed surfactants can. They did not test pure liquids to compare their results/examine if any other effects besides surface tension played a role.

The results of Tang *et al.* [74] support the findings of Chang *et al.*, i.e., sufficiently high concentrations of Silwet L-77 (a 'superspreading' surfactant) can drastically decrease contact angle on a lotus leaf surface. The main thrust of the work was spreading, rather than measuring contact angles [74]. Receding contact angles were not reported, nor was a comparison with pure liquids performed.

Ferrari *et al.* [75, 76], reported on wetting of SHS by mixtures of non-ionic, semi-polar, and ionic (SDS and hexadecyltrimethylammonium bromide (HTAB)) surfactant solutions (with or without salt). They created their SHS by roughening a glass slide, covering it with silica nanoparticles, and coating them with hydrophobic treatments. They tested a single concentration of each surfactant far below CMC and at $2 \times$ CMC, with no intermediate concentrations, with and without the addition of 20 mM NaCl. They found superhydrophobic behavior for low concentrations, and decreased but non-zero contact angles at $2 \times$ CMC. Salt was seen to have little effect at low surfactant concentration, and was somewhat detrimental to repellency at $2 \times$ CMC. Their explanation for the observed behavior suggested that self-limiting surfactant adsorption (decreased surface tension) explained the results. The mixed surfactant studies they performed showed more complicated behaviors that are beyond the scope of discussion in this chapter. Mixed surfactants still could not fully wet SHS, however.

If the hypothesis of Ferrari *et al.* and Chang *et al.* were correct (i.e., if the decreasing surface tension with surfactant concentration were the sole controlling factor of contact angle on SHS), then the same SHS would be expected to behave equally poorly with pure liquids of similar surface tension to the solutions. This was not seen by Mohammadi *et al.* [77], who studied a SHS produced by the natural formation of a rough microstructure on alkyl ketene dimmer (AKD), a wax that presents a saturated hydrocarbon surface. They found that pure liquids showed an abrupt drop in advancing[*] contact angle around a surface tension of 45 mN/m. This is similar to the work of Shibuichi *et al.* [36], who also studied AKD SHS, but with pure liquids

---

[*] The receding contact angle data reported in reference [77] has been discovered to be erroneous; using the traditional definition of receding contact angle (a sustained and constant value while the contact line is receding) reanalysis of the data showed that the receding contact angle for all liquids is zero. Shibuichi *et al.* [36] did not measure receding contact angle, and may be unaware if it was zero for their AKD surfaces.

only. Mohammadi *et al.* [77] also studied surfactant solutions of the same and lower surface tensions as the pure liquids. The surfactant solutions did not show the same drop in contact angles as the pure liquids did [77], and instead maintained high advancing CAs (above 90° in nearly all cases). Mohammadi *et al.* [77] were the first to propose that this was due to surfactants inhibiting penetration of the solution into the roughness; this hypothesis was supported by Milne [78] and will be examined in the work presented in this chapter.

Overall, it seems that SHS can show high advancing contact angles with surfactant solutions, but may not always do so. Receding contact angles vary between high and low values between and within studies. Considering the small number of experimental studies of surfactant wetting on SHS, more investigation is needed to study what effects (if any) various types of liquid impurities have on the wetting mode and contact angles of surfactant solution on topographically and chemically different SHS (i.e., complex surfaces). Theoretical models are also needed. As such, models based on modified forms of the Cassie and Wenzel equations for wetting of surfactant solutions on rough/heterogeneous surfaces is presented next in this chapter. The model predictions are then applied to understand and explain experimental results for wetting of 5 different SHS by various surfactant solutions. The surfactant results are also put into context by comparing them with wetting of SHS with pure liquids having similar surface tensions to the surfactant solutions.

C. Contact Angle Model Derivation

The goal in this section is to derive thermodynamic models for the contact angle of surfactant solutions on complex (i.e., heterogeneous and/or rough) surfaces. This derivation relies on the use of the Young, Wenzel, and Cassie equations for equilibrium contact angle on smooth and rough/heterogenous surfaces, leaving aside, for now, the question of contact angle hysteresis and the consequent advancing and receding contact angles present. The difference between advancing, receding, and equilibrium contact angles will be considered when the model is applied later in this chapter. The Wenzel and Cassie equations are used due to their broad application in literature.

Gibbs' adsorption equation for the differential change in surface energy with a differential change in surfactant concentration in an ideal dilute solution is:

$$d\gamma_{xy} = -\Gamma_{xy} RT \, d\ln(C_S), \tag{5}$$

where γ_{xy} is interfacial tension and Γ_{xy} is the surface excess per unit area, or coverage, of surfactants at the interface xy, where x and y can be any of solid (s), liquid (l), or vapor (v) phases (e.g., γ_{lv} is liquid–vapor interfacial tension), R is the universal gas constant, T is the absolute temperature, and C_S is surfactant concentration (throughout this chapter expressed non-dimensionally as the ratio of concentration over critical micelle concentration (i.e., C/C_{CMC}).

If a suitable isotherm equation relating Γ_{xy} to $\ln(C_S)$ is applied, then Eq. (5) is integrable for interfacial tension. The general isotherm equation proposed by Zhu and Gu [79] is used here since it allows for the expression of several 'types' of adsorption (e.g., Langmuir, S-type or 2 plateau). The isotherm of Eq. (6) can also be applied to all three interfaces.

$$\Gamma_{xy} = \Gamma_{xy}^{\infty} \frac{K_{xy} C_{\mathrm{S}}^{n_{xy}}}{1 + K_{xy} C_{\mathrm{S}}^{n_{xy}}}, \tag{6}$$

where Γ_{xy}^{∞} is the limiting surface coverage (i.e., the maximum possible concentration of surfactant on an interface) and K_{xy} is the product of the equilibrium constants for the first and second adsorption steps. The exponent, n_{xy}, is used as an empirical fitting parameter [80].

Combining Eqs (5) and (6) and integrating gives an expression for interfacial tension as a function of surfactant concentration.

$$\gamma_{xy}(C_{\mathrm{S}}) = \gamma_{xy}^{0} - \frac{\Gamma_{xy}^{\infty} RT}{n_{xy}} \ln(1 + K_{xy} C_{\mathrm{S}}^{n_{xy}}), \tag{7}$$

where γ_{xy}^{0} is the interfacial tension for the pure liquid, i.e., for $C_{\mathrm{S}} = 0$.

Proceeding, one can consider the Young, Wenzel and Cassie equations (Eqs (1)–(3)). Taking the derivative of Eq. (1) for a given material, m, with respect to $\ln(C_{\mathrm{S}})$, and applying Eq. (5), one obtains Eq. (8) describing the relationship between contact angle and surfactant concentration for a smooth surface.

$$\left(\frac{\gamma_{\mathrm{lv}}}{RT}\right) \times \left(\frac{\mathrm{d}\cos\theta_{\mathrm{y}}|_m}{\mathrm{d}\ln(C_{\mathrm{S}})}\right) - \Gamma_{\mathrm{lv}}\cos\theta_{\mathrm{y}}|_m + (\Gamma_{\mathrm{sv}} - \Gamma_{\mathrm{sl}})|_m = 0. \tag{8}$$

Equation (8) is identical to that obtained by El Ghzaoui [47]. Substituting the isotherm model of Eq. (6) into Eq. (8) yields:

$$
\begin{aligned}
\frac{\mathrm{d}\cos\theta_{\mathrm{y}}|_m}{\mathrm{d}\ln(C_{\mathrm{S}})} &- \frac{\Gamma_{\mathrm{lv}}^{\infty} RT}{\gamma_{\mathrm{lv}}}\left(\frac{K_{\mathrm{lv}} C_{\mathrm{S}}^{n_{\mathrm{lv}}}}{1 + K_{\mathrm{lv}} C_{\mathrm{S}}^{n_{\mathrm{lv}}}}\right)\cos\theta_{\mathrm{y}}|_m \\
&= \left(\frac{\Gamma_{\mathrm{sl}}^{\infty} RT}{\gamma_{\mathrm{lv}}}\left(\frac{K_{\mathrm{sl}} C_{\mathrm{S}}^{n_{\mathrm{sl}}}}{1 + K_{\mathrm{sl}} C_{\mathrm{S}}^{n_{\mathrm{sl}}}}\right) - \frac{\Gamma_{\mathrm{sv}}^{\infty} RT}{\gamma_{\mathrm{lv}}}\left(\frac{K_{\mathrm{sv}} C_{\mathrm{S}}^{n_{\mathrm{sv}}}}{1 + K_{\mathrm{sv}} C_{\mathrm{S}}^{n_{\mathrm{sl}}}}\right)\right)\Big|_m
\end{aligned}
\tag{9}
$$

with the γ_{lv} term in Eq. (9) given by Eq. (7) expressed at the liquid–vapor interface. In this chapter, solid–vapor adsorption of surfactant on hydrophilic surfaces is considered to model the autophobic effect. It is not needed (setting $\Gamma_{\mathrm{sv}}^{\infty} = 0$) for hydrophobic surfaces [27, 70].

Equation (9) is a linear, inhomogeneous, ordinary differential equation with variable coefficients. Applying the technique of variation of parameters results in Eq. (10a), a modified Young's equation relating changes in smooth surface contact angle with surfactant concentration:

$$
\begin{aligned}
&\theta_{\mathrm{y}}|_m(C_{\mathrm{S}}) \\
&= \cos^{-1}\left(\frac{\cos\theta_{\mathrm{y}}^{0}\gamma_{\mathrm{lv}}^{0} + \frac{\Gamma_{\mathrm{sl}}^{\infty} RT}{n_{\mathrm{sl}}}\ln(1 + K_{\mathrm{sl}} C_{\mathrm{S}}^{n_{\mathrm{sl}}}) - \frac{\Gamma_{\mathrm{sv}}^{\infty} RT}{n_{\mathrm{sv}}}\ln(1 + K_{\mathrm{sv}} C_{\mathrm{S}}^{n_{\mathrm{sv}}})}{\gamma_{\mathrm{lv}}^{0} - \frac{\Gamma_{\mathrm{lv}}^{\infty} RT}{n_{\mathrm{lv}}}\ln(1 + K_{\mathrm{lv}} C_{\mathrm{S}}^{n_{\mathrm{lv}}})}\right)\Big|_m,
\end{aligned}
\tag{10a}
$$

where θ_y^0 is the Young contact angle for pure liquid on a clean smooth surface. Following the same procedure but starting with Eq. (2) one arrives at:

$$
\theta_w(C_S) = \cos^{-1}\left(\left(\cos\theta_w^0\gamma_{lv}^0 + r\frac{\Gamma_{sl}^\infty RT}{n_{sl}}\ln(1 + K_{sl}C_S^{n_{sl}})\right.\right.
$$
$$
\left. - r\frac{\Gamma_{sv}^\infty RT}{n_{sv}}\ln(1 + K_{sv}C_S^{n_{sv}})\right)
$$
$$
\left. \Big/ \left(\gamma_{lv}^0 - \frac{\Gamma_{lv}^\infty RT}{n_{lv}}\ln(1 + K_{lv}C_S^{n_{lv}})\right)\right)
$$

(10b)

which describe the change of contact angles with surfactant concentration for rough surfaces wet in the Wenzel mode. Equation (10b) will be referred to as the modified Wenzel equation throughout this chapter. θ_w^0 is the Wenzel contact angle for a pure liquid on a clean surface.

Again following the same procedure but starting with Eq. (3), one arrives at:

$$
\theta_c(C_S) = \cos^{-1}\left(\left(\cos\theta_c^0\gamma_{lv}^0 - \sum_{\forall m}\left[f_m\left(\frac{\Gamma_{sv}^\infty RT}{n_{sv}}\ln(1 + K_{sv}C_S^{n_{sv}})\right.\right.\right.\right.
$$
$$
\left.\left.\left. - \frac{\Gamma_{sl}^\infty RT}{n_{sl}}\ln(1 + K_{sl}C_S^{n_{sl}})\right)\Big|_m\right]\right)
$$
$$
\left. \Big/ \left(\gamma_{lv}^0 - \frac{\Gamma_{lv}^\infty RT}{n_{lv}}\ln(1 + K_{lv}C_S^{n_{lv}})\right)\right),
$$

(10c)

where θ_c^0 is the Cassie contact angle for a pure liquid on a clean surface. Equation (10c) is for Cassie mode wetting of an arbitrary surface. As stated in the introduction, it is valid for any surface (smooth or rough, homogeneous or heterogeneous, with or without vapor remaining under the drop). Equation (10c) will be referred to as the modified *full* Cassie equation throughout this chapter. It can be simplified to the Eq. (10a), (10b), or (10d) by suitable choice of m and f_m. Considering a SHS with air remaining under the drop, one can simplify Eq. (10c) to:

$$
\theta_c(C_S) = \cos^{-1}\left(\left(\cos\theta_c^0\gamma_{lv}^0 + f_1\frac{\Gamma_{sl}^\infty RT}{n_{sl}}\ln(1 + K_{sl}C_S^{n_{sl}})\right.\right.
$$
$$
\left. + f_2\frac{\Gamma_{lv}^\infty RT}{n_{lv}}\ln(1 + K_{lv}C_S^{n_{lv}})\right)
$$
$$
\left. \Big/ \left(\gamma_{lv}^0 - \frac{\Gamma_{lv}^\infty RT}{n_{lv}}\ln(1 + K_{lv}C_S^{n_{lv}})\right)\right).
$$

(10d)

Equation (10d) is the modified version of the Cassie equation expressed for SHS (i.e., Eq. (3b)), and will be referred to as the modified Cassie equation throughout this chapter. The common (but not always allowable [31]) simplification of $f_1 = f$

and $f_2 = (1 - f)$ leads to:

$$
\theta_c(C_S) = \cos^{-1}\Bigg(\Bigg(\Big(\cos\theta_c^0\gamma_{lv}^0 + f\frac{\Gamma_{sl}^\infty RT}{n_{sl}}\ln(1 + K_{sl}C_S^{n_{sl}})
$$
$$
+ (1 - f)\frac{\Gamma_{lv}^\infty RT}{n_{lv}}\ln(1 + K_{lv}C_S^{n_{lv}})\Bigg) \tag{10e}
$$
$$
\Bigg/\Big(\gamma_{lv}^0 - \frac{\Gamma_{lv}^\infty RT}{n_{lv}}\ln(1 + K_{lv}C_S^{n_{lv}})\Big)\Bigg).
$$

which is only valid for the case of flat topped pillars with no penetration/curvature of the liquid–vapor interface into the crevices of the roughness (see Fig. 1). Equation (10b) through (10e) can also be derived by substituting Eq. (10a) (describing the effect of surfactant on the Young contact angle) into Eqs (2) and (3) directly.

Equations (10a)–(10c) are predictive tools to study the effect of surfactant adsorption at the solid–liquid and liquid–vapor interfaces of drops on smooth and rough surfaces. The above models are the first of their kind to describe the effect of surfactant adsorption on the wetting of complex surfaces. See references [27] and [52] for more details. For $r = f_1 = 1$ and $f_{2,\dots,\infty} = 0$, Eqs (10b) and (10c) both simplify to Eq. (10a), as expected.

Any integrable isotherm(s) could be used in place of Eq. (6), allowing for the modeling of different 'types' of adsorption for different surfactant-interface pairs. Having chosen the isotherm of Zhu and Gu [79], values must be determined for the constants (Γ_{xy}^∞, n_{xy}, K_{xy}, γ_{xy}^0, θ_y^0) specifying adsorption to each interface. This has been performed for solutions of Sodium Dodecyl Sulphate (SDS) on smooth Teflon AF coated silicon [27]; first Eq. (7) was fit to surface tension measurements to find l–v constants. These constants were input into Eq. (10a), which was then fit to contact angle measurements, to find s–l constants. Solid–vapor adsorption was neglected due to the hydrophobicity of the surface. In a separate work [52], Eq. (7) was fit to surface tension measurements take from reference [60] to find l–v constants for aqueous solutions of modified inulin. These were then used with Eq. (10a) to fit contact angle data of the same inulin surfactant on quartz glass, a system displaying the autophobic effect. In this way, some plausible values for s–l and s–v constants were found to allow a demonstration of the utility of the models explained here. Others sets of constants were also chosen on an *ad hoc* basis, based on characteristics described in the literature [52].

In this chapter, all of the above mentioned constants are used. The constants for SDS on Teflon AF are used throughout the rest of the chapter both to explore the model and to apply it to the wetting of SHS by surfactant solutions. The l–v constants for modified inulin, and some of the other constants from reference [52] are only used in this section to explore the model.

1. Model Results and Theoretical Behavior

Using the constants for all these systems, the surface plots in Fig. 2 were created. Figure 2a–d show predicted contact angle for SDS solutions from 0 to 1 CMC on

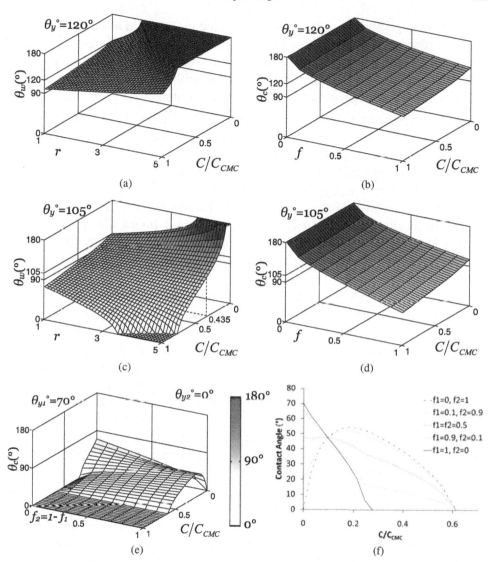

Figure 2. Surface plots of predicted contact angle for penetrated (Wenzel) wetting (a), (c) (using Eq. (10b)) and unpenetrated (Cassie) wetting (b), (d) (using Eq. (10e)) on rough surfaces (variable r and f) for various surfactant concentrations. Figure 2a, b and c, d correspond to intrinsic water contact angles of 120° and 105°, respectively, with adsorption coefficients held constant at those for SDS on Teflon AF. It is likely that the solid–liquid adsorption parameters would change on a surface with intrinsic contact angle of 105° compared to 120°. As such, Figure 2c, d should be seen as describing general trends. Subplots 'e' and 'f' are for two material smooth heterogeneous surfaces for which $\theta_y^0|_1 = 70°$ (with no autophobic effect), and $\theta_y^0 = 0°$ (with an autophobic effect present).

rough surfaces with topological parameters varying from $r = 1$–5 and $f = 0$–1, and intrinsic water contact angle varying from 120° to 105° (applying the adsorption constants for the SDS–Teflon AF system irrespective of intrinsic water contact angle). For simplicity, the surfaces being modeled in Fig. 2a–d are assumed to have flat pillar tops and flat liquid–vapor interfaces, allowing one to express $f_1 = f$ and $f_2 = (1 - f)$ (i.e., we consider surfaces of the type shown in Fig. 1a only here). Figure 2e–f show predicted contact angle for an arbitrary surfactant solution wetting a two material heterogeneous surface. In both plots, $\theta_y^0|_1 = 70°$ and material 1 does not experience the autophobic effect, while $\theta_y^0|_2 = 0°$ with material 2 experiencing an autophobic effect similar to that in reference [47]. Figure 2e shows predictions for a range of solid fractions from $f_2 = 0$ (all material 1) to $f_2 = 1$ (all material 2). Figure 2f shows five traces across the surface plot of Fig. 2e (at $f_2 = 0$, $f_2 = 0.1$, $f_2 = 0.5$, $f_2 = 0.9$ and $f_2 = 1$).

It is important to note that, in reality, if the intrinsic contact angle were to change drastically from 120° it is likely that the solid–liquid adsorption parameters would also change. As such, for surface chemistries which are not fluorinated, i.e., Fig. 2c, d, the results should be seen as describing general trends rather than exact values.

As Fig. 2a and c show, increasing surfactant concentration decreases contact angle for a surface with a given value of r being wet in the Wenzel mode. The magnitude of this decrease changes for different values of r. For example, for $r = 1$ (a smooth surface), the decrease in the Wenzel mode contact angle is $\sim21°$ as concentration changes from 0 to the C_{CMC} for a surface with $\theta_y^0 = 120°$ (Fig. 2a). For $r = 3$, the decrease in contact angle is $\sim62°$ for concentrations from 0 to the C_{CMC}. In this way, r can be thought of as amplifying the effects of surfactant adsorption. Since r is a measure of the increased area available under the drop and around the contact line, the increased area leads to increased surfactant adsorption and a greater change in contact angle as concentration changes. The roughness factor also acts to increase smooth surface contact angles, but only for those above 90° as discussed below.

The combined effects of intrinsic contact angle, concentration and roughness factor on Wenzel mode wetting can be understood by considering 'travel' along the r and concentration 'directions' in Fig. 2a, c. For $\theta_y^0 > 90°$ there are initially two competing influences, since in this case, as r increases the contact angle increases, but as concentration increases, contact angle decreases. For lower values of θ_y^0 (but still $>90°$), sufficiently high concentrations of surfactant allow the downward influence of concentration to overcome the influence of r. This can be seen in Fig. 2c compared to Fig. 2a. Specifically, it is seen in Fig. 2c for $C = 0.435C_{CMC}$, where θ_y (the smooth surface contact angle for a given surfactant concentration) decreases to 90°. When $\theta_y = 90°$ (or if $\theta_y^0 = 90$), r has no effect until surfactant adsorption to one or both interfaces decreases smooth surface contact angle, which then allows r to amplify the effect. This can be seen in the rapid decrease of contact angle with surfactant concentration in Fig. 2c once $\theta_y < 90°$. This also explains why it is much easier to use surfactants to fully wet surfaces which are already hydrophilic, since any roughness of the surface will act doubly to promote wetting.

Similarly to the Wenzel mode predictions, in Fig. 2b and d (showing Cassie mode wetting) surfactants are seen to decrease contact angle for any given f. In the Cassie mode however, it is seen, e.g., for $f = 0.5$ and $\theta_y^0 = 120°$ the decrease in Cassie mode contact angles is only ~13° over the entire range of C_S (Fig. 2a). This can be compared to a decrease of ~21° for a smooth surface ($f = 1$). In this way, the f parameter can be seen as attenuating the effect of surfactant adsorption.

The effect of intrinsic contact angle is seen to be less complicated for Cassie mode wetting (Fig. 2b and d) compared to the case of Wenzel mode wetting (Fig. 2a and c). For the Cassie mode, intrinsic contact angle determines the starting contact angle for high values of f, and surfactant adsorption leads to a relatively gradual decrease in contact angle. However, at lower values of f the Cassie contact angle becomes largely independent of intrinsic contact angle and/or concentration, again explained by the attenuating effects of f.

In addition to this attenuation, decreasing f is seen to increase contact angle, due to the $(1 - f)$ term in the modified Cassie equation (Eq. (10e)), indicating that the drop is 'sitting' on more air. In this way, the f parameter can also be thought of as acting in opposition to the effects of surfactant adsorption. Both the attenuation and opposition are expected intuitively, since there is less solid–liquid area under the drop for surfactants to adsorb to. This means that for low (but not zero) f planes (i.e., a superhydrophobic surface) surfactant concentration has little total effect; although adsorption still decreases the Cassie mode contact angle. This is not a full explanation for previous works [26, 73–78], that have found surfactant solutions often fail to wet SHS. A full explanation requires an explanation of how the Cassie mode (or perhaps a mode of partial penetration) is maintained, because the unpentrated Cassie mode is thermodynamically unlikely as surfactant concentration increases [32, 33]. This will be discussed in the Section E.

Considering Fig. 2a–d, one should note that the wetting mode (Wenzel or Cassie) of a drop, and thus which set of predictions to use (Fig. 2a and c or b and d), depends upon the topography and intrinsic contact angle of a surface. Predicting the wetting mode requires a thermodynamic analysis of the free energy. However, it is generally seen [32, 33] higher contact angles are less thermodynamically favorable. As such the predicted 180° contact angles in Fig. 2a and c for the Wenzel mode are likely thermodynamically unfavorable. Since determination of the favored mode requires [32, 33] *a priori* knowledge of both r and f, and since there is no one-to-one pairing of these parameters until a specific surface texture is chosen, the favored wetting mode could not be determined in Fig. 2a–d. In a recent paper these issues were investigated in a more theoretical way [27]. In this chapter, the developed model will be used to investigate and understand the experimental data presented in Section E.

Before considering the experimental data, the model for surfactant adsorption on heterogeneous surfaces will be briefly explored. A two material smooth ($f_1 + f_2 = 1$) heterogeneous surface is considered in Fig. 2e. As stated, $\theta_y^0|_1 = 70°$ (with material 1 not experiencing the autophobic effect), while $\theta_y^0|_2 = 0°$ (with material 2 experiencing an autophobic effect similar to that in reference [47]). Physically, this could be thought of as a glass surface ($\theta_y^0|_2 = 0°$) with, e.g., a thin PMMA coating ($\theta_y^0|_1 = 70°$) either completely covering the glass ($f_1 = 1$), or imperfectly coating the glass ($f_1 < 1$). Figure 2e shows the role of the f_1 and f_2 parameters in weighting the effect of surfactant adsorption on each portion of the heterogeneous surface. The effects of the combination on contact angle are non-linear (due to the cosine terms in Eq. (8)). As a result, the contact angles predicted by the modified Cassie model for $0 < f_1 < 1$ are more complex than for $f_1 = 0$ and $f_1 = 1$. Specifically, as concentration ranges from 0 to the CMC, multiple maxima and minima, and inflection points and 'kinks' (abrupt changes in slope) in the graph are seen for $0 < f_1 < 1$. To illustrate this, Fig. 2f shows five traces across the surface plot of Fig. 2e (at $f_1 = 0$, $f_1 = 0.1$, $f_1 = 0.5$, $f_1 = 0.9$ and $f_1 = 1$). The curves all intersect at $C/C_{CMC} \sim 0.1$, since at this concentration $\theta_y(C_s)|_1 = \theta_y(C_s)|_2$ and any weighting of the contact angles will have no net effect. As can be seen, even a heterogeneous surface that is a 9:1 combination of two materials can display different behaviors compared to $f_1 = 0$ and $f_1 = 1$ (all on Fig. 2f). The $f_1 = 0.9$ plot could correspond to a glass surface thinly coated in PMMA ($\theta_y^0|_1 = 70°$), with 10% bare patches ($\theta_y^0|_2 = 0°$, $f_2 = 0.1$) due to incomplete coating or coating damage. While the wetting of this surface with pure water would barely show a difference in contact angle between $f_1 = 1$ and $f_1 = 0.9$, surfactant solutions would behave very differently on a completely *versus* incompletely coated surface.

Looking at the $f_1 = 0.5$ plot in Fig. 2f, the behaviors are the most complex, showing local minima and maxima, inflection points, and abrupt changes in slope. The $f_1 = 0.1$ plot could correspond to a glass surface barely covered in PMMA, or to a hydrophilic surface with some hydrophobic patches due, e.g. to oil residue. In contrast to the $f_1 = 0.9$ plot, the $f_1 = 0.1$ plot shows little deviation from the $f_1 = 0$ ($f_2 = 1$) plot at high concentrations, and large deviations at low concentrations. This emphasizes the idea that it is the pure material contact angle *at a given concentration* that must be considered when judging the relative effect of small amounts of contamination on surfaces. This can be understood by considering that the f_1 and f_2 parameters acts on the cosine of the concentration dependent contact angles. As a result, a small change in the cosine of the heterogeneous surface contact angle at low contact angles results in a large change in contact angle. Conversely, an identical small change in the cosine at contact angles closer to 90° results in a relatively small change in contact angle.

Further discussion of surfactant solution wetting on smooth or rough heterogeneous surfaces are given in reference [52]. However, the great variations that are possible in surfactant solution wetting of heterogeneous surfaces point to the extreme importance in experimental work to either ensure the homogeneity of the surface, or to characterize the heterogeneity fully. Therefore, presented below is the setup and procedures (including surface preparation) for the experimental work discussed in this chapter. Following this, the experimental results are presented in Section E.

D. Experimental Procedure—Wetting of Smooth and Superhydrophobic Surfaces by Pure Liquids and Surfactant Solutions

Contact angles were measured on seven surfaces to investigate the role of differing topography and/or chemistry on contact angles. Surfaces were treated to present either a fluorinated or a saturated hydrocarbon chemistry to the liquid. Fluorinated chemistry was achieved either through exploiting the natural chemistry of the surface, or *via* spin coating with a 5:1 dilute solution of Teflon AF 1600 (DuPont Co.) in FC-75 (3M Co.). Saturated hydrocarbon chemistry was achieved either through exploiting the natural chemistry of the surface, or *via* immersion in a 12 mM solution of Octadecyltrichlorosilane (OTS) in extra dry toluene (both from Sigma Aldrich). See reference [78] for details of the coating and surface preparation. XPS analysis confirmed the chemical modification, and SEM and AFM were used to investigate surface texture. See Table 1 and Fig. 3 for the results of analysis.

Four different surface topographies were studied. Silicon wafers (referred to as control surfaces) were taken as received, diced, and coated with either Teflon AF or OTS (see reference [78] for details). They were seen to be relatively smooth (Table 1, Fig. 3); while smoother films are possible [81], the measured roughness of the control surfaces should not impact contact angle appreciably compared to the much rougher SHS. Plasma etched poly(tetrafluoroethylene) surfaces (here called PTFE) were prepared following the procedure outlined by Minko *et al.* [82]. The resulting topography is of a series of vertical spikes protruding from the surface (Table 1, Fig. 3). These surfaces are naturally fluorinated and could not be coated with a saturated hydrocarbon chemistry. Electrochemically etched aluminum samples (here called Teflon aluminum or OTS aluminum) were prepared following the process outlined by Hennig *et al.* [83]. The surfaces have a dual-scale, bumpy topography (Table 1, Fig. 3), and were coated with either Teflon AF or OTS. AKD surfaces were prepared following a previous study [77], and naturally develop a random topography of angled, plate-like structures (Fig. 3). AKD presents saturated hydrocarbon chemistry at its surface, with low quantities of hydrophilic heterogeneities, such as esters, confirmed with XPS analysis for coated and uncoated AKD. Teflon AF coating of AKD was shown (Table 1) to be incomplete; the chemistry is a mixture of saturated and fluorinated hydrocarbons; these surfaces will still be called Teflon AKD in this chapter. Wetting data for uncoated AKD (here called AKD) was taken from a previous study [77], but with corrected receding contact angle values.

SDS, HTAB and n-decanoyl-n-methylglucamine (MEGA 10), each at three or more submiceller concentrations, were used as probe liquids. Surfactants with different ionic properties were chosen to examine possible effects of charge interactions on the surface. Pure water, ethylene glycol (EG), bromonaphthalene (BN), and hexadecane (HD), were also tested to provide a comparison of wetting results for pure and impure liquids of similar surface tension/intrinsic contact angle.

Surface tension was measured by means of bubble tensiometry (SITA) and/or pendant drop analysis (First Ten Angstroms, Inc.) depending on the location of tests (Germany or Canada). Contact angle measurements are presented with respect to

Table 1.

AFM and XPS measurements. AFM measurements show roughness for smooth, coated silicon wafer, as well as plasma etched PTFE, etched aluminum and AKD superhydrophobic surfaces. XPS results are for fluorinated or saturated hydrocarbon surface chemistries for each surface where applicable. Value in parenthesis denotes standard deviation

	Silicon		PTFE	Aluminum		AKD	
	Teflon AF coated	OTS coated					
Ra (nm)	48 (26)	8 (0.02)	1328 (410)	400 (190)		1248 (170)	
Rq (nm)	57 (29)	14 (1)	1650 (780)	500 (210)		1530 (150)	
Rmax (nm)	276 (73)	437 (92)	8741 (350)	3800 (370)		7820 (180)	
Wenzel parameter r	—		2.78 (0.15)	1.77 (0.15)		2.85 (0.27)	
XPS Data	Teflon AF Coated	OTS Coated	Naturally Fluorinated	Teflon AF Coated	OTS Coated	Teflon AF Coated	Un-Coated
	[F]:[C]	[C–C] and [C–H]	[F]:[C] [C–C] and [C–H]	[F]:[C]	[C–C] and [C–H]	[F]:[C]	[C–C] and [C–H]
	~1.45	61.9%	~1.2 N/A	~1	58.8%	~0.3	>80%

Figure 3. SEM images of: (a) Teflon AF coated silicon, (b) plasma etched PTFE, (c) fluorinated etched aluminum, (d) uncoated AKD and (e) aluminum at higher magnification, showing fine features on top of larger features seen in (c), showing the dual-scale nature of the topography.

surface tension to allow for comparison between different liquids/solutions. Quasistatic advancing and receding contact angles tests were performed on a home built apparatus. Contact angles were measured using Axisymmetric Drop Shape Analysis (ADSA-P) operating in profile, sessile mode [84]. For more details on all experimental procedures, see reference [78].

E. Experimental Results and Discussion

Having derived a model describing contact angles on rough surfaces (wet in the Wenzel or Cassie mode) and smooth surfaces (Section C), these models are now applied to experimental data taken for Teflon AF and OTS coated silicon, as well as uncoated etched PTFE, Teflon AF and OTS coated etched aluminum, and Teflon AF coated and uncoated naturally rough AKD. For rough surfaces, the predictions should ideally use estimates/measurements of f_1 and f_2, capturing the roughness/corrugations of the solid–liquid and liquid–vapor interfaces under the drop. Due to the experimental difficulties in determining values of f_1 and f_2, we will make single estimates of f instead. Calculations [31] suggest that the errors inherent in this simplification are likely less than $5°$ for the low f_1 or f values seen for SHS (i.e., for $f \leqslant \sim 0.1$). The use of the correct equation would tend to cause the predictions of contact angle to decrease more slightly more quickly at higher values of concentration/lower values of surface tension.

1. Smooth Surfaces—Surfactant Solutions

Figure 4a and b show the contact angles of SDS, HTAB, and MEGA10 on Teflon AF and OTS coated silicon, respectively. Points are the average of advancing and receding contact angles (used as a proxy for equilibrium contact angle), and the end points of the bars mark the advancing and receding contact angles. Also shown (as lines in each graph) are the predicted contact angles using the model (Eq. (10a)) developed in Section C, with no solid–vapor adsorption since the surfaces are hydrophobic [70]. The adsorption constants used in the model are those for SDS on smooth Teflon AF, with the average water contact angle measured on each surface used as θ_y for each plot. As discussed in Section C, surfaces of different intrinsic contact angle might experience different solid–liquid adsorption behavior, causing differences in contact angle and surface tension. Further, different surfactants might result in different adsorption behavior, again leading to different contact angle and surface tension curves. Figure 4b shows, however, that the model predicts contact angles on the saturated hydrocarbon chemistry of OTS coated silicon similarly well to the Teflon AF coated silicon. Further, while individual plots of surface tension *versus* concentration and average contact angle *versus* concentration show differences for each different surfactant, when plotted as contact angle (Eq. (10a)) *versus* surface tension (Eq. (7)). Figure 4a and b both show that the model successfully predicts contact angle *versus* surface tension for HTAB and MEGA10 as well as SDS. The dotted regions of the lines are an extrapolation to super-CMC SDS concentrations, they predict the higher concentration (but still sub CMC) contact angles for HTAB and MEGA 10. In all, the model is seen to predict surfactant solution contact angles well in all cases studied here, and so it will be used to explore the wetting of SHS.

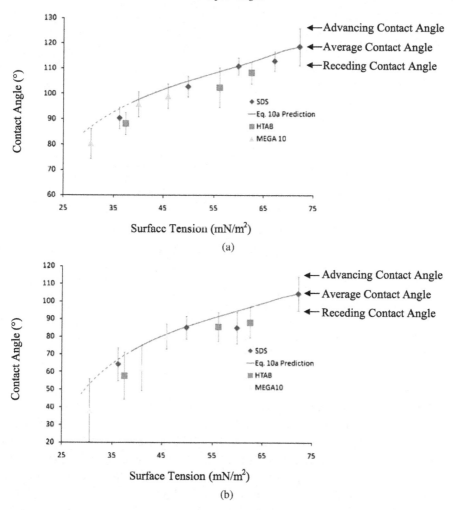

Figure 4. Average contact angle of surfactant solutions on (a) Teflon AF coated and (b) OTS coated smooth silicon wafer (symbols), and the predicted contact angles using Eq. (10a) with constants obtained from data of SDS on Teflon AF (line). The end points of the bars on symbols mark the advancing (top end) and receding (bottom end) contact angle. Standard deviation of advancing and receding contact angles are less than ~2.6° The dotted region of the line indicates super CMC concentrations of SDS.

2. PTFE Surfaces—Surfactant Solutions and Pure Liquids

Figure 5 shows advancing and receding contact angles of surfactant solutions and pure liquids on plasma etched PTFE, along with model predictions (made using Eq. (10b) and (10e)) for Wenzel and Cassie mode wetting of surfactants on the

Model and Experimental Study

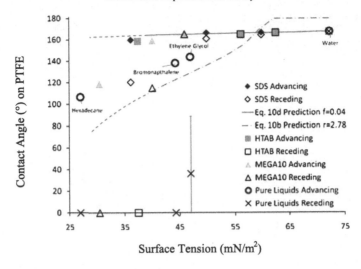

Figure 5. Advancing (closed symbol) and receding (open symbol) contact angles of surfactant solutions on naturally fluorinated plasma etched PTFE. Pure liquid contact angles are given as open circles (advancing) and '×'s (receding). Lines show the model prediction of contact angles (using Eqs (10b) and (10d) for Wenzel and Cassie mode wetting, respectively) using parameters found for aqueous solutions of SDS on Teflon AF. Standard deviations are generally of the same order as symbol size, and are not shown except for the large deviation of the receding contact angle of ethylene glycol (discussed in text).

SHS, respectively. The model requires values of r and f^* for the surface. The r value was measured using AFM (see Table 1). The f value was estimated by fitting the advancing contact angle for water on the PTFE surface to Eq. (3). The advancing contact angle was used instead of the average contact angle since the AKD surfaces (studied later) have zero receding contact angle (which makes using an average contact angle problematic). For the case of the PTFE and aluminum SHS, the advancing and receding contact angles for water are nearly identical, meaning that an estimate of f based on the advancing contact angle is essentially no different from an estimate based on average contact angle.

Using the measured r value, the model predictions of the modified Wenzel equation (10b) fails to predict the advancing contact angles of surfactant solutions on the PTFE surface considered in Fig. 5 (dashed-dotted line). Using an r value fit to the water contact angle *via* Eq. (2) produces similarly poor results (not shown), and results in an r value that does not reconcile with the measured value. The failure of the modified Wenzel model is not surprising since the PTFE surface has shown SHS behavior (high advancing and receding contact angle, therefore likely wetting in the Cassie mode). The modified Cassie equation (10e) predicts the advancing contact

---

* Predictions should use values of f_1 and f_2 in Eq. (10d). Single estimates of f, and Eq. (10e) are used instead for the sake of simplicity, with errors likely less than 5° and with the correct equation (Eq. (10d)) resulting in contact angle decreasing slightly more quickly [31].

angles much better, indicating that surfactant solutions wet the PTFE surface in the Cassie mode.

It is interesting that surfactant solutions display higher advancing contact angles on the PTFE surface compared to non-aqueous pure liquids with similar surface tensions. This supports the hypothesis put forth by Mohammadi *et al.* [77] that surfactant solutions are inhibiting penetration of surfactant solutions into the surface, either by forming a film over the pores/crevices, or by some other means. The highest concentration of MEGA10 (displaying the lowest surface tension) runs counter to this trend. It may indicate a mode of partial penetration for this solution (since it still is not matched by the Wenzel prediction). MEGA10 is a non-ionic surfactant, and thus has the lowest affinity for the surface compared to HTAB and SDS. This suggests that the highest concentration of MEGA10 does not interact with the surface sufficiently to prevent penetration into the SHS.

That the modified Cassie model predictions (dealing with equilibrium contact angle) predicts the advancing contact angle so well is an interesting point, and suggests that on SHS, advancing contact angle is a good proxy of equilibrium contact angle. This is not surprising when CAH is low (e.g., for water), since the high mobility of the drop on the surface means that the advancing and equilibrium contact angles are similar (i.e., there is a low energy barrier between the advancing and equilibrium state). As CAH increases however (e.g., as concentration increases and receding contact angle (discussed below), decreases) the model still predicts advancing contact angle well. This is a surprising result and suggests that surfactant solutions, in preventing penetration, maintain the Cassie mode and also preserve this similarity between the advancing and equilibrium contact angles.

The receding behavior of both surfactant solutions and pure liquids on PTFE is more complicated than the advancing. For higher surface tensions (lower concentrations of surfactants), the receding angle of surfactant solutions is seen to be high. At lower surface tensions however, the receding angle declines, eventually to zero degrees (in an abrupt transition). This is similar to the results seen by Shirtcliffe *et al.* [26] and counter to the findings of Ferrari [75, 76]. Shirtcliffe *et al.* proposed that the low receding angle could be due to stretching of a surfactant film across the pores/crevices of a SHS. It is also possible that re-organization of the surfactants across the receding contact line could lead to pinning and decreased receding contact angle [72]. Finally, the mechanism responsible for inhibiting the penetration of surfactant solutions into the roughness may also inhibit depinning on the recede at higher concentraiton. Regardless of cause, the result suggests that the effect of surfactants in increasing advancing contact angles compared to pure liquids is not necessarily carried over to receding contact angles.

Regardless of the decrease in receding contact angle, surfactant solutions still show higher receding contact angles than non-aqueous pure liquids of similar surface tension. It is worth noting that the high standard deviation on the receding contact angle measurement for ethylene glycol results from one high measurement and two measurements of zero receding angle. This suggests that this surface tension (for pure liquids) is near the Cassie to Wenzel transition point, at least during the receding stage. However, near this surface tension value, surfactant solutions still show high receding contact angles, supporting Mohammadi *et al.*'s hypothesis [77].

3. Aluminum Surfaces—Surfactant Solutions and Pure Liquids

Figure 6a and b show advancing and receding contact angles of surfactant solutions and pure liquids on Teflon AF coated and OTS coated electrochemically etched aluminum. For model predictions (*via* Eqs (10b) and (10e)), r values were measured by AFM (see Table 1) and f values were estimated for the aluminum surfaces in the same way as described for the PTFE surface. In addition, another f value was found for OTS Aluminum by fitting the highest concentration data point for SDS to Eq. (10c). As will be discussed later, this was done to allow a full explanation of the data.

In Fig. 6a, the successful prediction of the surfactant solution data by Eq. (10e) indicates that the Teflon AF coated aluminum surface is wet in the Cassie mode for all surfactant solutions. This surface also shows extremely high advancing contact angles with all pure liquids. Despite this, surfactant solutions are still seen to display slightly higher contact angles in comparison, lending further support to the hypothesis of surfactant solutions inhibiting penetration [77].

The receding contact angles for Teflon AF coated aluminum are seen to be similar to PTFE, and can be understood to occur due to the same mechanisms (stretching of a surfactant film across the pores [26], re-organization at the contact line [72], and/or pinning of the solution at the tops of the roughness). However, there are no high concentration solutions showing high receding contact angle on the aluminum surface, suggesting that that topography can play an important role in the receding behavior of surfactant solutions. A possible explanation is that the dual-scale topography of the aluminum (see Fig. 3c and e) could give a greater number of pinning points compared to the sharper topography of the PTFE (see Fig. 3b). Interestingly, Teflon AF coated aluminum and PTFE otherwise show little difference in terms of surfactant solution contact angles. On the other hand, the aluminum surface shows noticeably higher advancing and receding contact angles for pure liquids compared to the PTFE surface. The main difference between PTFE and Teflon AF coated aluminum is topography since liquid chemistry is unchanged and solid chemistry is similar, with PTFE expected to have a slightly lower intrinsic contact angle compared to Teflon AF [85]. Taking the observations for pure liquids and surfactant solutions together suggests that the inhibition of penetration caused by surfactant solutions can override the major differences in topography and minor differences in chemistry between the 'spiky' PTFE and 'bumpy' aluminum.

OTS coated aluminum (Fig. 6b) shows generally lower contact angles than Teflon AF coated aluminum (Fig. 6a), with a transition to full wetting on the recede at higher surface tension values. Individual wetting tests also showed that the receding contact angle was generally less stable (i.e., exhibited more instances of stick–slip)

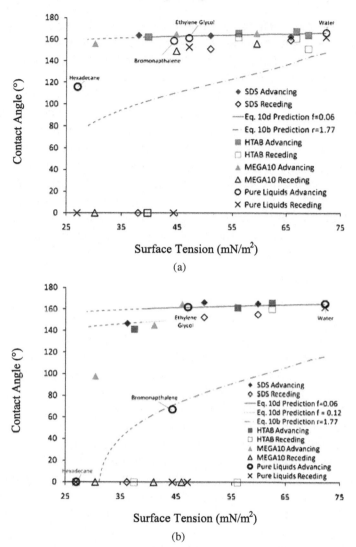

Figure 6. Advancing (closed symbol) and receding (open symbol) contact angles of surfactant solutions on (a) Teflon AF coated and (b) OTS coated, etched aluminum. Pure liquid contact angles are given as open circles (advancing) and 'x's (receding). Lines show the model prediction of contact angles (using Eqs (10b) and (10d) for Wenzel and Cassie mode wetting, respectively) using parameters found for aqueous solutions of SDS on Teflon AF. Standard deviations are generally of the same order as symbol size, and are not shown.

on OTS coated aluminum compared to Teflon AF coated aluminum. All of this is expected since the OTS coated silicon (Fig. 4b) shows lower intrinsic contact angle compared to Teflon AF coated silicon (Fig. 4a). Surfactant solution contact angles are still well predicted by the modified Cassie model, and are still higher than corresponding non-aqueous pure liquid contact angles. It seems, however, that OTS coated aluminum shows a state of partial penetration for the highest concentrations of all three surfactants. The dotted line on Fig. 6b corresponds to the modified Cassie model predictions when an f value of 0.12 is used (instead of $f = 0.06$ for the solid curve). This value of $f = 0.12$ was found by fitting the highest concentration data point for SDS to Eq. (10e). It is found to predict the high concentration data better than the modified Cassie prediction for $f = 0.06$. A similar idea was put forth by Ferrari *et al.* [76], where repeated advances of surfactant solution drops across a SHS were seen to result in contact angle decreasing by $\sim$5–15° over three cycles. They attributed the decreasing contact angle to an increase in solid fraction (i.e., increasing f, indicating a partial penetration of solution into the roughness). In Fig. 6b, hexadecane and bromonapthalene demonstrate extremely low contact angles indicating either an advanced state of partial penetration or the fully penetrated Wenzel mode. Thus, even if the surfactant solutions are partially penetrating the surface, the surfactants still inhibit penetration compared to non-aqueous pure liquids of similar surface tension.

4. AKD Surfaces–Surfactant Solutions and Pure Liquids

Figure 7a and b show contact angles of surfactant solutions and pure liquids on Teflon AF coated AKD and uncoated AKD, respectively. Uncoated AKD naturally presents a saturated hydrocarbon chemistry, with hydrophilic heterogeneities. It is clear from the data that surfactant solutions again show contact angles higher than pure liquids of similar surface tension. It is also clear that the topography of the AKD is less repellent than that of PTFE (see Fig. 5) or coated aluminum (see Fig. 6) since the AKD surfaces show generally lower contact angles.

Also shown on Fig. 7a and b are modified Cassie and Wenzel predictions of surfactant solution contact angles. Values of r were measured (see Table 1) and f values were found, as described for PTFE. The discrepancy in f values for the two surfaces calculated with water contact angle is due to the incomplete coating of Teflon AKD. This incomplete coating would result in a decreased intrinsic contact angle (compared to a completely Teflon AF coated surface) and an overestimation of f, but the f value for Teflon AKD will be used in order to allow progress. Again, additional values of f were found by fitting higher concentration contact angle data for SDS to Eq. (10e). Both AKD surfaces are heterogeneous *and* rough complex surfaces. While the full modified Cassie equation (10c) could be used to model both surfaces, it would require additional data regarding the distribution/adsorption properties of the heterogeneities that is not available. For now, Eq. (10e) is used to make progress as described below.

Regardless of the limitations posed by the heterogeneities of the AKD surfaces, certain observations can still be made. Examining Fig. 7a and b shows that neither a single modified Cassie prediction, nor a single modified Wenzel prediction, can successfully model the data. On Fig. 7b, one could instead attempt to model the data using the modified Cassie model at low concentrations and the modified Wenzel model at higher concentration, with r as an adjustable paramter. Modifying the fit in this way gives even larger r values, resulting in even worse predictions. It

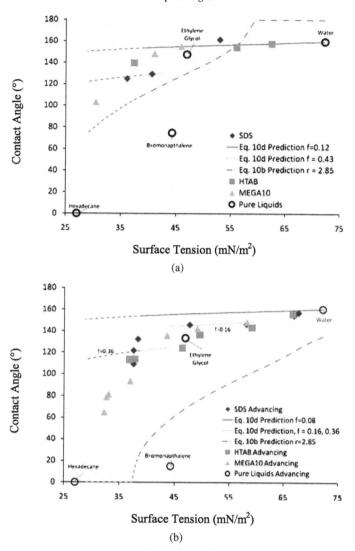

Figure 7. Advancing (closed symbol) contact angles of surfactant solutions and pure liquids on (a) Teflon AF coated and (b) naturally saturated hydrocarbon chemistry AKD. Pure liquid contact angles are given as open circles (advancing). Lines show the model prediction of contact angles (using Eqs (10b) and (10d) for Wenzel and Cassie mode wetting, respectively) using parameters found for aqueous solutions of SDS on Teflon AF. Standard deviations are generally of the same order as symbol size, and are not shown.

is possible to fit the high concentration data on Fig. 7a with the modified Wenzel equation using $r \approx 5$. This is incongruous with the measured value of $r = 2.85$ (see Table 1). Further, since uncoated AKD is seen not to be wetting in the Wenzel mode for high concentration surfactants, it is unlikely that Teflon AF coated AKD (with a higher intrinsic contact angle) would wet in the Wenzel mode. The Wenzel mode predictions for $r \approx 5$ are therefore not shown on Fig. 7a. Multiple Cassie mode predictions (at progressively higher values of f, i.e., increased amounts of penetration into the pores of the roughness) are thus necessary to model both surfaces. In contrast, the contact angles of bromonapthalene and hexadecane strongly suggest Wenzel mode wetting, supporting the hypothesis [77] that surfactant solutions are inhibiting penetration of surfactant solutions into the roughness of the SHS.

Considering that both coated and uncoated AKD surfaces seem to be wet in the Cassie mode by surfactant solutions, it is counterintuitive that they would display zero receding contact angle, both with surfactant solutions and pure liquids. This is expected to be due to the hydrophilic heterogeneities in the structure of the AKD pinning the contact line, and/or due to pinning on the microstructure of the AKD.

5. Investigating the Role of Intrinsic Contact Angle

Summarizing Fig. 5 through Fig. 7, one can see that surfactant solutions wet SHS differently than non-aqueous pure liquids of similar surface tension. Teflon AF coated aluminum shows higher contact angles than the PTFE surface for pure liquids, but both surfaces show roughly similar behavior with surfactant solutions. This indicates that on rough surfaces, surfactants can make up for a less repellent topography by preventing penetration (thereby maintaining a low f value). Considering the low contact angles on AKD, however, shows that surfactants cannot always do so. Another, more important, difference is that surfactant solutions show higher contact angles, indicating a preserved Cassie mode, compared to pure liquids of similar surface tension. In literature, the surface tension of a liquid is often taken as the controlling factor determining if said liquid will penetrate a given rough surface. However, intrinsic contact angle is a better measure, since as Eq. (4) suggests, a critical intrinsic contact angle must be surpassed to transition between the Cassie and Wenzel modes. As the analysis of experimental results in this chapter has, so far, relied upon surface tension, the question remains if the preserved Cassie mode for surfactant solutions on SHS is due to differences in intrinsic contact angle, or due to an additional effect of the surfactants. To answer this question, contact angles of pure liquids on smooth Teflon AF coated and OTS coated silicon surfaces were also measured. The results comparing surfactant solutions and pure liquids on smooth surface are presented in Fig. 8a and b.

For both Teflon AF and OTS coated silicon surfaces, pure liquids are seen to show generally similar contact angles to surfactant solutions of similar surface tension. At low surface tensions however, surfactant solutions show a slightly higher contact angle than the corresponding pure liquids (bromonapthalene and hexadecane). To investigate if this increase in intrinsic (smooth surface) contact angle explains the higher contact angles on SHS for surfactant solutions, one can plot contact angle on the SHS *versus* intrinsic contact angle. The range of intrinsic contact angles considered is shown by the dotted boxes in Fig. 8a and b.

Plotting the contact angle on the SHS *versus* intrinsic contact angle in Figs 9–11, one can see that advancing contact angle for pure liquids on SHS are generally significantly lower than the range of advancing contact angle for surfactants solu-

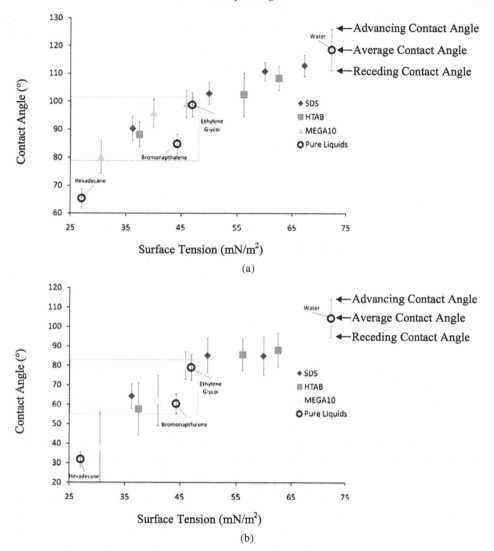

Figure 8. Average contact angle of surfactant solutions and pure liquids on (a) Teflon AF coated and (b) OTS coated, smooth silicon wafer. The end points of the bars on symbols mark the advancing (top end) and receding (bottom end) contact angle. Standard deviation of advancing and receding contact angles are less than ~2.6°.

tions for the same range of intrinsic (smooth surface) contact angles. On Figs 9–11, the modified Cassie prediction has been plotted for comparison with results, with the dashed regions of the lines an extrapolation to super-CMC SDS concentrations, they predict the higher concentration (but still sub CMC) contact angles for HTAB and MEGA 10 well, as argued at the end of Section F.1.

The increase in surfactant solution wetting compared to pure liquid wetting is most apparent in Fig. 9 for PTFE. Here two distinct regions have been drawn de-

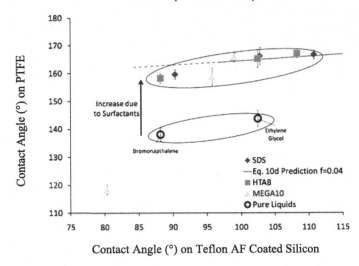

Figure 9. Advancing contact angle on PTFE *versus* advancing contact angle on Teflon AF coated, smooth silicon for pure liquids and surfactant solutions of similar intrinsic contact angle. Surfactant solutions show generally higher contact angles than pure liquids. For the modified Cassie prediction (Eq. (10e)), f is 0.04. Error bars show one standard deviation.

lineating the apparent contact angle range for pure liquids and the higher intrinsic contact angle range for surfactant solutions. The difference can also be seen clearly in Fig. 10b and Fig. 11a and b for OTS aluminum, Teflon AF coated AKD and uncoated AKD surfaces, respectively. In each of these three graphs, while solutions and pure liquids show similar apparent contact angles near the data point for ethylene glycol, for lower intrinsic contact angles the pure liquid contact angle drops to a lower value and with a steeper close than the surfactant solution contact angles. In Fig. 10a, for Teflon AF coated aluminum, the results are quite close to each other, and standard deviations of the data point overlap. The Teflon AF coated aluminum surface gave very high contact angles with pure liquids however, so while the surfactant solutions do show slightly higher advancing contact angles, if pure liquids are already not penetrating the SHS, surfactant solutions could not be expected to further inhibit penetration.

With the possible exception of Teflon AF coated aluminum, the increases seen cannot be explained by differences in intrinsic contact angle since the data in Figs 9–11 are normalized for this effect. Therefore, it is proposed that in addition to the effect of increased intrinsic contact angle, the surfactants are preventing penetration of the solution into the roughness, either by forming a film over the crevices [26, 77] or by some other means. The specifics of the exact mechanism remain to be determined, but investigations should be aided though the use of the newly developed modified Cassie and Wenzel equations, which have been shown capable of both predicting contact angles on SHS, and interpreting complex situations such as partial penetration.

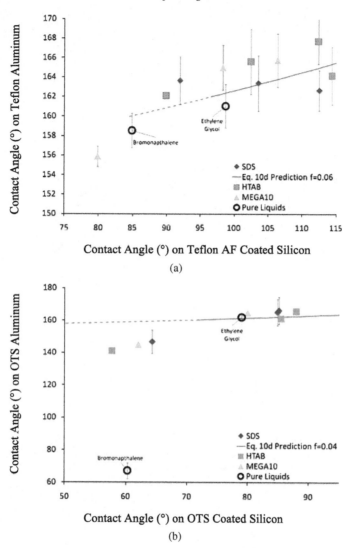

Figure 10. Advancing contact angle on (a) Teflon AF and (b) OTS coated aluminum *versus* advancing contact angle on (a) Teflon AF and (b) OTS coated, smooth silicon for pure liquids and surfactant solutions of similar intrinsic contact angle. Surfactant solutions show generally higher contact angles than the pure liquids. For the modified Cassie prediction, f is (a) 0.06 and (b) 0.04, the dashed line shows super CMC concentrations of SDS. In (a), the Teflon AF coated silicon contact angles are linear interpolations or extrapolations from the closest measured data to match measured values of concentration/surface tension of surfactant solutions on aluminum to Teflon AF coated silicon. Error bars show one standard deviation.

F. Closing Remarks

A model has been derived, from first principles, describing the effect of surfactant adsorption on equilibrium contact angle on complex surfaces (i.e., rough and/or heterogeneous) wetting in the Wenzel or Cassie modes. Experimental data for aqueous

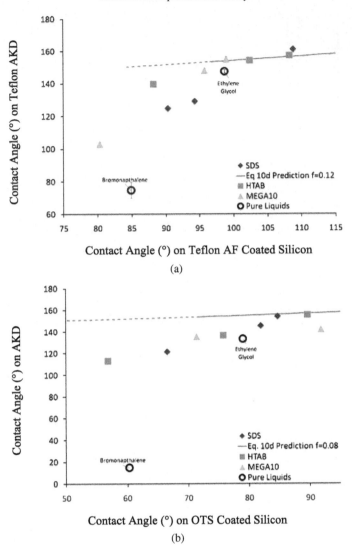

Figure 11. Advancing contact angle on (a) Teflon AF coated and (b) uncoated AKD *versus* advancing contact angle on (a) Teflon AF coated and (b) OTS coated silicon for pure liquids and surfactant solutions of similar intrinsic contact angle. Surfactant solutions show generally higher contact angles than the pure liquids. For the modified Cassie prediction, f is (a) 0.12 and (b) 0.08, the dashed line shows super CMC concentrations of SDS. In (b) (and for two data points in (a)), the coated silicon contact angles are linear interpolations or extrapolations from the closest measured data to match measured values of concentration/surface tension of surfactant solutions on AKD to coated silicon. Error bars show one standard deviation. Standard deviations are unavailable for uncoated AKD (data taken from reference [77]).

solutions of SDS contact angle on smooth Teflon AF surfaces was fit to estimate values for the adsorption coefficients in the model.

Using the model with these coefficients, it is seen that in the Wenzel mode, increasing r (increasing s–l area under the drop and around the contact line) leads to an increasing effect of surfactant adsorption on contact angle, as there is more area to adsorb to. The Wenzel roughness parameter, r, also has its usual effect of increasing contact angle for intrinsic contact angles greater than 90°, and decreasing contact angles otherwise. This means that on rough, hydrophilic surfaces, there is an amplification both of the intrinsic contact angle and the effect of surfactants, leading to enhanced wetting.

In Cassie mode wetting (on rough but homogeneous surfaces), it is instead seen that decreasing f_1 (decreasing s–l area) leads to a decreasing effect of surfactant adsorption as there is less area to adsorb to. For the flat-topped pillared surfaces considered in exploring the model, decreasing Cassie solid fraction, f_1, also leads to an increasing contact angle due to the increasing f_2 (or $(1 - f)$ for flat-topped surface texture) term in Eq. (10d), which quickly overwhelms the other decreasing terms in the model. Because of this, wetting in the Cassie mode is less affected by the intrinsic contact angle of the surface. The dominance of the f_2 (or $(1 - f)$ for flat-topped surface texture) term explains the high contact angles seen in literature for SDS solutions on SHS. This key finding is the first of its kind based on theoretical reasoning.

For Cassie mode wetting of smooth but heterogeneous surfaces, the f_m parameters are seen as weighting factors; determining how the behaviors of each material in a heterogeneous surface are combined. It is seen that small inclusions with high contact angles at a given concentration have a large effect on surfaces predominantly composed of a low contact angle material at the same given concentration. Conversely, low contact angle inclusion have a smaller effect on surfaces predominantly composed materials with a high contact angle at a given concentration. This can be understood by considering that the f_m parameters act on the cosine of the concentration dependent contact angles. As a result, a small change in the cosine of the heterogeneous surface contact angle at low contact angles results in a large change in contact angle. Conversely, an identical small change in the cosine at contact angles closer to 90° results in a relatively small change in contact angle. Even more complex behaviors are seen as the f_m parameters for each material approach each other.

The model was also used to investigate advancing and receding contact angles of nine surfactant solutions of various polarities. Four pure liquids with a similar range of surface tensions to the surfactant solutions (~27–73 mN/m) were also tested. Five superhydrophobic surfaces (SHS) with different topology ('spiky' etched PTFE, 'bumpy' etched aluminum and 'plate like' AKD) and chemistry (fluorinated or saturated hydrocarbon) combinations were probed along with two smooth control surfaces.

Based upon advancing contact angle measurements on the SHS, it was determined that the unpenetrated Cassie mode was the most commonly seen, with occasionally a mode that appears to be partial penetration at higher surfactant concentrations. The fully penetrated Wenzel mode was not seen. Pure liquids, on the other hand, generally showed lower contact angles than the surfactant solutions, indicating greater (often full) penetration of the pure liquids into the SHS.

Receding contact angle is high for high surface tension/intrinsic contact angle pure liquids and surfactant solutions. It likewise decreases as surface tension/intrinsic contact angle decreases. However, the receding contact angles of surfactant solutions are seen to decrease less than those for pure liquids. In addition, the PTFE surface shows receding contact angles above 90° for some high concentration surfactant solutions while the coated aluminum surface did not. The AKD surface showed zero receding contact angle for all liquids/solutions. This suggests that topography can play an important role in the receding behavior of surfactant solutions. In addition to the effects of intrinsic contact angle, for surfactant solutions, the low receding contact angle can be understood by considering stretching of a surfactant film across the pores [26], re-organization at the contact line [72], and/or pinning of the solution at the tops of the roughness. Regardless of cause the results suggests that the effect of surfactants in increasing advancing contact angles (compared to pure liquids) is not necessarily carried over to receding contact angles.

Taking advancing and receding contact angle data together it is clear that surfactant solutions show equal or higher apparent contact angles than pure liquids of similar surface tension/intrinsic contact angle on all five SHS. Since the analysis corrects for the slightly higher intrinsic contact angle of surfactant solutions, it is proposed that surfactant are inhibiting penetration of solutions into SHS, either by surfactant film formation or by pinning of the contact line at the top of the roughness.

Surface chemistry is seen to affect wetting mainly by means of the intrinsic contact angle. It is seen that for a given topography saturated hydrocarbon chemistry results in similar behavior to fluorinated chemistry in terms of general wetting trends and mechanisms describing them. However, the contact angles are generally lower for saturated hydrocarbon chemistry and drop to lower values more abruptly, and at higher surface tensions. In this way, an unsuitable chemistry can overrule the beneficial effects of topography.

G. Acknowledgements

The authors acknowledge Nicole Petong, Stefan Michel, Frank Simon and others of the Leibniz Institute of Polymer Research, Dresden, Germany, for sample preparation and characterization. Further acknowledgements to C. Blank at the Institut für Werkstoffwissenschaft at the Technical University of Dresden, for the high quality, high magnification SEM images and to Pedro J. Ramón-Torregrosa for sample characterization. Surface characterization was also performed at the Alberta Centre for Surface Engineering and Science. Funding was provided by the National Science and Engineering Research Council of Canada, the Leibniz Institute of Polymer

Research, Alberta Innovates—Technology Futures, and the University of Alberta International Office Research in Germany Award.

H. References

1. Carman, M., Estes, T., Feinberg, A., Schumacher, J., Wilkerson, W., Wilson, L., Callow, M., Callow, J. and Brennan, A., Biofouling, 22 (2006) 11–21.
2. Cheng, Y. T., Rodak, D. E., Wong, C. A. and Hayden, C. A., Nanotechnology, 17 (2006) 1359–1362.
3. Hoipkemeier-Wilson, L., Schumacher, J., Carman, M., Gibson, A., Feinberg, A., Callow, M., Finlay, J., Callow, J. and Brennan, A., Biofouling, 20 (2004) 53–63.
4. Shang, H. M., Wang, Y., Limmer, S. J., Chou, T. P., Takahashi, K. and Cao, G. Z., Thin Solid Films, 472 (2005) 37–43.
5. Shang, H. M., Wang, Y., Takahashi, K., Cao, G. Z., Li, D. and Xia, Y. N., Journal of Materials Science, 40 (2005) 3587–3591.
6. Di Mundo, R., Nardulli, M., Milella, A., Favia, P., d'Agostino, R. and Gristina, R., Langmuir, 27 8 (2011) 4914–4921.
7. Venkateswara Rao, A., Hegde, N. D. and Hirashima, H., Journal of Colloid and Interface Science, 305 (2007) 124–132.
8. Aussillous, P. and Quere, D., Nature, 411 (2001) 924–927.
9. Shikida, M., Ando, M., Ishihara, Y., Ando, T., Sato, K. and Asaumi, K., Journal of Micromechanics and Microengineering, 14 (2004) 1462–1467.
10. Ren, S., Yang, S., Zhao, Y., Yu, T. and Xiao, X., Surface Science, 546 (2003) 64–74.
11. Ren, S., Yang, S. and Zhao, Y., Acta Mechanica Sinica, 20 (2004) 159–164.
12. Zhai, L., Berg, M. C., Cebeci, F. Ç., Kim, Y., Milwid, J. M., Rubner, M. F. and Cohen, R. E., Nano Letters, 6 (2006) 1213–1217.
13. Meyyappan, S., Shadnam, M. R. and Amirfazli, A., Langmuir, 24 (2008) 2892–2899.
14. Choi, C.-H., Ulmanella, U., Kim, J., Ho, C.-M. and Kim, C.-J., Physics of Fluids, 18 (2006) 087105-1-8.
15. Fukagata, K., Kasagi, N. and Koumoutsakos, P., Physics of Fluids, 18 (2006) 051703-1-4.
16. Gogte, S., Vorobieff, P., Truesdell, R., Mammoli, A., van Swol, F. Shah, P. and Brinker, C. J., Physics of Fluids, 17 (2005) 051701-1-4.
17. Marmur, A., Langmuir, 22 (2006) 1400–1402.
18. Ou, J. and Rothstein, J. P., Physics of Fluids, 17 (2005) 103606-1-10.
19. McHale, G., Shirtcliffe, N. J., Evans, C. R. and Newton, M. I., Applied Physics Letters, 94 (2009) 064104-1-3.
20. McHale, G., Newton, M. I. and Shirtcliffe, N. J., Soft Matter, 6 (2010) 714–719.
21. Martell, M. B., Rothstein, J. P. and Perot, J. B., Physics of Fluids, 22 (2010) 065102-1-13.
22. Nakajima, A., Hashimoto, K. and Watanabe, T., Chemical Monthly, 132 (2001) 31–41.
23. Kako, T., Nakajima, A., Irie, H., Kato, Z., Uematsu, K., Watanabe, T. and Hashimoto, K., Journal of Materials Science, 39 (2004) 547–555.
24. Antonini, C., Innocenti, M., Horn, T., Marengo, M. and Amirfazli, A., Cold Regions Science and Technology, 67 (2011) 58–67.
25. Milne, A. J. B. and Amirfazli, A., Langmuir, 25 (2009) 14155–14164.
26. Shirtcliffe, N. J., McHale, G., Newton, M. I., Perry, C. C. and Roach, P., Materials Chemistry and Physics, 103 (2007) 112–117.

27. Milne, A. J. B., Elliott, J. A. W., Zabeti, P., Zhou, J. and Amirfazli, A., Model and Experimental Studies for Contact Angles of Surfactant Solution on Rough and Smooth Hydrophobic Surfaces, Phys. Chem. Chem. Phys., 13 (2011) 16208–16219.

28. Nishino, T., Meguro, M., Nakamae, K., Matsushita, M. and Ueda, Y., Langmuir, 15 (1999) 4321–4323.

29. Wenzel, R. N., Industrial & Engineering Chemistry, 28 (1936) 988–994.

30. Cassie, A. B. D. and Baxter, S., Transactions of the Faraday Society, 40 (1944) 546.

31. Milne, A. J. B. and Amirfazli, A., How Cassie's Equation is Meant to be Used, 10.1016/j.cis.2011.12.001.

32. Bico, J., Thiele, U. and Quéré, D., Colloids and Surfaces A: Physicochemical and Engineering Aspects, 206 (2002) 41–46.

33. Li, W. and Amirfazli, A., Advances in Colloid and Interface Science, 132 (2007) 51–68.

34. Kwok, D. Y. and Neumann, A. W., Advances in Colloid and Interface Science, 81 (1999) 167–249.

35. Fujita, M., Muramatsu, H. and Fujihira, M., Japanese Journal of Applied Physics 1, 44 (2005) 6726–6730.

36. Shibuichi, S., Onda, T., Satoh, N. and Tsujii, K., The Journal of Physical Chemistry, 100 (1996) 19512–19517.

37. Jeong, H. E., Lee, S. H., Kim, J. K. and Suh, K. Y., Langmuir, 22 (2006) 1640–1645.

38. Herminghaus, S., Europhysics Letters, 52 (2000) 165–170.

39. Barbieri, L., Wagner, E. and Hoffmann, P., Langmuir, 23 (2007) 1723–1734.

40. Shibuichi, S., Yamamoto, T., Onda, T. and Tsujii, K., Journal of Colloid and Interface Science, 208 (1998) 287–294.

41. Chen, W., Fadeev, A. Y., Hsieh, M. C., Öner, D., Youngblood, J. and McCarthy, T. J., Langmuir, 15 (1999) 3395–3399.

42. Neumann, A. and Spelt, J. K., Applied Surface Thermodynamics, 1st ed. Surfactant Science Series, Vol. 63. Marcel Dekker Inc., New York, 1996.

43. Extrand, C. W., Langmuir, 18 (2002) 7991–7999.

44. Gao, L. and McCarthy, T. J., Langmuir, 22 (2006) 6234–6237.

45. Marmur, A., Advances in Colloid and Interface Science, 50 (1994) 121–141.

46. Lucassen-Reynders, E. H., The Journal of Physical Chemistry, 70 (1966) 1777–1785.

47. El Ghzaoui, A., Journal of Colloid and Interface Science, 216 (1999) 432–435.

48. Starov, V., Journal of Colloid and Interface Science, 227 (2000) 185–190.

49. Starov, V. M., Journal of Colloid and Interface Science, 270 (2004) 180–186.

50. Starov, V. M., Advances in Colloid and Interface Science, 111 (2004) 3–27.

51. Kumar, N., Varanasi, K., Tilton, R. D. and Garoff, S., Langmuir, 19 (2003) 5366–5373.

52. Milne, A. J. B., Elliott, J. A. W. and Amirfazli, A., Contact Angles of Surfactant Solutions on Heterogeneous Surfaces, Submitted (2011).

53. Craster, R. V. and Matar, O. K., Langmuir, 23 (2007) 2588–2601.

54. Afsar-Siddiqui, A. B., Luckham, P. F. and Matar, O. K., Langmuir, 20 (2004) 7575–7582.

55. Frank, B. and Garoff, S., Langmuir, 11 (1995) 4333–4340.

56. Gerdes, S. and Tiberg, F., Langmuir, 15 (1999) 4916–4921.

57. Birch, W. R., Knewtson, M. A., Garoff, S., Suter, R. M. and Satija, S., Colloids and Surfaces A: Physicochemical and Engineering Aspects, 89 (1994) 145–155.

58. Qu, D., Suter, R. and Garoff, S., Langmuir, 18 (2002) 1649–1654.

59. Princen, H. M., Cazabat, A. M., Cohen Stuart, M. A., Heslot, F. and Nicolet, S., Journal of Colloid and Interface Science, 126 (1988) 84–92.

60. Nedyalkov, M., Alexandrova, L., Platikanov, D., Levecke, B. and Tadros, T. F., Colloid and Polymer Science, 286 (2008) 713–719.
61. Vogler, E. A., Langmuir, 8 (1992) 2005–2012.
62. Vogler, E. A., Langmuir, 8 (1992) 2013–2020.
63. Zisman, W. A., Advances in Chemistry Series, 43 (1964) 1.
64. Marmur, A. and Lelah, M., Chemical Engineering Communications, 13 (1981) 133–143.
65. Frank, B. and Garoff, S., Colloids and Surfaces A: Physicochemical and Engineering Aspects, 116 (1996) 31–42.
66. Afsar-Siddiqui, A. B., Luckham, P. F. and Matar, O. K., Langmuir, 19 (2003) 703–708.
67. Frank, B. and Garoff, S., Langmuir, 11 (1995) 87–93.
68. Troian, S., Wu, X. and Safran, S., Physical Review Letters, 62 (1989) 1496–1499.
69. Troian, S., Herbolzheimer, E. and Safran, S., Physical Review Letters, 65 (1990) 333–336.
70. Milne, A. J. B. and Amirfazli, A., Langmuir, 26 (2010) 4668–4674.
71. Dutschk, V., Sabbatovskiy, K. G., Stolz, M., Grundke, K. and Rudoy, V. M., Journal of Colloid and Interface Science, 267 (2003) 456–462.
72. Varanasi, K. S. and Garoff, S., Langmuir, 21 (2005) 9932–9937.
73. Chang, F.-M., Sheng, Y.-J., Chen, H. and Tsao, H.-K., Applied Physics Letters, 91 (2007) 094108.
74. Tang, X., Dong, J. and Li, X., Journal of Colloid and Interface Science, 325 (2008) 223–227.
75. Ferrari, M., Ravera, F., Rao, S. and Liggieri, L., Applied Physics Letters, 89 (2006) 053104.
76. Ferrari, M., Ravera, F. and Liggieri, L., Journal of Adhesion Science and Technology, 23 (2009) 483–492.
77. Mohammadi, R., Wassink, J. and Amirfazli, A., Langmuir, 20 (2004) 9657–9662.
78. Milne, A. J. B., Clean surfaces, dirty water: topography and chemistry in the wetting of superhydrophobic surfaces by pure liquids and surfactant solutions, MSc Thesis, University of Alberta, 2008.
79. Zhu, B.-Y. and Gu, T., Journal of the Chemical Society, Faraday Transactions 1, 85 (1989) 3813.
80. Vale, H. M. and McKenna, T. F., Colloids and Surfaces A: Physicochemical and Engineering Aspects, 268 (2005) 68–72.
81. Tavana, H., Petong, N., Hennig, A., Grundke, K. and Neumann, A. W., The Journal of Adhesion, 81 (2005) 29–39.
82. Minko, S., Müller, M., Motornov, M., Nitschke, M., Grundke, K. and Stamm, M., Journal of the American Chemical Society, 125 (2003) 3896–3900.
83. Hennig, A., Grundke, K., Frenzel, R. and Stamm, M., Tenside, Surfactants, Detergents, 39 (2002) 243–246.
84. Hoorfar, M. and Neumann, A. W., The Journal of Adhesion, 80 (2004) 727–743.
85. Grundke, K., Nitschke, M., Minko, S., Stamm, M., Froeck, C., Uhlmann, S., Pöschel, K. and Motornov, M., in: Contact Angle, Wettability and Adhesion, Part III, Vol. 3. VSP—An imprint of BRILL, 2003, pp. 1–25.

Collision and Attachment Interactions of Single Air Bubbles with Flat Surfaces in Aqueous Solutions

Anh V. Nguyen and Mahshid Firouzi

School of Chemical Engineering, The University of Queensland, Brisbane, Queensland, 4072, Australia. E-mail: anh.nguyen@eng.uq.edu.au

Abstract

Interaction of air bubbles with solid surfaces is important in many chemical engineering applications with mass and heat transfer, and separation processes. This review focuses on the collision and attachment interactions between a rising bubble and a flat surface in water and surfactant solutions. The analysis involves a numbers of important aspects of the bubble motion and contact interactions which are governed by both long-range and short-range (thin film) hydrodynamics of the bubble rise, surfactant adsorption and desorption, wetting film drainage and rupture, and spreading and relaxation of three-phase contact lines. Featured properties of air–water interface deformation, interfacial forces produced by adsorbed surfactants and interfacial mobility have a significant influence on the attachment and dewetting interactions.

Keywords

Bubble rise, surfactant adsorption and desorption, bubble deformation, surface forces, liquid film, dewetting

Contents

A. Introduction 212
B. Motion of Bubbles Before Interacting with the Surface 213
   1. Physiochemical Hydrodynamics and Interfacial Rheology of Bubble Rise 213
   2. terminal velocity]Terminal Velocity of bubble rise]Bubble Rise 216
C. deformation of bubble]Deformation of Bubble Interacting with the Surface 217
   1. Deceleration and Oscillation of Bubbles Interacting with the Surface 218
   2. Augmented Young–Laplace equation]Young–Laplace Equation 221
D. interfacial interaction forces]Interfacial Interaction Forces 222
   1. DLVO disjoining pressures]Disjoining Pressures 222
   2. Non-DLVO Disjoining Pressures 224
E. Drainage and Rupture of Intervening liquid films]Liquid Films 225
   1. Theories on drainage and rupture of wetting films]Drainage and Rupture of Wetting Films . 225
   2. Effect of surfactants]Surfactants on Wetting Film Drainage and Rupture 229
F. Spreading and Relaxation of contact lines]Contact Lines 231
   1. Hydrodynamic Theories 232
   2. molecular-kinetic theories]Molecular-Kinetic Theories 233

Drops and Bubbles in Contact with Solid Surfaces
© Koninklijke Brill NV, Leiden, 2012

3. Crossovers of Scales 234
G. Summary and Conclusions 235
H. Acknowledgements 236
 I. References 236

A. Introduction

Interaction between a gas bubble and a solid surface in a liquid has been the focus of scientific interest and industrial applications for many years [10, 17, 22, 33]. It plays a vital role in many applications, ranging from industrial multiphase applications such as bubble columns, chemical and biological reactors, mass and heat exchangers to medical applications such as echography and novel applications such as microfluidics devices and drag reductions of swimming clothes and ships using injected microbubbles. One of the important industrial applications of gas bubbles is the flotation separation of hydrophobic particles from a suspension of hydrophobic and hydrophilic particles [47]. Flotation is a well-known process that was originally developed around the turn of last century for recovering fine valuable mineral particles from the rock [53]. Today, more than ten of billion tons of ores are reportedly concentrated by flotation globally each year. At the heart of flotation is the bubble–particle interaction which has intensively been investigated for many years. Many important steps involved in the bubble interaction with the particle surface in flotation has been identified and quantified, and can analogically be applied to the collision and attachment of bubbles with a flat surface in water and surfactant solutions. However, by no means is the interaction of air bubbles with solid surfaces a simple process. On the contrary, it is a complicated process, which can be divided into a number steps as illustrated in Fig. 1. These steps can be grouped into the bubble rise before interacting with the surface and the bubble–surface interactions. The free rise of bubbles before making contact with the surface is considered when the bubble is far away from the surface and is strongly affected by the so-called long-range hydrodynamics. When the bubble and surface are at close proximity, the contact interaction is considered and is controlled by interfacial physics, e.g., interfacial hydrodynamics of the intervening liquid film, capillarity of bubble surface deformation, de-wetting dynamics, and colloidal interactions governed by intermolecular forces.

This chapter aims to review the steps involved in the interaction between an air bubble and flat solid surface in water and surfactant solutions. Specifically, the chapter focuses on single air bubbles rising in an aqueous solution towards a large planar solid which is placed horizontally to the direction of gravity, i.e., perpendicularly to the direction of the bubble rise.

This chapter is organised as follows. In Section B, physiochemical hydrodynamics and interfacial rheology of rising bubbles in an aqueous solution is briefly reviewed. Since this topic is reviewed in books, only the key aspects of the bubble rise and its velocity are described here for reference in the subsequent sections. On approaching the solid surface, the rising bubble first loses its momentum by

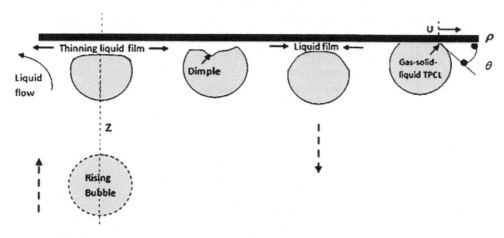

Figure 1. Sequence of events for collision and attachment of a bubble rising in an aqueous solution onto a large planar solid surface placed horizontally to gravity.

dissipating the kinetic energy into the potential energy of the bubble surface deformation. The bubble–surface collision may undergo sequential oscillatory motion until the momentum of the bubble rise is sufficiently lost. Finally, the bubble rests at the surface to allow for the liquid drainage and rupture. The bubble oscillatory motion and the deformation of the bubble surface due to collision and capillarity and intermolecular interactions are described in Section C. Interfacial forces due to intermolecular interactions are specifically described in Section D. The film drainage and rupture are described in Section E. After the film rupture, a three-phase contact (TPC) area is formed, expands and relaxes. The dynamics of the TPC spreading and relaxation is described in Section F. A summary and conclusions of the review are given in Section G.

B. Motion of Bubbles Before Interacting with the Surface

As briefly highlighted in the Introduction, collision and attachment of a rising bubble and a solid surface can be effectively described by splitting the interaction zone into two small zones which are dominantly governed by the (long-range) hydrodynamic forces or (short-range) interfacial forces. In the long-range hydrodynamic zone, which is far from the solid surface, the bubble is considered to freely rise in the aqueous solution. At steady state, the bubble motion is governed by the balance between the fluid resistance and gravitational forces which leads to various drag laws and predictions for the bubble rise velocity. Summarised below are many important aspects of physiochemical hydrodynamics, interfacial rheology and the terminal velocity of bubble rise.

1. Physiochemical Hydrodynamics and Interfacial Rheology of Bubble Rise

Bubble rise velocity is a complicated function of the bubble geometry, the physical properties of the medium and the dynamic physicochemical properties of the gas–

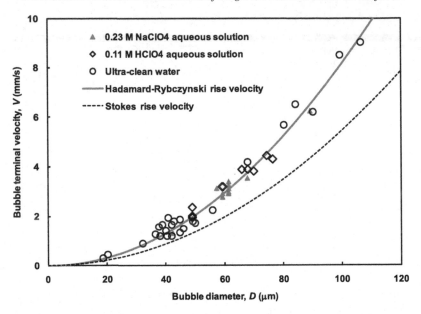

Figure 2. Experimental (symbols) [24, 55] and model (lines) results for terminal rise velocity of N_2 bubbles as a function of bubble size in ultra-clean water and salt solutions.

liquid interface. The shape that the bubble assumes is in turn a complex function of the hydrodynamic, viscous and interfacial forces. The studies have shown the importance of the local water flow passing air bubbles in determining the bubble rise velocity. In particular, the mobility of the air bubble surface appears to be critical. The rise velocity of bubbles with an immobile surface can be determined by the method for solid spheres. The rise velocity of bubbles with a mobile surface in clean water was investigated by a number of researchers [53]. The Hadamard–Rybczynski–Boussinesq equation for the terminal bubble rise velocity, U, is described as follows [53]:

$$U = \frac{2R_{\rm b}^2 \delta g}{9\mu} \frac{3 + 2Bou}{2 + 2Bou},$$ (1)

where $R_{\rm b}$ is the bubble radius, δ and μ are the water density and viscosity, respectively, and g is the acceleration due to gravity. The Boussinesq number, $Bou = \mu_{\rm s}/(\mu R_{\rm b})$, where $\mu_{\rm s}$ is the surface shear viscosity, is determined by ratio of the surface to bulk shear stresses, based on the hypothesis that a thin liquid layer of higher viscosity exists near the water–air interface. If the bubble surface is rigid, Eq. (1) reduces to the Stokes equation for the bubble terminal velocity: $U_{\rm Stokes} = 2R_{\rm b}^2 \delta g/9\mu$. If the bubble surface is fully mobile, Eq. (1) reduces to the Hadamard–Rybczynski equation: $U_{\rm HR} = R_{\rm b}^2 \delta g/3\mu$. Experiments have confirmed that the rise velocity of small bubbles in clean water follows the Hadamard–Rybczynski prediction (Fig. 2) and the bubble surface in clean (surfactant-free) aqueous solutions should be mobile.

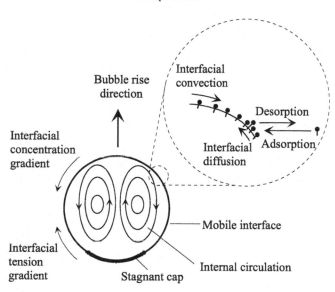

Figure 3. Cross-sectional meridional-plane view of a bubble rising in a surfactant solution and the interfacial surfactant transport processes: adsorption and desorption of surfactants, diffusion in the continuous phase, and surface diffusion at the interface [53].

The motion of bubbles is significantly impeded by the presence of impurities such as surfactants [18]. Levich, unlike Boussinesq, considered the presence of a surface tension gradient and suggested various transport mechanisms by which the surfactant molecules control the surface tension gradient (Fig. 3), and the bubble surface rheology and rise velocity. Surfactant concentration varies along the bubble surface taking the maximum and minimum at the rear and front stagnation points, respectively. The concentration distribution is attributed to the relative bubble–water motion and disappears if the relative motion is not present or negligible. Owing to the surface concentration gradient, there is an interfacial gradient along the bubble surface which retards the bubble surface and strongly affects the local stress balance on the boundary between a bubble and the liquid media. The Frumkin–Levich approach explains the fore-and-aft symmetrical liquid flow and circulation of the gas phase inside a bubble. The surface contamination (surfactant molecules or other impurities) often forms an immobile cap, termed the stagnant cap, on the rear surface around the stagnation point (Fig. 3). The forward part of the bubble surface remains mobile. The leading part of the mobile surface is stretched, the lowest part is compressed. The low surface concentration in the stretched surface provides a continuous supply of adsorbed substance from the bulk phase. High concentrations in the compressed part initiate desorption of surfactants. Stagnant cap is formed if the interfacial convection is faster than the interfacial and bulk diffusion and the adsorption and desorption of surfactants. In the case of low Reynolds numbers, it is theoretically possible to relate the geometry of the stagnant cap angle with the terminal bubble rise velocity [10].

2. Terminal Velocity of Bubble Rise

Developments during the last few decades in the physico-chemical hydrodynamics of bubble rise are summarized by Clift *et al.* [10] and Dukhin *et al.* [17]. The bubble terminal velocity in a surfactant solution is still difficult to predict. Coupling between the surfactant and mass transfer causes difficulties in the quantifying the role of the surfactant concentration in bubble motion.

Experimental data for the terminal velocity of the bubble rise are well reported in the literature [10] and shown in Fig. 4. Three special systems, namely, pure water, grossly contaminated water and transition between pure water and grossly pure

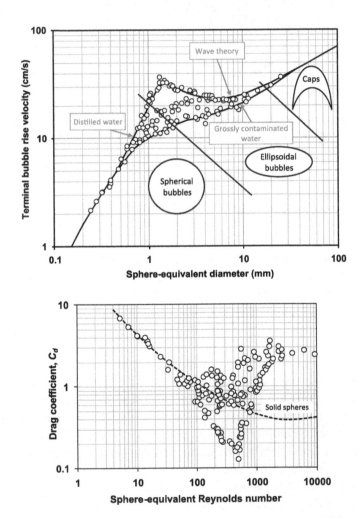

Figure 4. Experimental domain of terminal rise velocity of single bubbles in water [53] and the corresponding drag coefficient, C_d, *versus* sphere-equivalent Reynolds number. The bubble shape domains are estimated on the basis of equivalent rise velocity. Dash line shows the drag coefficient for solid spheres [53].

water can be identified. Many surfactants added to water in a very small (ppm) concentration, sufficiently render constant bubble terminal velocity at a given size. A further increase in concentration produces no effect on the bubble terminal velocity. Grossly contaminated water represents the case of minimum constant terminal velocity at a given size. The terminal velocity of small bubbles follows that of solid spheres until the volume-equivalent Reynolds number exceeds a critical value. For bubbles in pure and grossly contaminated water, the critical value is 25 and 60 respectively. However, the bubble deformation is insignificant and the bubble shape remains spherical up to a much higher Reynolds number which is about 200 [10]. For small bubbles (with $Re < 130$) in contaminated water, the shape of the bubble is spherical and the bubble rise velocity follows that of solid particles, i.e.,

$$U = U_{Stokes} \left\{ 1 + \frac{Ar/96}{(1 + 0.079Ar^{0.749})^{0.755}} \right\}^{-1}, \tag{2}$$

where $Ar = 8R_b^3\delta^2 g/\mu^2$ is the Archimedes number for bubble rise.

The drag coefficient of bubble rise in contaminated water starts to deviate from the standard curve of solid spheres at approximately $Re = 130$. At this regime, the bubbles are non-spherical. The shape factor has to be considered in predicting the drag coefficient of the bubble rise in contaminated water and the bubble terminal rise velocity becomes a function of the Morton number, $Mo = g\mu^4/(\delta\sigma^3)$ where σ is the surface tension, as follows:

$$U = 18U_{Stokes} \left(\frac{4a^2 Ar^{2b-1} Mo^{0.46b}}{2.85} \right)^{\frac{1}{2-2b}}, \tag{3}$$

where a and b are the model parameters [53]. The contaminated water is practically justified by the simple criterion of Clift *et al.* [10]: the velocity–volume curve of bubbles should not pass through any maximum peak. This condition is usually satisfied by the bubble rise in many surfactant solutions.

C. Deformation of Bubble Interacting with the Surface

When subjected to external forces, gas bubbles can be deformed and undergo different shapes. For freely rising bubbles before interacting with the solid surface, the extensive experimental dada for the radius, L, of the bubble cross section area projected perpendicularly to its path can be related to the bubble sphere-equivalent radius, R_b, as follows [53]:

$$L = R_b \frac{1.488 + 0.042Ta^{1.908}}{1.488 + 0.025Ta^{1.908}}, \tag{4}$$

where $Ta = Re(Mo)^{0.23}$ is the Tadaki number given as a function of the bubble Reynolds number, $Re = 2R_b U\delta/\mu$, determined by the sphere-equivalent bubble radius, R_b, and the Morton number, Mo.

When approaching and interacting with the solid surface, the bubble surface is further deformed by (1) the liquid pressures produced by the flows confined between the bubble and solid surfaces, and/or (2) the intermolecular forces within the intervening liquid film. The dynamic deformation and static deformation of a bubble interacting with the solid surface are described below.

1. Deceleration and Oscillation of Bubbles Interacting with the Surface

When a bubble rises towards a horizontal surface, the velocity of the bubble centroid decelerates from the terminal rise velocity to zero. The bubble can rebound and oscillates at the solid surface. For slowly rising bubbles, no significant rebound and oscillation are expected. For fast rising bubbles with sufficiently high values of the Reynolds and Weber numbers, the bubble rebound and oscillation are significant. Quantitatively, air bubbles rising in water can undergo shape oscillation upon collision without a clearly visible rebound if their Weber numbers based on their terminal rise velocity are less than about 0.3 [71].

Evidently, the bubble oscillation and rebound result from the interchange between the bubble kinetic energy, surface energy and potential energy. The bubble dynamic behaviour at the solid surface can be analyzed using the equation of motion written for the bubble centroid which gives [34, 59, 82]:

$$
\rho_G V \frac{dU}{dt} = V\rho_L g - C_d(\pi L^2)\frac{\rho_L U^2}{2} - C_{AM}\rho_L V\frac{dU}{dt} \\
- 6\pi\mu R_b \int_{-\infty}^{t} K(t,s;Re)\dot{U}(s)\,ds - \int_0^{r_{max}} 2\pi r p\,dr.
$$

(5)

Equation (5) accounts for various forces acting on a bubble rising towards a solid surface in a still aqueous solution. The left hand side of Eq. (5) describes the bubble inertia. The terms on the right hand side describe the bubble buoyancy, liquid resistance, added-mass force, history force and bubble–surface interaction force, respectively. The hydrodynamic equation for the bubble–surface interaction force can be inferred and modelled from the pressure field obtained from the 2D Navier–Stokes equations simplified for flows within the liquid films. The bubble weight is neglected in the motion equation due to the big difference between the air and water densities. Equation (5) is featured by various factors, such as the drag coefficient C_d, the added mass coefficient C_{AM} and the kernel $K(t,s;Re)$ of the history force, associated with the surface deformation of bubbles interacting with the solid surfaces. These factors are known for motions of bubbles and particles far away from the solid surface [10, 53] and are less known for bubbles interacting with the solid surface. The factors for the bubble motion are generally complicated as it can be appreciated by the dependence of the drag coefficient on bubble Reynolds number shown in Fig. 4. Nonetheless, progress has been made and good agreement between the theory and experiments has been established. Figure 5 shows an example available in the literature, where the model prediction uses the lubrication theory in determining the bubble–surface interaction force in Eq. (5).

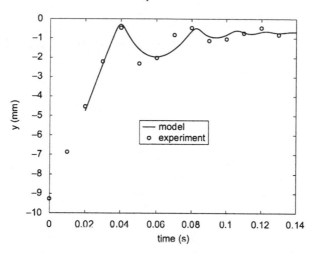

Figure 5. Perpendicular position the centroid of a bubble with 0.85 mm radius ($Re = 522$ and $We = 0.866$) rising toward and oscillating at a horizontal solid surface [59].

The energy balance for the deformable bubbles bouncing from the solid surface has also been analysed [19, 29, 38, 81]. A few convenient locations (or times) on the bubble trajectory can be considered in assessing the energy components, namely, (1) the bubble position at its terminal velocity at a sufficient distance from the wall so that the effect of the solid surface on the bubble shape and liquid velocity can be neglected, (2) the position bubble arrested upon collision with the surface, and (3) the lowest position of the bubble bounce. The bubble velocity is equal to zero at the last two positions. The experimental data [71] show that 95% of the kinetic and surface energy of bubbles colliding with the solid surface at their closest proximity is stored as surface energy and only 41% of the initial energy is converted into the gravitational and surface energy at the lowest bubble position on its bounce trajectory. Furthermore, the energy dissipated during the rebound is significantly larger than the energy loss predicted based on the drag on a spherical bubble in an unbounded fluid with the same initial energy. The shape of bubbles is nearly spherical after the rebound. The additional energy losses may be due to the acoustic radiation by the bubble oscillation and the energy dissipation into the thinning of the intervening liquid films. The interchange between the kinetic and surface energies during the bubble–surface interaction has been quantitatively analysed [82]. The collision-rebound process is also analysed by considering the restitution coefficient (defined as the ratio of rebound to impact velocity) in a similar way to analysing the solid particle–solid surface collision [51] which predicts the collision and rebound distance as a function of the Stokes number. There is a critical Stokes number below which the viscous effect on the bubble collision overcomes the inertial effect and no bubble rebound at the solid surface occurs [19, 29, 38]. However, the bubble–surface interaction is significantly different from that of the solid particle in one aspect, that is, the bubble approach distance and velocity are similar

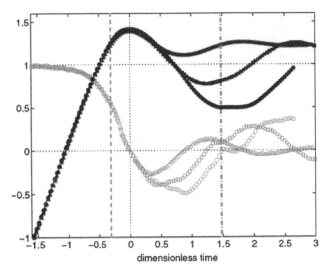

Figure 6. Normalised position of bubble centroid (filled symbols), $2z/d_{eq} + 2$, and normalised velocity (open symbols), $(dz/dt)/U$, *versus* the normalised time, tU/d_{eq}, for different modified Stokes numbers, $St^* = 3.9$ (triangles), 10.8 (squares) and 21.1 (circles) [82]. The zero time is off-set to the zero bubble velocity. The dash lines describe the bubble positions away from the solid surface by one equivalent diameter.

when normalised to account for deformation (Fig. 6). This aspect is an important consideration for modelling the drainage process.

The surface wettability and microscopic roughness have been found to significantly influence the bubble rebound and oscillation [37, 81]. In particular, there is a significant difference between the bubble oscillation at the hydrophobic and hydrophilic surfaces at long time. At the hydrophobic surface (such as Teflon), the intervening liquid film was ruptured and the three phase contact was formed. However, at the hydrophilic surface (such as mica) the bubble arrested beneath the surface and no direct contact was formed. The interchange of the bubble kinetic to surface energy during the collision with the hydrophobic Teflon surface is also lower than that with the hydrophilic (mica and glass) surfaces. The difference is due to the presence of sub-microscopic air pockets present on the rough hydrophobic Teflon surfaces which cause an earlier film rupture during the bubble rebound and oscillation. Surfactants such as simple alcohols effectively reduce not only the bubble rise velocity [35, 46] and hence the kinetic energy of the bubble rise, but also significantly change the bubble rebound and oscillation at the hydrophobic and hydrophilic surfaces [36, 37]. Specifically, surfactant mixtures can synergistically adsorb at both the bubble surface and the solid surface, and change the surface properties, such as the surface charge and hydrophobicity, of both the bubble and solid surfaces [39]. These surface properties ultimately change the surface forces and the liquid film drainage and rupture, as described in Sections D and E.

2. Augmented Young–Laplace Equation

After rebound and oscillation, the bubble can rest at the solid surface with or without an intervening liquid film. In addition to the capillary pressure and hydrostatic pressure, the bubble surface can subjected to one more pressure, called the disjoining pressure, Π, which arises from the intermolecular interactions within the liquid film [16, 28]. The components of the disjoining pressure are briefly described in Section D. Shown below is a modification of the classical Young–Laplace equation for the bubble deformation in the presence of the disjoining pressure.

When the bubble rests at the solid surface, all the transient terms in Eq. (5) become zero and only the buoyancy and bubble–surface interaction force on the right hand side remain. These two terms can be further refined to give the Young–Laplace equation for a sessile bubble. Specifically, the equation can be directly established by minimizing the surface energy, the gravitational potential energy, and the surface free energy due to intermolecular interactions between the bubble and the surfaces at close proximity. The minimization [1, 6] yields the augmented Young–Laplace equation described as

$$\frac{d}{r\,dr}\left\{\frac{r(dh/dr)}{\sqrt{1+(dh/dr)^2}}[\gamma+\sigma(r,h)]\right\} = \Delta p + g\rho h + \sqrt{1+(dh/dr)^2}\frac{\partial\sigma}{\partial h}, \quad (6)$$

where r is the radial coordinate measured from the axis of symmetry, h describes the bubble surface coordinate measured from the solid surface as a function of the radial coordinate r, Δp is the pressure at the lower bubble apex and γ is the surface tension. $\sigma(r,h)$ is the surface interaction free energy per unit area between the bubble and solid surfaces separated by the distance, s, directed from the bubble centroid.

By definition, the surface interaction energy is related to the disjoining pressure by $\Pi(s) = -\partial\sigma/\partial s$, where the partial derivative can be linked with the bubble coordinate on the basis of the distance balance. It gives $[R_b+s(r)]^2 = (\sqrt{s^2-h^2}+r)^2+[h(0)+R_b]^2$.

Assuming that air is incompressible under the normal condition and therefore the bubble volume remains constant during the interaction, giving $V_b = 2\pi\int rh\,dr = const$. This volume constancy has to be considered when integrating Eq. (6). The initial conditions for integrating Eq. (6) include: $h = h(0)$ and $dh/dr = 0$ at $r = 0$ (at the upper bubble apex). The disjoining pressure as a function of separation is described in the next section.

Differential Eq. (6) with implicitly unknown pressure Δp presents an initial-boundary problem which can successfully be solved by the shooting technique. The integration starts at $r = 0$ with a guess for Δp and ends at the lower bubble apex. The guess for Δp is updated until bubble volume constancy is met. The fourth-order Runge–Kutta method can be used to carry out the integration. In particular, following the traditional numerical integration of the Young–Laplace equation [23], the integration of Eq. (6) is improved by integrating along the arc length of the bubble surface. Figure 7 show the numerical results.

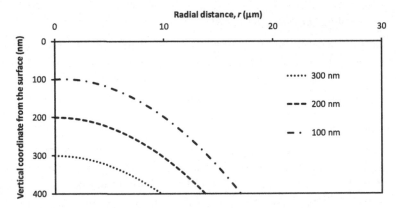

Figure 7. Numerical results of Eq. (6) for the bubble surface profiles arrested underneath of a solid surface in water by the electrical double layer and van der Waals repulsions given by Eqs (11) and (7). The legend shows the shortest separation distance at the upper bubble apex, $h(0)$. See the text for additional information about the model parameters.

D. Interfacial Interaction Forces

When a bubble approaches a solid surface, interfacial forces become significant at small separation distances. These forces arise from molecular interactions between charged and uncharged atoms and molecules of the interacting bodies and the surrounding medium, which include (1) van der Waals (electrodynamic) interactions, (2) electrostatic double-layer interactions, and (3) non-DLVO interactions. The first two interactions form the key components of the celebrated DLVO (Derjaguin–Landau–Verwey–Overbeek) theory of colloid stability. Summarised below are the disjoining pressures of the interfacial interactions relevant to the bubble–surface collision and attachment.

1. DLVO Disjoining Pressures

For bubble–solid surface interaction, the van der Waals interaction is repulsive. It can be determined using the macroscopic (Hamaker) and the microscopic (Lifshitz) theories. The combined Hamaker–Lifshitz approach is useful. For two planar parallel surface elements of a water film confined between the gas phase and the solid phase, the combined theory gives:

$$\Pi_{\text{vdw}}(H) = -\frac{A}{6\pi H^3} + \frac{1}{12\pi H^2}\frac{dA}{dH},\tag{7}$$

where H is the thickness and A is the Hamaker–Lifshitz function, which is a weak function of the thickness H. The function covers both the short- and long-range van der Waals interactions and can be described as follows:

$$A = A^0(1 + 2\kappa H)e^{-2\kappa H} + A^{\xi}(H),\tag{8}$$

where κ is the Debye constant. The zero-frequency, A^0, and non-zero frequency,

A^ξ, terms are described by Nguyen and Schulze [53]:

$$A^0 = \frac{3k_BT}{4} \sum_{m=1}^{\infty} \left\{ \frac{77}{79} \frac{78-\varepsilon}{78+\varepsilon} \right\}^m / m^3, \tag{9}$$

$$A^\xi(H) = -0.235\hbar\omega \frac{n_p^2 - 1.887}{n_p^2 - 1}$$
$$\times \left\{ \frac{0.588}{[1+(H/5.59)^q]^{1/q}} - \frac{(n_p^2+1.887)^{-1/2}}{[1+(H/\lambda_p)^q]^{1/q}} \right\}, \tag{10}$$

where ε is the solid dielectric constant, $\hbar$ is the Planck constant divided by 2π, n_p is the solid refractive index, $\omega = 2 \times 10^{16}$ rad/s, $\lambda_p = 9.499/\sqrt{n_p^2+1.887}$ is a modified London wavelength accounting for the effect of electromagnetic retardation on the van der Waals interaction, and H is in nm.

The bubble surface is negatively charged (-55 mV) in deionised water. The solid surface is also negative in water at neutral pH. The electrostatic double-layer interaction between the air–water and water–solid surfaces is usually repulsive. The double-layer disjoining pressure between two planar parallel surface elements of a water film confined by the gas phase and the solid phase can be obtained from the solution of the Poisson–Boltzmann equation. The calculation of the double-layer force as a function of the separation distance is recently reviewed [50, 53]. For low surface (zeta) potentials (<50 mV), the Hogg–Healy–Fuerstenau approximation for the double-layer interaction at constant surface potentials gives

$$\Pi_{edl} = \frac{\varepsilon_w\varepsilon_0\kappa^2}{2} \frac{2\psi_b\psi_s\cosh(\kappa H) - \psi_b^2 - \psi_s^2}{\sinh^2(\kappa H)}, \tag{11}$$

where ψ_s and ψ_b are the solid and bubble surface potentials, ε_0 is the dielectric constant of vacuum and $\varepsilon_w = 78$ for water. For the double-layer interaction at constant surface charge one obtains

$$\Pi_{edl} = \frac{\varepsilon_w\varepsilon_0\kappa^2}{2} \frac{2\psi_b\psi_s\cosh(\kappa H) + \psi_b^2 + \psi_s^2}{\sinh^2(\kappa H)}. \tag{12}$$

Another useful approximation can be obtained by superposition which is valid for high surface potentials and long separation distance and gives

$$\Pi_{edl} = 32\varepsilon_w\varepsilon_0\kappa^2 \left(\frac{k_BT}{ez}\right)^2 \tanh\left(\frac{ez\psi_b}{4k_BT}\right) \tanh\left(\frac{ez\psi_s}{4k_BT}\right) \exp(-\kappa H), \tag{13}$$

where k_B is the Boltzmann constant, T is the absolute temperature, z is the valence of the symmetric $z:z$ electrolyte.

The surface (zeta or streaming) potential of the air–water and solid–water interfaces can be measured by microelectrophoresis and streaming potential method. Useful results for the bubble and solid surfaces are summarised in the books [53].

2. Non-DLVO Disjoining Pressures

The non-DLVO forces can be the (repulsive) hydration forces (between hydrophilic surfaces) or the (attractive) hydrophobic forces (between hydrophobic surfaces). Another important non-DLVO force is the steric force between adsorption layers of macromolecular reagents used as depressants or flocculants.

Hydration forces have been extensively studied between clay, mica and silica surfaces in water [28, 56]. In these systems, the surfaces and particles would remain in strong adhesion or coagulate in salt solutions in a primary minimum only if the forces were the DLVO forces. Indeed, water molecules are strongly hydrated by the salt ions and/or strongly bounded to the hydrophilic surfaces. Theory [45] and experiments [28] show the exponential decay of the hydration disjoining pressure *versus* separation distance as follows:

$$\Pi_h(H) = K \exp(-H/\lambda), \tag{14}$$

where λ is the decay length and K is a constant. The hydration force between two mica surfaces is monotonically repulsive below $H = 5$ nm. An oscillation with a mean periodicity of about 0.25 nm ($\sim$water diameter) is observed below $H = 1.5$ nm [28].

Hydrophobic surfaces are inert to water as they are unable to interact or bind with water either by electrostatic means or *via* hydrogen bonds. Hydrophobic forces between microscopic hydrophobic surfaces have generally been found to increase with the hydrophobicity of the surfaces, as conventionally defined by the water contact angle. The first direct measurements of the hydrophobic force show that the attraction has a very long range and decays exponentially as predicted by Eq. (14) with $K < 0$ [27]. Since the first measurements, experimental data have shown that the attraction between hydrophobic surfaces is strong and long-ranged and can be empirically described by a double exponential function of two decay lengths as

$$\Pi_h = K \exp(-H/\lambda) + K^* \exp(-H/\lambda^*). \tag{15}$$

Alternatively, the measured hydrophobic attraction can be described by a power law similarly to the van der Waals interaction [9, 80] as

$$\Pi_h = \frac{K_{132}}{h^3}, \tag{16}$$

where the empirical constant can be correlated with the surface hydrophobicity measured by the advancing and receding contact angle as $K_{132} = -\exp[a(\cos\theta_R + \cos\theta_A)/2 + b]$. The data for K, K^*, a and b are summarised in the book [53]. The constant in the front of the single exponential term can also be determined using the surface thermodynamics of acid–base interactions [49, 77].

There is no consensus on explaining the strong hydrophobic attraction between hydrophobic surfaces despite a number of proposed mechanisms [52]. Since the hydrophobic attraction is difficult to model at present, the empirical correlation fitted with double exponentials is often used. The double exponential dependence has no real physical basis—it only describes a difference between DLVO and experimental data for surface forces. Indeed, the double exponential reflects the presence of surface nanobubbles [2, 8, 21, 52, 68, 79]. The nanobubbles have been found significant for coagulation of solid particles [26, 67], and drainage and rupture of liquid films [54]. However, removing nanobubbles or degassing aqueous solution of wetting films and foam films is difficult to experimentally carry out because the gas bubbles are always present in the systems. Therefore, the empirical correlations presented for the non-DLVO interfacial interaction in the above are useful.

E. Drainage and Rupture of Intervening Liquid Films

The wetting film formed between the rising bubble and the solid surface initially thins under the influence of gravity and capillary suction at the film periphery. When the film thickness reduces to 300–200 nm, the interfacial intermolecular forces start affecting the film drainage. Then the films normally become unstable and rupture at a critical thickness within the range of 200–10 nm, depending on concentration of surfactant and surface impurities. Theories on the film drainage and rupture are summarised below.

1. Theories on Drainage and Rupture of Wetting Films

The Navier–Stokes equations can be simplified and solved for the 2D fluid flow within the film. The pressure field can be inferred from the flow solution and, hence, the hydrodynamic equation for film drainage and rupture can be established. Such a theoretical description for the film drainage rate was obtained long time ago and is referred to as the Stefan–Reynolds equation as follows [47]:

$$\frac{dh}{dt} = -\frac{2h^3}{3\mu R^2}(P_\sigma - \Pi), \tag{17}$$

where R is the (constant) film radius and P_σ is the capillary pressure at the film periphery. The Stefan–Reynolds (lubrication) theory is restricted to planar parallel and tangentially immobile film surfaces. For films with non-planar parallel film surfaces, the film thickness is also a function of the film radial coordinate, r, and the lubrication theory gives

$$\frac{\partial h}{\partial t} = \frac{1}{3r\mu}\frac{\partial}{\partial r}\left(h^3 r \frac{\partial \Delta P}{\partial r}\right). \tag{18}$$

The dynamic pressure inside the film, ΔP, in Eq. (18) depends on the local deformation of the gas–liquid interface and is described as $\Delta P(h) = P_g - \frac{\gamma}{r}\frac{\partial}{\partial r}(rh'/\sqrt{1+h'^2}) - \Pi(h)$, where the first term is the (constant) gas pressure inside the bubble and the prime describes the derivative with respect to r. Equation (18) can be numerically solved to validate and predict the spatial and temporal film profiles influenced by the van der Waals and double layer disjoining pressures [66]. An example is shown in Fig. 8.

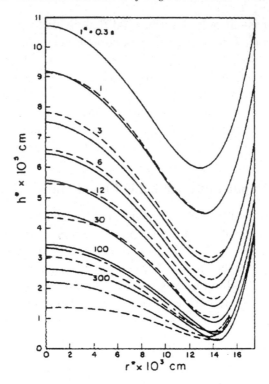

Figure 8. Numerical results (solid lines) of Eq. (18) and experimental results (dash lines) for wetting film thickness profiles between a bubble and a silica surface in an aqueous solution at different times [7].

Due to the radial pressure gradient by the out-flow (the hydrodynamic pressure) in draining films, a variation in film thickness is expected and when the approach speed is sufficiently high and the separation is small, the hydrodynamic pressure can be large enough to invert the curvature of the fluid interface to form a dimple at the film centre (Fig. 8). Recently, a more complex interface shape was observed by experiments [74] which can be referred to as a wimple (with two local minima and one local maximum on the film profile) and a pimple (the film profile has two inflection points but just one local minimum at the centre).

Films with diameter smaller than 100 μm are known to follow the Stefan–Reynolds theory [44]. The smallness of the film radius is quantitatively described by $R_{rim} \leqslant \sqrt{12\gamma \bar{h}/\Delta P}$, where R_{rim} is the film radius at the rim, $\bar{h}$ is the profile-average film thickness and ΔP is the pressure on the meniscus outside the film minus the disjoining pressure calculated with the average thickness. For films with larger radius, a dimple is often formed during the film drainage. If the film thickness fluctuation is small, averaging Eq. (18) gives the $R^{4/5}$-prediction for the fast drainage of dimpled thin films [72]:

$$\frac{d\bar{h}}{dt} = -\frac{1}{6\mu} \sqrt[5]{\frac{\bar{h}^{12}\Delta P^8}{4\gamma^3 R^4}}. \tag{19}$$

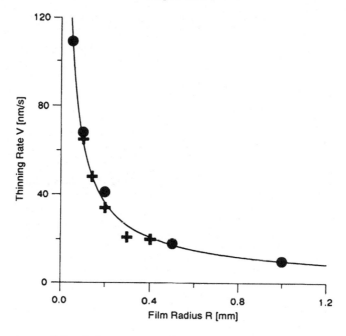

Figure 9. Dependence of film drainage rate on film radius as predicted by Eq. (19). The points describes the experimental data [72].

Comparison of Eq. (19) with experimental results is shown in Fig. 9.

If the spatial correlation between the film surface domains is described by a fractal dimension, α, having a value between 0 and 2, Eqs (17) and (19) can be generalised as [31]

$$\frac{d\bar{h}}{dt} = -\frac{2\bar{h}^3 \Delta P}{3\mu R^2} \left\{ \frac{R^2 \Delta P}{16\gamma \bar{h}} \right\}^{\frac{2-\alpha}{2+\alpha}}. \tag{20}$$

For $\alpha = 2$, Eq. (20) recovers the Stefan–Reynolds drainage law for planar parallel films. If $\alpha = 1$, the film contains an axisymmetric dimple causing faster drainage. If $\alpha = 1/2$, the film exhibits numbers of asymmetric dimples and the draining rate is even faster [72]. For $\alpha = 0$, the film contains spatially uncorrelated domains causing the fastest possible drainage.

In the case of bubbles in ultra-clean water, the boundary condition of tangential immobility employed in the Stefan–Reynolds theory is no longer valid and has to be replaced by the boundary condition of tangential mobility at the air–water film surface. Since the air–water film surface is fully mobile, the water velocity on the air–water surface of the film is of the order of the bubble rise velocity and the inertial terms in the Navier–Stokes equations can become significantly large and cannot be neglected as in the case of the lubrication approximation. This is particularly the case of strong-slip regime at the water–solid interface, which renders a plug flow within the film. If the slip regime is weak (i.e., the slip length is of the order of

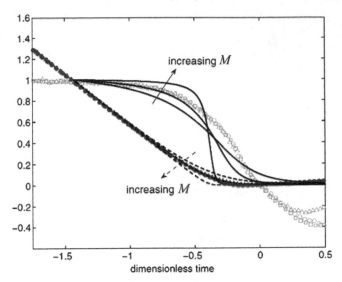

Figure 10. Comparison between the experimental (points) and the inertial drainage model described by Eq. (22) for the dimensionless drainage rate $(dh/dt)/U$ (dash lines) and film thickness h/d_{eq} (solid lines) *versus* time tU/d_{eq} for varying $M = 5$, 10 and 50 [82].

the film thickness or smaller) and the inertial terms can be ignored, the lubrication approximation with zero stress at the air–water surface gives

$$\frac{\partial h}{\partial t} = \frac{1}{3r\mu}\frac{\partial}{\partial r}\left[(4h^3 + 12\beta h^2)r\frac{\partial \Delta P}{\partial r}\right], \tag{21}$$

where β is the slip length of the water–solid surface. In the case of strong-slip regime, the inertial terms should be retained. The viscous terms can be ignored in the modelling of the inertial drainage. The water velocity at the air–water film surface is of the order of the bubble rise velocity. The inertial drainage can be described as follows [82]:

$$\frac{dh}{dt} = U\left(\frac{2M + 1}{2M + d_{eq}/h}\right)^{3/2}, \tag{22}$$

where h is the film thickness, i.e., the distance between the bubble and solid surfaces, at the upper bubble apex, the initial thickness at $t = 0$ is equal to the bubble equivalent (geometric mean) diameter. The other model parameter is described as $M \approx 32C_{AM}/3$, which is O (10). Figure 10 shows a comparison between Eq. (22) and experimental data. The inertial drainage mechanism fits the experimental more closely at the beginning and the deviation between the model and the theory indicates that viscous effects can be dominated for very thin films.

It should be noticed that as in reality there are usually contaminants or surfactants in the liquid medium, the surfaces are between these two extreme Stefan–Reynolds and Scheludko models for fully immobile and mobile surfaces respectively. In this case the film thickness can be obtained by numerical solutions or experimental methods.

When the film becomes thinner, the repulsive forces decelerate the drainage rate and the attractive forces can cause the film to rupture. There are two traditional approaches used to describe the film rupture: (1) the thermal fluctuation of the gas–liquid interface in the presence of the attractive forces, which amplifies the fluctuation, and (2) the interfacial capillary waves (also called spinoidal dewetting) which increase the interfacial area and the interfacial energy of the film, causing the film rupture to minimise the system free energy [53]. The film rupture process can be established using the linear or nonlinear hydrodynamic stability theories. The theories can predict the critical film thickness and wavelength, and the time of the film rupture.

The third theory proposed for film rupture is based on the density fluctuations inside the film in the vicinity of a hydrophobic solid–liquid interface [53]. Specifically, the film rupture can occurs by a gas nucleation process. The theory was further developed by considering the stability of 'holes' in the liquid layer on the solid surface. The film rupture is first triggered by a single hole and then, depending on the capillary and gravitational forces, the liquid may fill the hole or the hole may continue expanding. Holes of dimensions smaller than critical ones close and bigger than that expand. Sharma and Ruckenstein [64] developed a model for critical hole diameter which is a function of film thickness and contact angle.

Recently, a number of authors have related the liquid film rupture on hydrophobic surfaces to the coalescence of the interfacial nanobubbles [21]. After rupturing the local foam films between the nanobubbles and the big bubbles, a hole with three phase contact (TPC) in place of nanobubbles is formed which can be followed by the TPC expansion if the hole is bigger than the critical dimension [69, 70].

Many authors confirmed the existence of the surface nanobubbles using IR spectroscopy, force measurement and AFM [20, 41, 75]. Also in a recent study by Karakashev and Nguyen [32], it has been shown that migration of dissolved gasses has a strong effect on the film rupture. Therefore, it is likely that the film rupture is significantly influenced by the surface nanobubbles and/or submicroscopic and nanobubbles of dissolved gases in the bulk solution.

2. Effect of Surfactants on Wetting Film Drainage and Rupture

Surfactants significantly influence the dynamic properties of thin liquid films and the film life time. Figure 11 shows the effect of sodium dodecyl sulphate (SDS) on the lifetime and rupture of wetting films on hydrophobised silica.

In the presence of surfactants the tangential liquid velocity at the film surfaces may be substantially reduced by an opposing gradient of the surface tension, the so-called dynamic elasticity or Marangoni effect [40]. Many works have been focused on the influence of the mechanical properties of the film surfaces (e.g., the Marangoni stress), the type and concentration of surfactants on the film drainage and stability [30, 60, 63, 76]. It is shown that surfactant adsorption at the film sur-

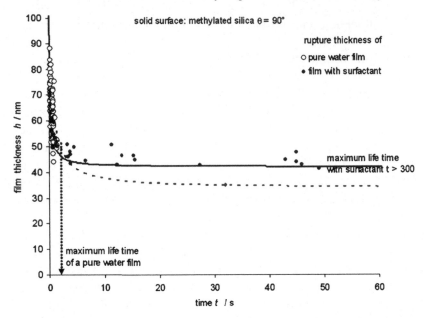

Figure 11. Influence of surfactant on the lifetime and rupture of wetting films on hydrophobised silica with a contact angle of 90°. The maximum lifetime without SDS is 3 s. Addition of 1×10^{-5} M SDS increases the lifetime of the wetting film up to 300 s [70].

face plays a critical role in film drainage. Although most of both theoretical and experimental results are primarily related to the foam films that form between gas bubbles, in many respects they can be principally generalised for wetting films [43].

The results indicated that the film drainage in the presence of surfactants critically depends on two factors:

- The film radius which increases the thickness inhomogeneities and the film drainage rate. The formation of thickness inhomogeneity starts at a film radius of about 50 micron [61].

- The Gibbs elasticity, E_G, which controls the film surface mobility. It expresses the resistance of the film to local deformation. Films become practically rigid (non-deformable) at $E_G = 1$ mN/m.

The effect of surfactant adsorption on foam film drainage can be adequately described by a number of models [30]: the model of Scheludko [65], the model of Radoev, Dimitov and Ivanov [60], the model of Ruckenstein and Sharma [63] and the model of Manev, Tsekov and Radoev [42]. However, the model description for the drainage of wetting films in the presence of surfactants is very limited. Nonetheless, Tsekov *et al.* [72] explained the deviation of wetting film drainage from the

Stefan–Reynolds theory by a factor, which is expressed as an effective viscosity and replaces the dynamic viscosity in the Stefan–Reynolds Eq. (17), as follows:

$$\mu_e = \mu \frac{Ma + Na + Ap + MaNa + ApNa/2}{12 + 4Ma + 4Na + MaNa}, \tag{23}$$

where $Ma = \frac{ahE_g}{\mu(aD_s+Dh)}$ is the Marangoni number, $Na = \frac{h\beta}{\mu}$ is the Navier number, $Ap = -\frac{ah^2\Gamma}{\mu(aD_s+Dh)}\frac{\partial\Pi}{\partial\Gamma}$ is the adsorption number. Here $a = \partial\Gamma/\partial c$ is the adsorption length (Γ is the adsorption density and c is the surfactant concentration), $E_G = -\partial\gamma/\partial\ln\Gamma$ is the surface Gibbs elasticity, D_s and D are the surface and bulk diffusion coefficients of the surfactants. The Marangoni number and Navier number account for the mobility of the air–water film surface and water solid film surface, respectively. In the limit as $Ma \gg 1$ and $Na \gg 1$ which corresponds to the tangentially immobile film surfaces, Eq. (23) gives $\mu_e = \mu$ and the Stefan–Reynolds theory is recovered. The adsorption–pressure number, Ap, takes into account the dependence of the disjoining pressure on the adsorption on the film surfaces. Unlike the other dimensionless numbers in Eq. (23), the adsorption number can be positive and negative.

The above models do not account for the dynamic effects produced by the electrical double layer, which can arise with the ionic surfactants present in the film solution. Therefore, an electrical term can be added into the bulk pressure stress tensor and the electrostatic potential into the film to make the theory valid for ionic surfactants [73, 76]. Depending on the adsorption layer at the film interface, the interfacial charge density can vary with the film thickness, the so-called charge regulation. Two particular models are well known in the literature, namely constant interfacial charge density and constant interfacial potential. According to the numerical analysis of the governing equations, ionic surfactants can influence the film drainage in two ways: at high surfactant or salt concentration, the interfaces become tangentially immobile and then dynamic changes in the concentration, adsorption, electric charge and middle plane potential affect the film thinning due to the change in the non-equilibrium part of the electrostatic disjoining pressure. At small surfactant concentration, it can influence the film drainage by reducing the surface mobility.

F. Spreading and Relaxation of Contact Lines

The classical hydrodynamics of the spreading and relaxation of the three-phase contact (TPC) lines leads to a singularity due to a non-integratable stress balance and an infinite curvature of the interface when the contact line is approached. A slip length of nanometre scales is introduced to describe the unique hydrodynamic mechanisms operating in a very small neighbourhood of the contact line which is called the inner region. To ease the modelling the TPC motion, the governing equations are solved in three different domains (Fig. 12): (1) the outer region, which is far

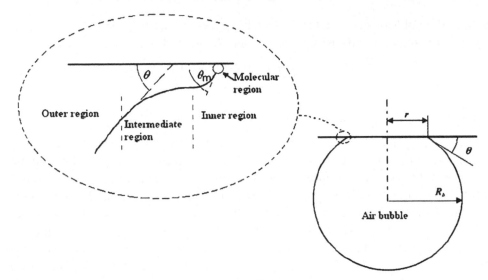

Figure 12. Magnified view of the molecular, inner, intermediate and outer regions of a dynamic meniscus with the outer (macroscopic), θ and inner (microscopic), θ_m contact angles.

from the contact line, (2) the inner region, which is close to the contact line and (3) the intermediate region which is between the inner and outer regions.

The molecular-kinetic theory and the hydrodynamic theory can be used to establish the relationship between the velocity of the three phase contact line and the dynamic contact angle. These two models differ in the mode of energy dissipation. The molecular-kinetic approach emphasizes dissipation due to displacements of molecules in the immediate vicinity of the contact line while the hydrodynamic approach focuses on dissipation due to viscous flow within the wedge of liquid near the moving contact line.

1. Hydrodynamic Theories

The hydrodynamic theory links the molecular region (described by the dynamic contact angle, θ_m and the molecular characteristic length scale, λ) to the static-like (outer) region described by the macroscopic contact angle, θ and the macroscopic (capillary) length, L.

Voinov [78] and Cox [11, 12] were among the first who developed useful hydrodynamic models for the dependence of the moving line velocity, U, on the macroscopic dynamic contact angle, θ. The Voinov model was derived from the continuity and Stokes equations near the contact line and can be described as follows:

$$\theta = \{\theta_m^3 + 9Ca[\ln(L/\lambda)]\}^{1/3}, \tag{24}$$

where $Ca = \mu U/\sigma$ is the capillary number and U is the TPC line velocity.

When the inertia is significant, the convective term in the Navier–Stokes equation is expected to affect the fluid flow and gas–liquid interface shape. The model given by Cox [13] considers the effect of flow inertia on the TPC motion. In this case, the hydrodynamic zone is also controlled by the flow Reynolds number which can be described as

$$g(\theta) = g\left\{\theta_m^3 + 9Ca\left[\ln\left(\frac{2L}{\lambda}Re\right)\right]\right\}^{1/3} + Ca\ln Re, \qquad (25)$$

where g is defined by $g(x) = 1.532(x - \sin x)$ and $Re = \rho U L/\mu$.

In the above two models, a local coordinate system with the origin located at (and moving with) the TPC point is used. In reality, the solution has to be corrected for the relative motion of the origin *via* the Galilean transformations which is a drawback of this model. Moreover, the macroscopic length scale, L, is usually not known precisely. Phan *et al.* [58] followed the special approach developed for bubbles and drops by Hocking and Rivers [25] who solved the governing equations relative to a fixed coordinate system with the origin located at the centre of the contact area and is stationary during the TPC expansion. In this approach, the unknown macroscopic length is not needed. Solving the motion equations are accompanied by a set of integration constants. The determination of the unknown constants is complicated since the deformed interface deviates from the spherical profile. Moreover, due to the singularity, the motion equations have been solved in three separate domains as discussed previously. Matching the results gives

$$Ca = \frac{G(\theta) - G(\theta_m)}{-\ln(\lambda) + \ln(r) - Q_0(\theta) + Q_1(\theta_m)} \qquad (26)$$

where $Q_0(\theta) = 1 + \ln(1 + \cos\theta) + J(\theta)/g(\theta)$ and $Q_1(\theta_m) = j(\theta_m)/g(\theta_m)$. Functions $G(x)$, $g(x)$, $J(x)$ and $j(x)$ can be found from the solution of motion equations in outer and inner regions. The details of the analysis are given in [58].

2. Molecular-Kinetic Theories

The motion of TPC line is determined by the overall statistics of the individual gas-liquid molecular displacement on the solid substrate. This molecular displacement is made by molecular jumping with a mean distance, λ, on the solid substrate. The molecular-kinetic theory gives [4, 53]

$$U = 2K\lambda \sinh\left\{\frac{\sigma\lambda^2}{2k_B T}(\cos\theta_0 - \cos\theta_m)\right\}, \qquad (27)$$

where k_B is the Boltzmann constant, T is the absolute temperature and θ_0 is the equilibrium (or the so-called Young contact angle).

Phan *et al.* [57] examined both hydrodynamic and molecular-kinetic models accuracy by comparing the experimental results of observing the motion (the radial position) of the TPC line for a small rising bubble ruptured by a submerged horizontal glass plate. It was shown that both models were not able to describe the experimental data using the physically consistent values for the model parameters. As shown in Fig. 13, the models fit only the first few experimental data. The difference between the model and the experimental data at long time of the TPC motion

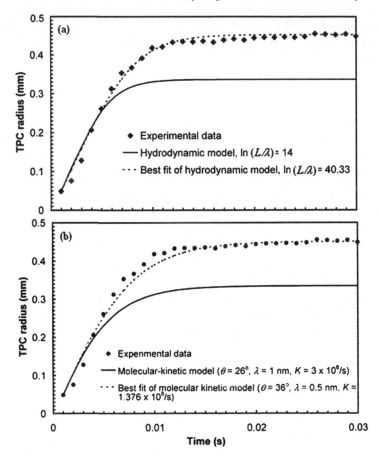

Figure 13. Experimental data (points) and predictions (line) by hydrodynamic and molecular-kinetic theories for TPC radius *versus* time [57].

is caused by the significant difference between the Young and macroscopic contact angles. The least-squares regression analysis was used to obtain the best fit values for the model parameters, but the fitted parameters were inconsistent with the physics involved in the models. Since neither of these models is able to describe the experimental data, a combined model of the hydrodynamics and molecular-kinetics is able to consistently describe the data [57] and is discussed below.

3. Crossovers of Scales

It is noted that both types of dissipation (the viscous loss in the thin liquid wedge and the wetting line friction) do exist simultaneously [15]. Brochard-Wyart and de Gennes [5] pointed out that non-hydrodynamic dissipation dominates at relatively large value of contact angle, while the hydrodynamic dissipation prevails at smaller contact angles. To overcome the inconsistency between these two cases several attempts have been made for a combined theory which considers both types of dissipation at the same time.

Blake [3] studied the formulation of the combined theory based on the molecular-kinetic approach by simple adding the viscous flows effect to the barriers created by the solid–liquid attractions.

Another type of dissipation energy in precursor film was considered by de Gennes [14]. The capillary force is balanced by the total energy dissipation which comprises of the viscous dissipation, the dissipation at the advancing contact line and in a precursor film.

Using a slightly different method, de Ruijter *et al.* [15], derived closed-form equations for the evaluation of the droplet's base radius and specify regimes at which different dissipation energies dominate. They derived the following equation for the TPC velocity by combining these two models into one:

$$Ca = \frac{\cos\theta_0 - \cos\theta}{\zeta_0 + 6\phi \ln[R(t)/a]}, \tag{28}$$

where $\zeta_0 = k_B T / K \lambda \mu$ is the (dimensionless) friction coefficient of the liquid molecules in the vicinity of the contact line, parameter a is the radius of the core region where the dissipation is negligible and $\phi = \frac{\sin^3\theta}{2-3\cos\theta+\cos^3\theta}$ is the geometrical factor. If the frictional forces are dominant, Eq. (28) reduces to molecular-kinetic model and if the viscous forces control the spreading, it reduces to hydrodynamic model. It was also shown that both energies of dissipation can have a dominant effect on the wetting process at different time scale, namely, non-hydrodynamic dissipation prevails at relatively short times and hydrodynamic dissipation dominates at long time. The authors also found explicitly the crossover time, t_2, separating these two regimes. For spreading times longer than t_2, most of the work is dissipated due to viscous flow, and therefore a hydrodynamic regime is predicted and for spreading time below t_2 the molecular-kinetic model is the governing channel of dissipation. The value of t_2 depends naturally on both the friction coefficient and the viscosity.

Recently, Roques-Carmes *et al.* [62] studied these different regimes experimentally. They confirmed the existence of several spreading regimes which, depending on the fluid particle volume and viscosity, the regimes sequence can be different (Fig. 14). They also showed that the nature of the regimes is strongly influenced by surfactant concentration. They estimated a critical contact angle for the crossover between molecular-kinetic regime and hydrodynamic regime. A critical contact angle of 35° was found. The critical value was also independent of the presence of surfactant and the drop volume.

G. Summary and Conclusions

In this chapter the interaction of air bubbles with solid surfaces has been reviewed. It focuses on the rising bubbles interacting with a planar solid surface placed horizontally relative to gravity in aqueous solutions. Bubbles motion in the solutions is influenced by the bubble size and shape, and surfactants and the other impurities

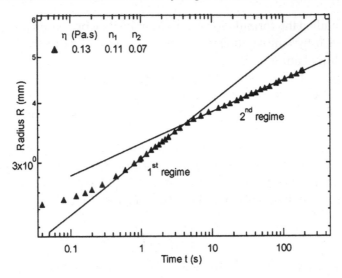

Figure 14. Influence of the viscosity on the time evaluation of the drop radius [62].

in water. There is a domain of the bubble rise velocity and drag coefficient *versus* the bubble equivalent diameter. The interaction of the rising bubbles with the solid surface is influenced by the bubble kinetic energy and surface energy, the intervening liquid film and the surface roughness and wettability. Interestingly, the recent studies show that the deformed bubble profile and velocity during the approach can be normalised to obtain the master curves. The wetting films formed between the rising bubble and the solid surface can be analysed within the framework of lubrication approximation when the inertial effects can be neglected. When the bubble surface is clean (free from surfactants), the water velocity at the air–water film surface is of the order of the bubble rise velocity and the inertial effects on the film drainage are significant. Both the DLVO and non-DLVO colloidal forces are relevant to the drainage and rupture of the intervening liquid films. The strong hydrophobic attraction is influenced by the nanobubbles of dissolved gases in the solution and at the water–solid surface. The hydrodynamic and molecular-kinetic models for the spreading and relaxation of the three-phase contact lines have been reviewed. A combination of the two models can provide a better agreement with the experimental results.

H. Acknowledgements

This research is supported under Australian Research Council's Discovery Projects funding scheme (Grants DP0663688 and DP0985079).

I. References

1. Attard, P. and Miklavcic, S. J., Effective spring constant of bubbles and droplets. Langmuir, 17(26) (2001) 8217–8223.

2. Attard, P., Moody, M. P. and Tyrrell, J. W. G., Nanobubbles: the big picture. Phys. A: Stat. Mech. Appl., 314(1–4) (2002) 696–705.

3. Blake, T. D., Dynamic contact angle and wetting kinetics, in: J. C. Berg (Ed.), Wettability. Marcel Dekker, New York, 1993.

4. Blake, T. D. and Haynes, J. M., Kinetics of liquid/liquid displacement. J. Colloid Interface Sci., 30(3) (1969) 421–423.

5. Brochard-Wyart, F. and De Gennes, P. G., Dynamics of partial wetting. Adv. Colloid Interface Sci., 39 (1992) 1–11.

6. Chan, D. Y. C., Dagastine, R. R. and White, L. R., Forces between a rigid probe particle and a liquid interface. I. The repulsive case. J. Colloid Interface Sci., 236(1) (2001) 141–154.

7. Chen, J.-D. and Slattery, J. C., Effects of London–van der Waals forces on the thinning of a dimpled liquid film as a small drop or bubble approaches a horizontal solid plane. AIChE Journal, 28(6) (1982) 955–963.

8. Christenson, H. K. and Claesson, P. M., Direct measurements of the force between hydrophobic surfaces in water. Adv. Colloid Interface Sci., 91(3) (2001) 391–436.

9. Claesson, P. M., Blom, C. E., Herder, P. C. and Ninham, B. W., Interactions between water-stable hydrophobic Langmuir–Blodgett monolayers on mica. J. Colloid Interface Sci., 114(1) (1986) 234–242.

10. Clift, R., Grace, J. R. and Weber, M. E., Bubbles, Drops and Particles. Academic Press, New York, 1978, 380 pp.

11. Cox, R. G., The dynamics of the spreading of liquids on a solid surface. Part 1. Viscous flow. J. Fluid Mech., 168 (1986) 169–194.

12. Cox, R. G., The dynamics of the spreading of liquids on a solid surface. Part 2. Surfactants. J. Fluid Mech., 168 (1986) 195–220.

13. Cox, R. G., Inertial and viscous effects on dynamic contact angles. J. Fluid Mech., 357 (1998) 249–278.

14. de Gennes, P. G., Wetting: Statics and dynamics. Rev. Mod. Phys., 57(3) (1985) 827–863.

15. de Ruijter, M. J., de Coninck, J. and Oshanin, G., Droplet spreading: Partial wetting regime revisited. Langmuir, 15(6) (1999) 2209–2216.

16. Derjaguin, B. V., Churaev, N. V. and Muller, V. M., Surface Forces. Kluwer Academic Pub., New York, 1987.

17. Dukhin, S. S., Kretzschmar, G. and Miller, R., Dynamics of Adsorption at Liquid Interfaces: Theory, Experiment and Application. Elsevier, Amsterdam, 1995, 581 pp.

18. Edwards, D. A., Brenner, H. and Wasan, D. T., Interfacial Transport Processes and Rheology. Butterworth-Heinemann, Boston, 1991, 558 pp.

19. Gondret, P., Lance, M. and Petit, L., Bouncing motion of spherical particles in fluids. Phys. Fluids, 14(2) (2002) 643–652.

20. Gong, W., Stearnes, J., Fornasiero, D., Hayes, R. A. and Ralston, J., The influence of dissolved gas on the interactions between surfaces of different hydrophobicity in aqueous media. Part II. A spectroscopic study. Phys. Chem. Chem. Phys., 1(11) (1999) 2799–2803.

21. Hampton, M. A. and Nguyen, A. V., Nanobubbles and the nanobubble bridging capillary force. Adv. Colloid Interface Sci., 154(1–2) (2010) 30–55.

22. Happel, J. and Brenner, H., Low Reynolds Number Hydrodynamics. Prentice Hall, Englewood Cliffs, NJ, 1965, 438 pp.

23. Hartland, S. and Hartley, R. W., Axisymmetric Fluid–Liquid Interfaces: Tables Giving the Shape of Sessile and Pendant Drops and External Menisci, with Examples of Their Use. Elsevier, Amsterdam, 1976, 782 pp.

24. Henry, C. L., Parkinson, L., Ralston, J. R. and Craig, V. S. J., A mobile gas–water interface in electrolyte solutions. J. Phys. Chem. C, 112(39) (2008) 15094–15097.
25. Hocking, L. M. and Rivers, A. D., Spreading of a drop by capillary action. J. Fluid Mech., 121 (1982) 425–442.
26. Holuszko, M. E. et al., The effect of surface treatment and slime coatings on ZnS hydrophobicity. Miner. Eng., 21(12–14) (2008) 958–966.
27. Israelachvili, J. and Pashley, R., The hydrophobic interaction is long range, decaying exponentially with distance. Nature, 300(5890) (1982) 341–342.
28. Israelachvili, J. N., Intermolecular and Surface Forces. Academic Press, London, 2005, 291 pp.
29. Joseph, G. G., Zenit, R., Hunt, M. L. and Rosenwinkel, A. M., Particle wall collisions in a viscous fluid. J. Fluid Mech., 433 (2001) 329–346.
30. Karakashev, S. I. et al., Comparative validation of the analytical models for the Marangoni effect on foam film drainage. Colloids Surf., A, 365(1–3) (2010) 122–136.
31. Karakashev, S. I., Ivanova, D. S., Manev, E. D., Kirilova, R. and Tsekov, R., An experimental test of the fractal model for drainage of foam films. J. Colloid Interface Sci., 353(1) (2011) 206–209.
32. Karakashev, S. I. and Nguyen, A. V., Do liquid films rupture due to the so-called hydrophobic force or migration of dissolved gases? Langmuir, 25(6) (2009) 3363–3368.
33. Kim, S. and Karrila, S. J., Microhydrodynamics: Principles and Selected Applications. Butterworth-Heinemann, Boston, 1991, 507 pp.
34. Klaseboer, E., Chevaillier, J. P., Mate, A., Masbernat, O. and Gourdon, C., Model and experiments of a drop impinging on an immersed wall. Phys. Fluids, 13(1) (2001) 45–57.
35. Kracht, W. and Finch, J. A., Effect of frother on initial bubble shape and velocity. Int. J. Miner. Process., 94(3–4) (2010) 115–120.
36. Krasowska, M., Krzan, M. and Małysa, K., Bubble collisions with hydrophobic and hydrophilic surfaces in alpha-terpineol solutions. Physicochem. Probl. Miner. Process., 37 (2003) 37–50.
37. Krasowska, M. and Malysa, K., Kinetics of bubble collision and attachment to hydrophobic solids: I. Effect of surface roughness. Int. J. Miner. Process., 81(4) (2007) 205–216.
38. Legendre, D., Daniel, C. and Guiraud, P., Experimental study of a drop bouncing on a wall in a liquid. Phys. Fluids, 17(9) (2005) 1–13.
39. Leja, J., Surface Chemistry of Froth Flotation. Plenum Press, New York, NY, 1982, 758 pp.
40. Levich, V., Physicochemical Hydrodynamics. Prentice Hall, Englewood, NJ, 1962.
41. Mahnke, J., Stearnes, J., Hayes, R. A., Fornasiero, D. and Ralston, J., The influence of dissolved gas on the interactions between surfaces of different hydrophobicity in aqueous media. Part I. Measurement of interaction forces. Phys. Chem. Chem. Phys., 1(11) (1999) 2793–2798.
42. Manev, E., Tsekov, R. and Radoev, B., Effect of thickness non-homogeneity on the kinetic behaviour of microscopic foam films. J. Dispersion Sci. Technol., 18(6–7) (1997) 769–788.
43. Manev, E. D. and Nguyen, A. V., Critical thickness of microscopic thin liquid films. Adv. Colloid Interface Sci., 114–115 (2005) 133–146.
44. Manev, E. D. and Nguyen, A. V., Effects of surfactant adsorption and surface forces on thinning and rupture of foam liquid films. Int. J. Miner. Process., 77(1) (2005) 1–45.
45. Marcelja, S. and Radic, N., Repulsion of interfaces due to boundary water. Chem. Phys. Lett., 42(1) (1976) 129–130.
46. Navarra, A., Acuña, C. and Finch, J. A., Impact of frother on the terminal velocity of small bubbles. Int. J. Miner. Process., 91(3–4) (2009) 68–73.
47. Nguyen, A. V., Historical note on the Stefan–Reynolds equations. J. Colloid Interface Sci., 231(1) (2000) 195.

48. Nguyen, A. V., Flotation, in: I. D. Wilson (Ed.), Encyclopedia of Surface and Colloid Science, 2nd Edn. Elsevier, Amsterdam, 2007, pp. 1–27.

49. Nguyen, A. V., Drelich, J., Colic, M., Nalaskowski, J. and Miller, J. D., Bubble: Interaction with solid surfaces, in: P. Somasundaran (Ed.), Encyclopedia of Surface and Colloid Science, 2nd Edn. CRC Press, New York, NY, 2007, pp. 1–29.

50. Nguyen, A. V., Evans, G. M. and Jameson, G. J., Electrical double-layer interaction between spheres: approximate expressions, in: P. Somasundaran (Ed.), Encyclopedia of Surface and Colloid Science, 2nd Edn. CRC Press, New York, 2006, pp. 1971–1981.

51. Nguyen, A. V. and Fletcher, C. A. J., Particle interaction with the wall surface in two-phase gas-solid particle flow. Int. J. Multiphase Flow, 25(1) (1999) 139–154.

52. Nguyen, A. V., Nalaskowski, J., Miller, J. D. and Butt, H. J., Attraction between hydrophobic surfaces studied by atomic force microscopy. Inter. J. Miner. Process., 72(1–4) (2003) 215–225.

53. Nguyen, A. V. and Schulze, H. J., Colloidal Science of Flotation. Marcel Dekker, New York, 2004, 840 pp.

54. Nguyen, P. T. and Nguyen, A. V., Drainage, rupture, and lifetime of deionized water films: Effect of dissolved gases. Langmuir, 26(5) (2010) 3356–3363.

55. Parkinson, L., Sedev, R., Fornasiero, D. and Ralston, J., The terminal rise velocity of 10–100 mm diameter bubbles in water. J. Coll. Interface Sci., 322(1) (2008) 168–172.

56. Pashley, R. M., DLVO and hydration forces between mica surfaces in lithium, sodium, potassium, and cesium ions electrolyte solutions: a correlation of double-layer and hydration forces with surface cation exchange properties. J. Colloid Interface Sci., 83(2) (1981) 531–546.

57. Phan, C. M., Nguyen, A. V. and Evans, G. M., Assessment of hydrodynamic and molecular-kinetic models applied to the motion of the dewetting contact line between a small bubble and a solid surface. Langmuir, 19(17) (2003) 6796–6801.

58. Phan, C. M., Nguyen, A. V. and Evans, G. M., Combining hydrodynamics and molecular kinetics to predict dewetting between a small bubble and a solid surface. J. Colloid Interface Sci., 296(2) (2006) 669–676.

59. Podvin, B., Khoja, S., Moraga, F. and Attinger, D., Model and experimental visualizations of the interaction of a bubble with an inclined wall. Chem. Eng. Sci., 63(7) (2008) 1914–1928.

60. Radoev, B. P., Dimitrov, D. S. and Ivanov, I. B., Hydrodynamics of thin liquid films effect of the surfactant on the rate of thinning. Colloid Polym. Sci., 252(1) (1974) 50–55.

61. Radoev, B. P., Scheludko, A. D. and Manev, E. D., Critical thickness of thin liquid films: Theory and experiment. J. Colloid Interface Sci., 95(1) (1983) 254–265.

62. Roques-Carmes, T., Mathieu, V. and Gigante, A., Experimental contribution to the understanding of the dynamics of spreading of Newtonian fluids: Effect of volume, viscosity and surfactant. J. Colloid Interface Sci., 344(1) (2010) 180–197.

63. Ruckenstein, E. and Sharma, A., A new mechanism of film thinning: enhancement of Reynolds' velocity by surface waves. J. Colloid Interface Sci., 119(1) (1987) 1–13.

64. Sharma, A. and Ruckenstein, E., Dewetting of solids by the formation of holes in macroscopic liquid films. J. Colloid Interface Sci., 133(2) (1989) 358–368.

65. Sheludko, A., Thin liquid films. Adv. Colloid Interface Sci., 1(4) (1967) 391–464.

66. Slattery, J. C., Interfacial Transport Phenomena. Spinger-Verlag, New York, NY, 1990, 1162 pp.

67. Snoswell, D. R. E., Duan, J., Fornasiero, D. and Ralston, J., Colloid stability and the influence of dissolved gas. J. Phys. Chem. B, 107(13) (2003) 2986–2994.

68. Steitz, R. et al., Nanobubbles and their precursor layer at the interface of water against a hydrophobic substrate. Langmuir, 19(6) (2003) 2409–2418.

69. Stöckelhuber, K. W., Stability and rupture of aqueous wetting films. Eur. Phys. J. E, 12(3) (2003) 431–435.

70. Stöckelhuber, K. W., Radoev, B., Wenger, A. and Schulze, H. J., Rupture of wetting films caused by nanobubbles. Langmuir, 20(1) (2004) 164–168.

71. Tsao, H.-K. and Koch, D. L., Observations of high Reynolds number bubbles interacting with a rigid wall. Phys. Fluids, 9(1) (1997) 44–56.

72. Tsekov, R., The R4/5-problem in the drainage of dimpled thin liquid films. Colloids Surf., A, 141(2) (1998) 161–164.

73. Tsekov, R. et al., Streaming potential effect on the drainage of thin liquid films stabilized by ionic surfactants. Langmuir, 26(7) (2010) 4703–4708.

74. Tsekov, R. and Vinogradova, O. I., A qualitative theory of wimples in wetting films. Langmuir, 21(26) (2005) 12090–12092.

75. Tyrrell, J. W. G. and Attard, P., Atomic force microscope images of nanobubbles on a hydrophobic surface and corresponding force-separation data. Langmuir, 18(1) (2002) 160–167.

76. Valkovska, D. S. and Danov, K. D., Influence of ionic surfactants on the drainage velocity of thin liquid films. J. Colloid Interface Sci., 241(2) (2001) 400–412.

77. van Oss, C. J., Interfacial Forces in Aqueous Media. CRC, New York, 2006, 456 pp.

78. Voinov, O. V., Hydrodynamics of wetting (in Russian). Mechanics of Liquids and Gas, 5 (1975) 76–84.

79. Yang, J., Duan, J., Fornasiero, D. and Ralston, J., Very small bubble formation at the solid–water interface. J. Phys. Chem. B, 107(25) (2003) 6139–6147.

80. Yoon, R. H., Flinn, D. H. and Rabinovich, Y. I., Hydrophobic interactions between dissimilar surfaces. J. Colloid Interface Sci., 185(2) (1997) 363–370.

81. Zawala, J., Krasowska, M., Dabros, T. and Malysa, K., Influence of bubble kinetic energy on its bouncing during collisions with various interfaces. Can. J. Chem. Eng., 85(5) (2007) 669–677.

82. Zenit, R. and Legendre, D., The coefficient of restitution for air bubbles colliding against solid walls in viscous liquids. Phys. Fluids, 21(8) (2009) 083306.

Electrowetting of Ionic Liquids in Solid–Liquid–Liquid Systems

Rossen Sedev, Craig Priest and John Ralston

Ian Wark Research Institute, ARC Special Research Centre for Particle and Material Interfaces,
University of South Australia, Mawson Lakes, SA 5095, Australia

Keywords

Electrowetting, ionic liquid, contact angle, spreading, wetting kinetics, energy dissipation

Contents

A. Introduction 241
    A.1. young–lippmann equation]Young–Lippmann Equation 242
    A.2. Electrowetting Saturation 244
    A.3. Solid–Liquid–Liquid Electrowetting 244
    A.4. conductive liquids]Conductive Liquids 245
    A.5. wetting dynamics]Wetting Dynamics 245
B. Materials and Methods 247
    B.1. Insulator 247
    B.2. Ionic Liquids 249
    B.3. Electrode 249
    B.4. Electrowetting 249
C. Results 250
D. Discussion 255
    D.1. Electrowetting Performance 255
    D.2. electrowetting saturation]Electrowetting Saturation 258
    D.3. dynamics of electrowetting]Dynamics of Electrowetting 260
E. Conclusion 263
F. Acknowledgements 263
G. References 264

A. Introduction

In electrowetting the contact angle is altered by applying an external voltage [1–3]. In the most common configuration (Fig. 1a), experiments are carried out with an electrically conductive droplet resting on a flat electrode insulated with a hydrophobic polymer layer. A parallel-plate capacitor is formed between the electrode and

Drops and Bubbles in Contact with Solid Surfaces
© Koninklijke Brill NV, Leiden, 2012

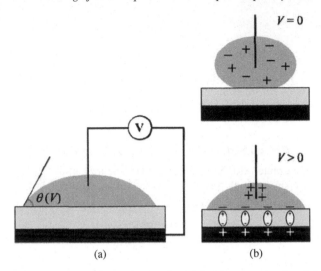

Figure 1. Electrowetting of a conductive liquid droplet on an insulated electrode.

the droplet. When an external voltage, V, is applied the capacitor is charged. Charge carriers from the liquid accumulate at the solid–liquid interface (Fig. 1b) and reduce the effective solid–liquid interfacial tension, γ_{SL}. The wettability of the solid surface is improved, the macroscopic contact angle, θ, decreases and is now a function of the applied voltage.

Substantial contact angle changes can be achieved through electrowetting which means that capillary forces can be varied widely in a controllable manner. This is especially important in miniaturized devices where wettability and capillary pressure play a leading role [4]. Electrowetting is fast, easily implemented and electronically controlled. A variety of devices have been developed, e.g., optical switches [5], microlenses [6], microvalves [7], triggers [8], pixels [9–11] and micropumps [12, 13]. A detailed review of what can be achieved through electrowetting in microfluidic platforms has been presented by Fair [14]. In all these applications frequent switching is involved and therefore the magnitude, reversibility, and robustness of the electrowetting effect are crucial.

A.1. Young–Lippmann Equation

The wettability in a solid–liquid–fluid system is most often characterized by the contact angle. For an ideal system, the equilibrium contact angle, θ_0, is given by the Young equation [15, 16]:

$$\cos\theta_0 = \frac{\gamma_{SV} - \gamma_{SL}}{\gamma}. \tag{1}$$

The three interfacial tensions, γ_{SV}, γ_{SL}, and γ, are macroscopic parameters accounting for the intermolecular interactions at the solid–vapour, solid–liquid and liquid–vapour interfaces. Wettability is directly related to adhesion, usually expressed as work of adhesion, W_A [15, 16]:

$$W_A = \gamma + \gamma_{SV} - \gamma_{SL} = \gamma(1 + \cos\theta_0). \tag{2}$$

The enhanced adhesion achieved under electrowetting conditions can be formally split into 'chemical' work of adhesion and 'electrical' work of adhesion (in analogy to the chemical and electric components of the electrochemical potential):

$$W_A = W_0 + W_{EL} = W_0 + \frac{1}{2}CV^2. \tag{3}$$

The first term is given by Young–Dupré's equation (2), while the second one is the electrostatic energy (per unit area) stored in the capacitor. From equations (2) and (3) one obtains the Young–Lippmann equation [1–3] which relates the contact angle, θ, as to the applied voltage, V:

$$\cos\theta = \cos\theta_0 + \frac{\varepsilon\varepsilon_0}{2\gamma h}V^2. \tag{4}$$

The dielectric constant of the insulating material is ε, ε_0 is the permittivity of vacuum, γ is the interfacial tension of the liquid/fluid interface, h is the thickness of the insulating layer, and θ_0 is the contact angle at zero voltage ($V = 0$).

The Young–Lippmann equation is a thermodynamic equation and therefore strictly applies only under equilibrium conditions [17]. Significant contact angle hysteresis is often observed in experiments and one should interpret such results carefully. This chapter is devoted to electrowetting in solid–liquid–liquid systems, where contact angle hysteresis is minimal and therefore should not be of major concern. It is assumed that the applied voltage affects the solid–liquid interface only and this seems to be a good approximation [2, 17, 18]. The electrowetting effect (the second term in equation (4)) is predicted to be identical for positive and negative voltages. This could be different if strong specific adsorption is present but such cases are rare. For instance, in solid–liquid–vapour electrowetting, we observed a systematic deviation from the Young–Lippmann equation at positive potentials. Due to the influence of ionic strength, pH and oxygen content in the fluoropolymer insulating coating the effect was attributed to specific adsorption of hydroxyl ions [18]. The electrowetting effect is larger for insulating layers with higher dielectric constant and smaller thickness [19–21]. Finally, the effect in solid–liquid–liquid systems should be more significant as the interfacial tension of a pure liquid is lower than its surface tension. Surfactants adsorbing at the liquid-fluid interface can be very effective in reducing the voltage needed to obtain a certain electrowetting effect [22–24].

The Young–Lippmann equation has been criticized but, although alternative or improved approaches should be explored, we see no reason to reject the simple and clear physical basis it provides [17].

A.2. Electrowetting Saturation

The Young–Lippmann equation describes adequately most experimentally obtained electrowetting curves (contact angle *vs.* voltage), provided that the voltage does not exceed a threshold value, V_S (saturation potential). At voltages beyond V_S, the macroscopic contact angle deviates from the Young–Lippmann equation and becomes effectively independent of the applied voltage. Contact angle saturation limits strongly the practical usefulness of electrowetting.

The physical mechanism of saturation is a matter of debate. Several mechanisms explaining contact angle saturation have been proposed: charge trapping at the solid surface [25], ionization of the ambient fluid close to the contact line [26], defects in the insulating layer [27], non-zero liquid resistance [28], and dielectric breakdown [29, 30]. We argue that the validity of the Young–Lippmann equation is limited from below by $\gamma_{SL} = 0$ which is achieved at $V = V_S$. This condition provides the following prediction [2] for the saturation contact angle, θ_S:

$$\cos\theta_S = \frac{\gamma_S}{\gamma},\tag{5}$$

where γ_S is the surface tension of the solid. According to this equation, θ_S is completely determined by the material properties of the system (namely γ_{SV} and γ). Thus the saturation contact angle is not expected to be zero and electrowetting saturation is not a defective phenomenon but imposed by the limit of validity of the thermodynamic model. Equation (5) provides a reasonable estimate for fluoropolymer surfaces and several other cases (using the critical surface tension of wetting, γ_C, as an approximation for γ_S) [2]. More recently, electrowetting measurements in the presence of surfactants [22, 23] have lent further support to the validity of the zero-interfacial-tension hypothesis. It should be noted that equation (5) estimates the point of deviation from the Young–Lippmann equation rather than the lowest contact angle achievable during electrowetting.

A.3. Solid–Liquid–Liquid Electrowetting

A major portion of the research on electrowetting has been carried out in solid–liquid-air systems, usually a drop of electrolyte in ambient air [3]. Replacing air with an immiscible oil, however, offers some advantages [14] (no evaporation, lesser contamination, small contact angle hysteresis and therefore easier actuation, and improved liquid transport) and solid–liquid–liquid systems are getting more and more popular. Janocha *et al.* [31] attempted electrowetting of a decane droplet immersed in water on several polymer surfaces with variable success. Berge and Peseux [6] used organic liquid droplets immersed in an aqueous solution of sodium sulphate. Quilliet and Berge [32] estimated theoretically that, under equilibrium conditions, a thin film of ambient oil ($\sim$20 nm, stabilized by van der Waals forces) could be present under the water droplet. This film of oil effectively lubricates the water droplet and hence is responsible for the very low contact angle hysteresis found in solid–liquid–liquid systems. This idea is encountered in microfluidic studies of a water droplet moving through an immiscible oil [33] or physiological fluids undergoing multiple-step manipulation on a single chip [34]. The ambient oil significantly reduces biofouling in microfluidics and is of major importance in this field [14, 34]. Static and transient capacitance measurements have demonstrated convincingly the presence a wetting film of oil [14]. Berry and co-workers studied the electrowetting of aqueous droplets containing salt and surfactant in alkanes on an amorphous fluoropolymer surface [22–24]. Electrowetting experiments have been performed with mercury in salty water on mica [35]. Staicu and Mugele [36] studied the dynamic entrapment of an oil film between an aqueous droplet and the insulating polymer layer. The film is unstable and breaks into droplets. The same effect provides opportunities to optimize the design of electrowetting display pixels [11].

A.4. Conductive Liquids

Ionic liquids are a new class of solvents made widely available in recent years [37–40]. These are organic salts with relatively low melting points (usually $\leqslant 100°C$) due to the different size of the ions. Their fluidity and good thermal stability are attractive properties. The volatility of ionic liquids is extremely low (they can be probed with vacuum-based techniques, e.g., X-ray photoelectron spectroscopy [41, 42]) and therefore exclusively suitable for applications where evaporation is undesirable. Many ionic liquids can be synthesized and this chemical diversity offers numerous options to tailor their properties [43]. On the other hand minor components, such as water or halides, and other additives can affect their properties considerably. The surface tension of ionic liquids is intermediate between alkanes and water, i.e., similar to that of polar molecular solvents [44]. The toxicity of ionic liquids is largely unexplored [45]. Ionic liquids are conductive though their conductivity is somewhat reduced by their viscosity (typically 50–500 times higher than the viscosity of water). The concentration of ions is naturally very high though ionicity can vary widely [46]. Ionic liquids are often stable within a wide range of potentials [47] and this makes them excellent electrowetting agents. We showed that ionic liquids can electrowet fluoropolymer surfaces in air, though not very efficiently [48]. More recently, we have focussed on the solid–liquid–liquid electrowetting of ionic liquids on a fluoropolymer surface in ambient hexadecane [49–51].

A.5. Wetting Dynamics

One major advantage of electrowetting over other actuation mechanisms is that it is fast. The kinetics of spreading and retraction of the conductive droplet is of key importance to achieve controlled manipulation. In the electrowetting experiment (Fig. 1), when voltage is switched on, the wettability of the solid substrate is improved and the droplet spreads until the macroscopic contact angle reaches its final value. When the voltage is switched off, the original wettability is restored, and the droplet retracts back to its initial position. During the transition between these two

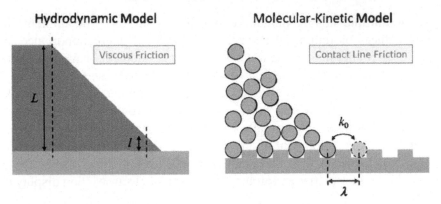

Figure 2. Hydrodynamic and molecular-kinetic approaches to the dynamics of wetting.

states, the dynamic contact angle, θ, changes in time and is a function of the speed, u, of the moving contact line.

The velocity dependence of the contact angle is most often described in terms of a hydrodynamic model or a molecular-kinetic model [52] (Fig. 2). In the hydrodynamic approach, a small zone in the vicinity of the contact line (characteristic size l) is effectively excluded from consideration as the no-slip condition is violated [53–55].

An elaborated hydrodynamic model has been provided by Cox [54] but a simplified version, derived by Voinov, performs well for dynamic contact angles smaller than 130°. Voinov's equation is [53]

$$\theta^3 = \theta_0^3 + 9\frac{\mu u}{\gamma}\ln\frac{L}{l}. \tag{6}$$

In this model the main dissipation mode is viscous friction and the viscosity of the liquid, μ, naturally appears in equation (6). The influence of the macroscopic length scale, L, and the microscopic length scale, l, is weak because of the logarithmic dependence. The viscous dissipation, D_{HD}, has been estimated for a liquid wedge contained between L (e.g., droplet size) and l (e.g., slip length) assuming a film flow under a fixed pressure gradient [55]

$$D_{\mathrm{HD}} = \int \mu\left(\frac{\partial u_y}{\partial x}\right)^2 dx\,dy\,dz \cong const\frac{\mu u^2}{\theta}\ln\frac{L}{l}. \tag{7}$$

The molecular-kinetic approach describes the macroscopic movement of the contact line as a series of molecular jumps. Each molecular jump is described as a rate-activated process and arrives at the following equation [56–58]

$$\cos\theta = \cos\theta_0 - \frac{2k_{\mathrm{B}}T}{\gamma\lambda^2}\sinh^{-1}\frac{u}{2k_0\lambda}, \tag{8}$$

where λ is the average size of the molecular jump, k_0 is the jump frequency at a static contact line ($u = 0$), k_{B} is the Boltzmann constant, and T is the absolute temperature. The dissipation mode in this case is molecular friction at the contact line, though the activation energy related to k_0 can be reinterpreted to account explicitly for the viscosity of the liquid ($k_0 = k_B T / \mu \lambda^3 \exp(-\Delta G / k_B T)$ [57]). By expressing the contact line friction coefficient, B, through a diffusion coefficient, D, the following estimate for the molecular-kinetic dissipation, D_{MK}, is obtained:

$$D_{MK} = fu = Bu^2 = \frac{k_B T}{D} \frac{1}{\lambda} u^2 = \frac{k_B T}{k_0 \lambda^3} u^2. \qquad (9)$$

It is physically clear that both viscous and molecular dissipation should be considered [58, 59], but most often the hydrodynamic and the molecular-kinetic approaches are treated as excluding alternatives. Models combining different dissipation channels tend to be unreliable or intractable [52, 58] and have not gained wider acceptance.

In this work we used a range of ionic liquids (1-alkyl-3-methyl-imidazolium tetrafluoro-borates, Rmim.BF$_4$, 1-butyl-3-methyl-imidazolium ionic liquids with various anions, bmim.X, and bmim.BF$_4$-water mixtures) to electrowet a Teflon AF1600-coated electrode, immersed in n-hexadecane. The solid–liquid–liquid systems showed robust electrowetting behaviour: very large contact angle variations (up to 130°), excellent reversibility and insignificant contact angle hysteresis. Electrowetting curves obtained with DC and AC voltages were similar and overlapping but lower contact angles could be achieved with AC voltage. The spreading and retracting of the ionic liquid droplet under electrowetting conditions was very fast and involved both viscous and molecular dissipation.

B. Materials and Methods

B.1. Insulator

Teflon AF1600 (an amorphous perfluorinated polymer available from DuPont) was dip-coated on ITO-coated glass slides. AF1600 is a random copolymer of 2,2-bis(trifluoromethyl)-4,5-difluoro-1,3-dioxole and tetrafluoroethylene in a ratio PDD:TFE = 2:1. It is soluble in perfluorinated solvents (e.g., 3M Fluoroinert FC-75) and therefore can be coated on various surfaces. It is very popular in electrowetting studies either as a thin hydrophobic layer on top of an insulating coating [1, 60] or as a hydrophobic insulator [2, 18, 27, 48–50].

In aqueous solutions, at normal pH, Teflon AF1600 acquires spontaneously a negative charge. The reasons for that are passionately debated but the mechanism appears to be specific adsorption of OH$^-$ ions at the hydrophobic interface [61, 62]. The zeta potential, ζ, as a function of bulk pH for both AF1600 and PTFE is shown in Fig. 3. The trend is very similar for both polymers and their isoelectric points ($\zeta = 0$) are identical (pH = 4). However, the magnitude of the negative potentials is significantly larger in the case of AF1600. This can be attributed to the oxygen present in the dioxole monomer (PDD). This oxygen has been detected by surface spectroscopies, such as Secondary Ion Mass Spectroscopy (SIMS) [18] and X-ray Photoelectron Spectroscopy (XPS) [63], and therefore it resides in close

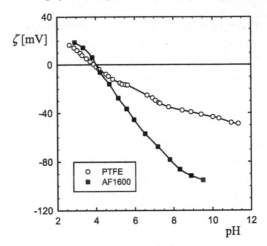

Figure 3. Zeta potential *vs.* pH (in aqueous 1 mM KCl) for PTFE (data from Werner *et al.* [65]) and Teflon AF1600 (data from Zimmermann *et al.* [61]). Reprinted with permission from Paneru *et al.* [51]. Copyright 2011 VSP.

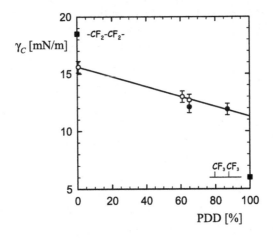

Figure 4. Critical surface tension of wetting, γ_C, for Teflon AF with various ratio between the PDD and TFE monomers. Reprinted with permission from Quinn *et al.* [18]. Copyright 2003 American Chemical Society.

proximity of the polymer surface. However, it does not affect the wettability of the Teflon AF1600 surface which is very hydrophobic. The critical surface tension of wetting, γ_C, obtained with alkanes using the Zisman approach [15, 64] for several amorphous fluoropolymers (Teflon AF) with different PDD:TFE ratios is shown in Fig. 4.

The PDD:TFE ratio has a minor effect on the critical surface tension of wetting. All values for amorphous fluoropolymers are intermediate between those obtained on $-(CF_2–CF_2)_n-$ (i.e., PTFE) and on a monolayer of closely packed $-CF_3$ groups. Therefore the hydrophobicity of Teflon AF is due to the predominant exposure of

Table 1.
Ionic liquids used in this study

Abbreviation	Cation	Anion
emim.BF$_4$	1-ethyl-3-methylimidazolium	tetrafluoroborate
bmim.BF$_4$	1-butyl-3-methylimidazolium	tetrafluoroborate
hmim.BF$_4$	1-hexyl-3-methylimidazolium	tetrafluoroborate
omim.BF$_4$	1-octyl-3-methylimidazolium	tetrafluoroborate
dmim.BF$_4$	1-decyl-3-methylimidazolium	tetrafluoroborate
bmim.PF$_6$	1-butyl-3-methylimidazolium	hexafluorophosphate
bmim.I	1-butyl-3-methylimidazolium	iodide
bmim.TFA	1-butyl-3-methylimidazolium	trifluoroacetate
bmim.NTf$_2$	1-butyl-3-methylimidazolium	bis(trifluorosulfonyl)imide

–CF$_3$ and –CF$_2$– groups and is unaffected by the oxygen from the PDD monomers. This apparent contradiction reflects the different sampling depths achieved with XPS and SIMS (6–8 nm and 1–2 nm, respectively [66]) as compared with contact angle measurements ($\sim$0.5 nm [64, 67]).

B.2. Ionic Liquids

The imidazolium-based ionic liquids used in this study are listed in Table 1. Two series of pure ionic liquids were considered: (i) constant anion (1-alkyl-3-methylimidazolum tetrafluoroborates, Rmim.BF$_4$) and (ii) constant cation (1-butyl-3-methylimidazolium with various anions, bmim.X). Mixtures of bmim.BF$_4$ and water with various concentrations were also included. All ionic liquids were purchased from Merck and purified by extraction with activated charcoal, filtering through a 0.2 µm Teflon filter, extraction with ethyl acetate, and evacuation under medium vacuum (0.1 mbar) for 24 h [68].

B.3. Electrode

Glass slides, coated with a 30 nm Indium Tin Oxide (ITO) layer, were obtained from Delta Technologies (Stillwater, MN). They were cleaned with isopropanol, dried with nitrogen, and dip-coated with Teflon AF1600 (6% AF1600 dissolved in Fluorinert FC-75, Derbyshire, UK). The AF1600 coating was dried in air (30 min in a laminar flow cabinet and 30 h in an oven at 100°C). The thickness of the AF1600 was determined by capacitance and stylus (Zeiss HandySurf profilometer) measurements. The average thickness of the insulating layer was (3.8 $\pm$ 0.2) µm.

B.4. Electrowetting

The insulated electrode was immersed in a cell (Fig. 5) filled with hexadecane (Sigma-Aldrich 99%, additionally purified by passing it through an aluminium oxide column). A small droplet of ionic liquid (typical volume 10 µl) was deposited on the electrode with an automated syringe. The droplet was then contacted with

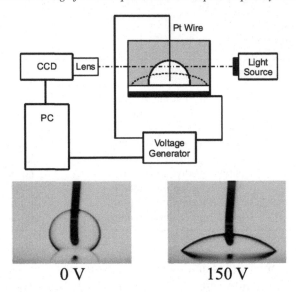

Figure 5. Diagram of the electrowetting setup. The pictures were taken with the system Teflon AF1600-bmim.BF$_4$-hexadecane subjected to DV voltage. Reprinted with permission from Paneru *et al.* [50]. Copyright 2010 American Chemical Society.

platinum needle (the thickness and exact position of the needle turned out to be unimportant) and external voltage was applied between the insulated electrode and the conductive droplet.

DC voltage was obtained from a power supply and an amplifier (Trek 610D, Medina, NY). AC voltage was generated with a signal generator (Kenwood, CR Oscillator, AG-203). The polarity of the voltage applied to the insulated electrode is reported. The cell was combined with a sessile drop apparatus (Fig. 5) and electrowetting curves (static contact angle *vs.* voltage) were recorded. Voltage was increased from zero to the maximum desired value and reduced back to zero in various increments. A fresh electrode surface was used in every experiment.

Digital images of static droplets (624 × 580 pixels, 256 grey levels) were captured with a progressive scan CCD camera (JAI CV-M10BX) with a resolution of 6.7 μm/pixel. The dynamics of spreading and retraction of ionic liquid droplets was followed with a high-speed camera (Olympus Encore MAC-2000) connected to the same optical setup and operating at 1000 frames per second. In all cases, the droplet silhouette was digitized and diameter and contact angle were determined by using image processing software (ImageJ [69]).

All experiments were carried out in a class 1000 clean room at room temperature (24°C) and normal humidity (45%).

C. Results

A typical electrowetting curve obtained with DC voltage for the Teflon AF1600-bmim.BF$_4$-hexadecane system is shown in Fig. 6. It is symmetric with respect to

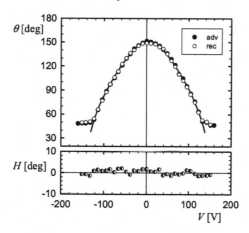

Figure 6. Static contact angle, θ, as a function of the applied DC voltage, V, for a droplet of bmim.BF$_4$ immersed in hexadecane on an electrode insulated with Teflon AF1600. The solid line is the best fit of the Young–Lippmann. The lower graph shows contact angle hysteresis, $H = \theta_A - \theta_R$, vs. applied voltage, V. Reprinted with permission from Paneru et al. [50]. Copyright 2010 American Chemical Society.

voltage polarity, i.e., around $V = 0$, and closely obeys the Young–Lippmann equation (shown with a solid line) all the way up to the saturation point ($V_S = \pm 120$ V). Beyond that voltage the contact angle remains effectively constant. The advancing and receding static contact angles (shown with open and filled symbols), obtained when increasing or decreasing the voltage so that the ionic liquid advances or recedes over the solid surface, practically coincide. In other words, the contact angle hysteresis, H ($= \theta_A - \theta_R$), is very small.

The electrowetting effect is very large and the value of the static contact angle can be quickly and easily varied between 145° (at $V = 0$) and 48° (at $|V| \geqslant |V_S|$). When a droplet of ionic liquid was repeatedly electrowet on the same location of the electrode, the electrowetting effect was robust and reproducible. Variations in the voltage increment used (5–100 V) did not affect the curve shown in Fig. 6.

Electrowetting in the same solid–liquid–liquid system was even more effective when using AC voltage—Fig. 7. This result was obtained with a square wave at a frequency of 500 Hz. Small variations in the shape or the frequency (100–1000 Hz) of the AC signal did not affect the outcome significantly. The electrowetting curve obtained with AC voltage is also reversible, robust, and covers an even wider range of static contact angles: from 145° down to about 15° (Fig. 7).

The electrowetting curves obtained with DC voltage for three different tetrafluoroborate ionic liquids are shown in Fig. 8. Another three 1-butyl-3-methylimidazolium ionic liquids with different anions were also tested and the results are plotted in Fig. 9. In all cases the electrowetting curves are symmetric, follow the Young–Lippmann equation (shown with a solid line), and saturate at contact angles of about 50°. The contact angles at zero voltage are very similar for all the BF$_4$ and

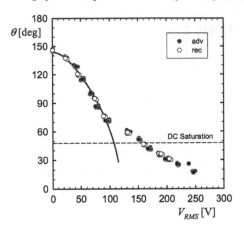

Figure 7. Electrowetting curve (contact angle *vs.* RMS voltage) for Teflon AF1600-bmim.BF_4-Hexadecane obtained with AC voltage (square wave, 500 Hz). Five consecutive experiments were carried out on the same location on the Teflon-coated electrode. The solid line is the best fit of the Young–Lippmann. The dotted line marks the saturation contact angle obtained with DC voltage. Reprinted with permission from Paneru *et al.* [50]. Copyright 2010 American Chemical Society.

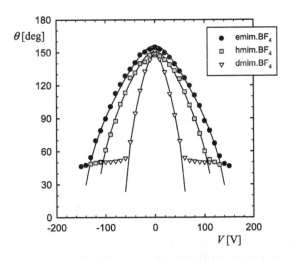

Figure 8. Static contact angle, θ, as a function of the applied DC voltage, V, for a droplet of ionic liquid immersed in hexadecane on an electrode insulated with Teflon AF1600. The solid lines are the best fit of the Young–Lippmann equation. Reprinted with permission from Paneru *et al.* [51]. Copyright 2011 VSP.

the PF_6 ionic liquids but markedly larger for the iodide and smaller for the NTf_2 anion. The saturation voltages are also very different.

The electrowetting curves obtained with DC voltage for various bmim.BF_4-water mixtures are shown in Fig. 10.

The curves are symmetric with respect to voltage polarity (only the positive branches are shown) and follow the Young–Lippmann equation (solid lines) prior

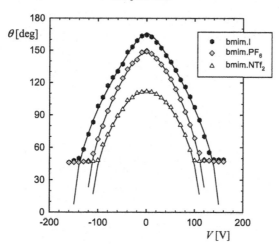

Figure 9. Static contact angle, θ, as a function of the applied DC voltage, V, for a droplet of ionic liquid immersed in hexadecane on an electrode insulated with Teflon AF1600. The solid lines are the best fit of the Young–Lippmann equation. Reprinted with permission from Paneru *et al.* [51]. Copyright 2011 VSP.

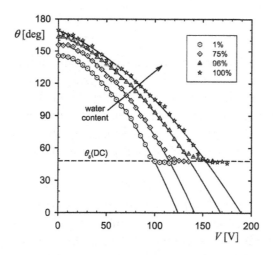

Figure 10. Static contact angle, θ, as a function of the applied DC voltage, V, for a droplet of a bmim.BF$_4$-water mixture immersed in hexadecane on an electrode insulated with Teflon AF1600. The solid lines are the best fit of the Young–Lippmann equation. The dashed line indicates the saturation contact angle, θ_S. Adapted with permission from Paneru *et al.* [49]. Copyright 2010 American Chemical Society.

to saturation. As the water content increases, both the contact angle at zero voltage, θ_0, and the saturation voltage, V_S, increase. The saturation contact angle is essentially constant ($\sim 50°$, dashed line). The electrowetting curves obtained with AC voltage (Fig. 11) are similar and again electrowetting performance is better though the improvement over DC voltage diminishes as water content increases.

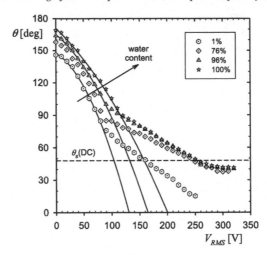

Figure 11. Static contact angle, θ, as a function of the applied AC voltage (sine wave, 700 Hz), V_{RMS}, for a droplet of a bmim.BF4-water mixture immersed in hexadecane on an electrode insulated with Teflon AF1600. The solid lines are the best fit of the Young–Lippmann equation. The dashed line indicates the saturation contact angle, θ_S, obtained with DC voltage. Adapted with permission from Paneru *et al.* [49]. Copyright 2010 American Chemical Society.

The dynamics of spreading and retraction of the ionic liquid is well illustrated by the time dependence of the base area of the droplet, $A \ (=\pi r^2)$, shown in Fig. 12. After switching on the DC voltage (at $t = 0$), the area A increases, initially very rapidly, then more gradually, and finally arrives asymptotically (not shown in the figure) at the static value prescribed by the electrowetting curve shown in Fig. 6. When the voltage is switched off the ionic liquid droplet quickly retracts back to its original shape.

As the applied DC voltage is increased one and the same droplet of ionic liquid spreads further (Fig. 12a). Positive and negative voltages of the same magnitude produce identical effects (open and filled symbols in Fig. 12). At $|V| \geqslant |V_S|$, a limiting spreading curve is obtained and a subtle difference is found—positive voltage is slightly less efficient than negative (open and filled triangles in Fig. 12a). During the retraction of the droplet (Fig. 12b, when voltage is switched off), polarity is irrelevant and the droplet contracts to one and the same position (and therefore base area) which is essentially identical to the position assumed before the electrowetting experiment had started.

The spreading kinetics for a series of Rmim.BF$_4$ ionic liquids is shown in Fig. 13. The behaviour is exponential as discussed above and the rate of spreading decreases as the length of the alkyl chain increases from 2 to 10 carbon atoms. The spreading of bmim.BF$_4$-water mixtures (not shown) is similar, though at very low viscosity (high water content) the sudden application of a DC voltage disrupts the shape of the conductive droplet and the $A(t)$ curve becomes non-monotonic [49].

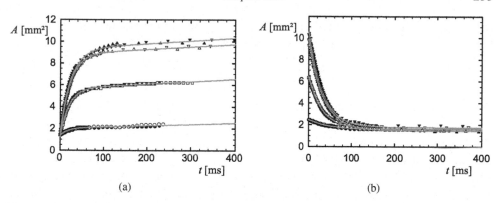

(a)   (b)

Figure 12. Base area, A, *vs.* time, t, for a bmim.BF$_4$ droplet immersed in hexadecane on a Teflon–coated electrode and subjected to a DC voltage (○ 50 V; ● −50 V; □ 100 V; ■ −100 V; △ 150 V; ▲ −150 V; ▽ 200 V; ▼ −200 V): (a) spreading droplet; (b) retracting droplet. The solid lines are the best fits of (a) $A = a + b[1 - \exp(-t/\tau)] + kt$ (exponential saturation with four parameters); (b) $A = a + b\exp(-t/\tau)$ (exponential decay with three parameters). Reprinted with permission from Paneru *et al.* [50]. Copyright 2010 American Chemical Society.

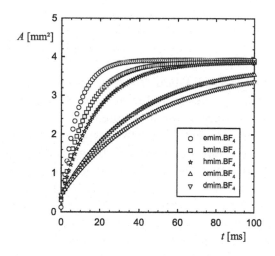

Figure 13. Base area, A, *vs.* time, t, for a Rmim.BF$_4$ droplet immersed in hexadecane on a Teflon–coated electrode, subjected to a DC saturation voltage.

D. Discussion

D.1. Electrowetting Performance

The range of static contact angles achievable with the system AF1600-ionic liquid-alkane is extremely wide: from about 150° down to 48° (when using DC voltage) or about 15° (when using AC voltage). The large values of the contact angles at zero voltage, θ_0, contribute to that range considerably. Since both alkane and Teflon are apolar materials [70] (i.e., they interact almost exclusively through dispersion forces), Fowkes approach [15] can be used to relate the interfacial tension, γ_{12}, to the surface tension of the two separate materials, γ_1 and γ_2, as:

$$\gamma_{12} = \gamma_1 + \gamma_2 - 2(\gamma_1 \gamma_2)^{1/2}. \tag{10}$$

With the surface tensions of hexadecane (27.6 mJ/m^2) and Teflon AF1600 (12.4 mJ/m^2) we estimate the interfacial tension of the AF1600/HD interface at 3.0 mJ/m^2. By using Young equation with $\theta_0 = 150°$ and $\gamma_{\mathrm{IL/HD}} = 21.7$ mJ/m^2, we estimate the interfacial tension of the AF1600/bmim.BF$_4$ to be 21.8 mJ/m^2. Because $\gamma_{\mathrm{AF1600/HD}}$ and $\gamma_{\mathrm{IL/HD}}$ are very close, the static contact angle at zero voltage is very large. In solid–liquid–vapour systems, such large contact angles are seen only on super-hydrophobic surfaces [71], where—due to the combination of hydrophobicity and surface roughness—γ_{SV} and the effective γ_{SL} are very similar (the liquid droplet is supported largely on a 'cushion' of air).

Another key feature of solid-ionic liquid-alkane electrowetting is the very low contact angle hysteresis—approximately 2° for both DC and AC voltages. This hysteresis is substantially lower than the lowest one found in low-hysteresis electrowetting systems: Teflon-aqueous salt-vapour ($\theta_A - \theta_R < 10°$) [2, 18]. Again, such low (effectively negligible) hysteresis is also observed in superhydrophobic systems [71]. Contact angle hysteresis is a macroscopic manifestation of the imperfections of the solid surface and can be very large [15]. One of the best behaved electrowetting curves for a solid–liquid–vapour system was reported by Verheijen and Prins [25]. The hysteresis on Teflon AF1600 surface was within 2° but, crucially, the dry solid surface was impregnated with silicone oil before carrying out the electrowetting experiment. Unusually low hysteresis during electrowetting in solid–liquid–liquid systems was found by Berge and Peseux [6] and they attributed the fact to the presence of a residual thin oil film trapped under the droplet. Janocha et al. [31] reported large contact angle changes and small hysteresis in solid–liquid–liquid electrowetting. Maillard et al. [72] also reported hysteresis of 2° or less in various solid–liquid–liquid systems exhibiting large static contact angles in the absence of applied voltage. Static and transient capacitance measurements convincingly showed the presence of an oil film intercalated between the conductive droplet and the solid surface [14]. Replacing air with oil in microfluidic electrowetting experiments significantly reduces evaporation, contamination, and biofouling, and enhances droplet movement [14, 34]. These are considerable practical advantages when designing devices (e.g., valves or actuators) where a maximum change in the capillary force is required.

The electrowetting curves obtained in this study were fully reversible when voltage was cycled. This was observed even in runs where the voltage exceeded V_S. Therefore charge injection into the insulator, which is widely considered as a major mechanism for electrowetting saturation [25], was not significant in our experiments. This may be tentatively related to the presence of a thin wetting film of hexadecane separating the ionic liquid from the insulator.

Table 2.
Maximum surface charge, σ_{max}, in various physical situations ($L = (e/\sigma_{max})^{1/2}$ is the average distance between two surface charges)

Phenomenon	System	σ_{max} mC/m^2	L nm
Air ionisation	Flat insulating surface in dry air	0.03 [75]	78
Electrowetting (SLV)	0.1 M aqueous KCl on AF1600 in air	0.25 [18]	25
Elcctrowetting (SLL)	bmim.BF$_4$ on AF1600 in hexadecane	0.5 [50]	18
Charge Injection	Polystyrene and Chlorinated Polyethylene in 10^{-2} M aqueous HCl	1.7 [74]	10
Electrokinetics (streaming potential)	Neutral polymers in aqueous salt solutions	4.8 [73]	6
Electrokinetics (streaming potential and streaming current)	Teflon AF1600 in 10^{-3} M aqueous KCl	5 [61]	6

All of our results are well-described by the Young–Lippmann equation (4). The solid lines shown in Figs 6–11, are the least-squares fits obtained by using the equation (4) in the form $\cos\theta = \cos\theta_0 + \alpha(\varepsilon\varepsilon_0/\gamma d)V^2$, with θ_0 and α taken as fitting parameters. The factor α is obtained since all quantities in the electrowetting term are known. The closeness of α to $\frac{1}{2}$ is a good indicator of the quality of the experimental data [2, 17, 18]. Thus the Young–Lippmann equation provides a consistent description of the electrowetting curve as long as $|V| \leqslant |V_S|$.

It is informative to estimate the surface charge density, σ, at the solid surface. The limiting value of σ is $\sigma_{max} = CV_S$, and is of the order of 0.5 mC/m^2 (Table 2). This value is close to the one estimated for Teflon AF1600-aqueous 0.1 M KCl-air systems [18]. It is about one order of magnitude smaller than values typically encountered in electrokinetic studies of polymers surfaces [61, 73] or charge injection measurements [74]. It is also one order of magnitude larger than the limiting surface charge possible in air [75].

This rather low surface charge density corresponds nicely with the fact that the electrowetting curves presented here do not show any unusual or asymmetric behaviour with respect to the Young–Lippmann equation [17]. The average distance between two surface charges, L (Table 2) is much larger than the size of the charge carrier. In other words, the charges accumulated on the surface can be considered as randomly distributed point charges.

The experiments with bmim.BF$_4$-water mixtures provide another insight into the mechanism of electrowetting. When changing the composition of the mixture, the conductive droplet varies from concentrated electrolyte (i.e., ionic liquid) to dilute electrolyte. The interfacial structure and electrochemical behaviour of dilute and solvent-free electrolytes are very different [76, 77]. Nonetheless the electrowetting

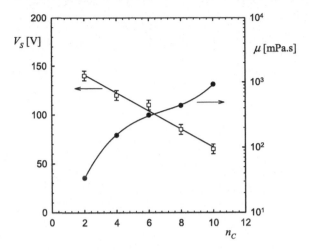

Figure 14. Saturation voltage, V_S, and viscosity, μ, as a function of the number of carbon atoms, n_C, in the alkyl chain of the imidazolium cation for tetrafluoroborate ionic liquids. Adapted with permission from Paneru *et al.* [51]. Copyright 2011 VSP.

curves retain their shape and behaviour—Figs 10 and 11, with minor changes reflecting the variation in material parameters.

D.2. Electrowetting Saturation

All ionic liquids studied electrowet in accordance with the Young–Lippmann equation and saturate—under DC voltage—at about the same contact angle ($\sim 50°$). Therefore the saturation voltage, V_S, is very different. For the homologous tetrafluoroborates (shown in Fig. 8), the saturation voltage, V_S, drops linearly with the length, n_C, of the alkyl chain of the imidazolium cation—Fig. 14.

As n_C increases, lower voltages produce the same contact angle change. However, the advantage of using imidazolium ionic liquids with longer side chains may be outweighed by the logarithmic increase in viscosity (Fig. 14).

We have shown [2] that the saturation contact angle, θ_S, for solid–liquid–vapour systems is often correctly estimated by equation (5). In the present case, for the solid-ionic liquid-hexadecane system, the zero-interfacial-tension hypothesis is written as:

$$\cos \theta_S = \frac{\gamma_{S/HD}}{\gamma_{IL/HD}}. \tag{11}$$

The values estimated through (11) range between 72° and 83° (with the exception of dmim.BF$_4$, 44°) and are very different from the values found in our experiments: $\theta_S = 48°$ (when using DC voltage) and $\theta_S \leqslant 15°$ (when using AC voltage). Therefore the zero-interfacial-tension hypothesis cannot be used directly for solid–liquid–liquid systems. Intriguingly, the value estimated through the above equation, approximates reasonably well the inflection point on the AC electrowetting curve [49] (Fig. 7). The inflection point may indicate a transition between different charging mechanisms and therefore would reinforce the status of equation (11) as estimating the limit of applicability of the simple thermodynamic model of a sessile drop [2, 17]. It does not predict contact angle behaviour beyond this point and these results may be useful in developing further the theory of electrowetting saturation.

A key difference between solid–liquid–vapour and solid–liquid–liquid systems is the much higher density and viscosity of the ambient fluid (hexadecane in our case). In early research on electrowetting in solid–liquid–liquid systems [6] it was already suggested that an alkane film is trapped between the droplet and the solid surface. Since then further evidence for that has been accumulated [6, 14, 36]. Staicu and Mugele [36] found a silicone oil film entrapped under the aqueous NaCl droplet on Teflon AF1600 surface. The oil film was initially thick ($\sim$400 nm), then became unstable and broke into small droplets (the same effect has been used to enhance the performance of electrowetting display pixels [11]). In their experiments, voltage was applied before the droplet had touched the insulated electrode and the viscosity ratio was very high ($m = \mu_{oil}/\mu_{water} \approx 100$). We therefore suspect that the oil film was hydrodynamically entrained, i.e., was transient in character. We estimate this was not the case in our experiments because of the specifically designed measurement protocol.

Quilliet and Berge [32] estimated theoretically that an equilibrium oil film may be present due to van der Waals interactions. This seems unlikely as, for instance, the refractive indices of bmim.BF$_4$ (1.421) and hexadecane (1.434) are very similar and therefore the Hamaker constants for the two materials should not be significantly different [78]. Thus we expect only weak van der Waals forces to operate. Electrostatic repulsion, as understood in traditional colloidal terms, should be screened very effectively by the large ionic strength of the ionic liquid [47]. An oil film would be even more unstable under electrowetting conditions [32, 79]. Recently strong structural forces with oscillatory decaying character have been reported in confined ionic liquids [80–83] and these may prove to be important. In view of these developments, the zero-interfacial-tension hypothesis should be reconsidered in relation to the Frumkin–Derjaguin model [84, 85] (the model in which the liquid meniscus contacts a thin wetting film rather than the non-wetted solid–fluid interface). Nonetheless, in the next section we provide a consistent interpretation of the dynamics of electrowetting within a macroscopic framework which does not consider thin wetting films.

Contact angle saturation in electrowetting is highly undesirable as it limits its usefulness in actuation applications. It is therefore significant that, when using AC voltage, saturation is delayed and very low contact angles are achievable— Fig. 7 and Fig. 11. We have previously seen the same effect in solid–liquid–vapour systems [2]. This superior performance of AC voltage in electrowetting has been explained by Hong *et al.* [86]. They modelled numerically the electromagnetic field inside a conductive droplet sitting on an insulator in air. When AC voltage (frequency 1–16 kHz) was used to achieve a contact angle change identical to that obtained with DC voltage, the local electric field near the contact line was weaker and did not trigger contact angle saturation. Regardless of the physical details, solid-ionic liquid-alkane systems perform very well in electrowetting and can be integrated in suitable applications.

D.3. Dynamics of Electrowetting

There is a continuing interest in electrowetting as a possible mechanism for actuation in microfluidics because it is electrically controlled, reversible and very fast. Under the influence of an applied DC voltage, the ionic liquid droplet spreads so that its base area, A (the area of the insulator-ionic liquid contact) expands exponentially. There is no generally accepted description of the evolution of the base area of a spreading droplet, but Lavi and Marmur [87], after reviewing extensively experimental results, proposed the following empirical equation:

$$A = A_0\{1 - \exp[-(t/\tau)^n]\}. \tag{12}$$

This equation has the advantage of not being related to a specific model of the dynamic contact angle. In practice, the exponent n is often close to unity and an exponential growth of the contact area has been recorded for a variety of systems: the spreading of a silicone oil in air on a commercial polyurethane paint [88], the spreading of liquid molybdenum on a glass surface in an argon-hydrogen atmosphere [89], and the dewetting of a hydrophobic solid surface when contacted by an air bubble [90, 91], to name a few. We did observe an exponential increase and decrease of the base area for all ionic liquids studied (Fig. 12 and Fig. 13) and this suggests that there is nothing exceptional in the spreading behaviour of ionic liquids under electrowetting conditions.

The spreading of liquids is directly related to their bulk viscosity because the liquid flows during spreading [85]. In electrowetting experiments, the droplet of ionic liquid is subjected to a driving force which takes it from a spherical cap (the influence of gravity is insignificant) with a small contact area (high contact angle) to a spherical cap with a large contact area (small contact angle). Since the contact angles at zero voltage and at saturation are rather similar for all the ionic liquids studied (Fig. 6, Fig. 8, Fig. 9) the following estimate of the average viscous stress, τ_{xy}, can be made:

$$\tau_{xy} = \mu\frac{\partial u_x}{\partial y} \cong \mu\frac{u_x}{x} \cong \mu\frac{x}{\tau x} \cong \frac{\mu}{\tau} \cong const. \tag{13}$$

In this equation Newton's law is used to relate the bulk viscosity, μ, with a characteristic spreading time, τ. If the stress τ_{xy} is approximately constant, a linear relation is predicted and, indeed, this was observed in our experiments—Fig. 15. A similarly strong correlation between spreading time and viscosity was found for the bmim.BF$_4$-water mixtures but it was non-linear probably due to the more complex hydrodynamic behaviour of less viscous droplets [49].

There is however, a distinct difference between the characteristic time during spreading and retraction. Receding is definitely slower than advancing and we speculate that the arrangement of the bulky ions of the ionic liquids at the solid–liquid

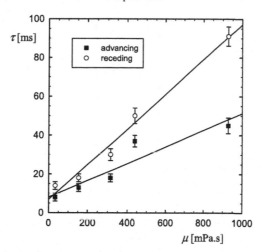

Figure 15. Characteristic spreading (■) and retraction (○) time, τ, as a function of the viscosity, μ, of the Rmim.BF$_4$ ionic liquid. Adapted with permission from Paneru *et al.* [51]. Copyright 2011 VSP.

interface, achieved under a given voltage, is easier to establish than to destroy. This was confirmed in experiments with a glycerol-water mixture of a viscosity 160 mPa s (similar to that of bmim.BF$_4$). The characteristic time was very similar ($\tau \approx 20$ ms) and identical during spreading and retraction. It should be noted that the behaviour of ionic liquids near a charged wall is non-trivial [76, 77] and rather poorly understood.

The above discussion suggests that spreading dynamics is dominated by viscous dissipation. We now consider the influence of the contact line friction. The dependence, usually considered in wetting dynamics, is the relation between dynamic contact angle, θ, and speed of the contact line, u. The instantaneous dynamic contact angle was calculated from the base area, A, and the fixed droplet volume assuming the droplet shape was spherical. The speed of the contact line was calculated from the derivative of the base area: $u = dx/dt = [2(\pi A)^{1/2}]^{-1}dA/dt$. The results obtained for a bmim.BF$_4$ droplet spreading under constant DC voltage, exceeding the saturation voltage, are presented in Fig. 16.

The format of the graphs is chosen so that a linear dependence should be seen if equations (6) and (8) are obeyed. In line with previous work [58, 91], we find that each of the models fits only a portion of the velocity dependence. Up to a speed of about 0.04 m/s, the dynamic contact angle is well described by the hydrodynamic description. At speeds higher than about 0.01 m/s, the molecular-kinetic description is more adequate. The slope of the hydrodynamic line is about four times smaller than the one estimated through equation (6) but this level of discrepancy is rather common [52]. From a hydrodynamic point of view, the viscosity of both advancing and receding liquids should be taken into account [54]. However, bmim.BF$_4$ is quite viscous and the viscosity ratio, m ($=\mu_{HD}/\mu_{IL} = 0.02$) is in effect the same as for a water droplet spreading in air ($m = \mu_{air}/\mu_{water} = 0.022$). The molecular

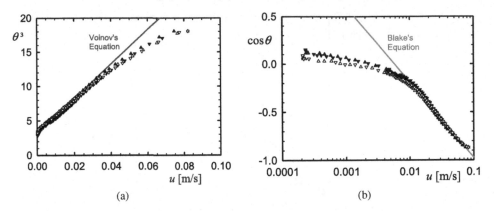

Figure 16. Dynamic contact angle, θ, *vs.* contact line speed, u (DC voltage: $\triangle$ 150 V; $\blacktriangle$ −150 V; $\triangledown$ 200 V; $\blacktriangledown$ −200 V): (a) hydrodynamic approach, the line is the best fit of Voinov's equation (6); (b) molecular-kinetic approach, the line is the best fit of the exponential approximation of Blake's equation (8). Adapted with permission from Paneru *et al.* [50]. Copyright 2010 American Chemical Society.

parameters derived from Fig. 16b (size of the average molecular jump $\lambda = 1.1$ nm and jump frequency at a static contact line $k_0 = 70$ MHz) are reasonable and in line with previous reports [52, 58]. Thus the parameters obtained for both equations are physically plausible.

The wetting dynamics, shown in Fig. 16, agrees with the concept given by Brochard-Wyart and de Gennes [59], which considers both viscous, D_{HD}, and molecular dissipation, D_{MK}:

$$\frac{D_{HD} + D_{MK}}{\mu u^2} = \frac{const}{\theta} \ln \frac{L}{l} + \exp\left(\frac{\lambda^2 W_A}{k_B T}\right). \tag{14}$$

The hydrodynamic term includes the dynamic contact angle, θ, and as the liquid wedge becomes thinner during spreading, θ decreases, and the importance of the viscous term increases [55]. At the same time, molecular dissipation (the second term; we have used the hypothesis [92] $\Delta G = W_A \lambda^2$) is not affected by the contact angle [58, 92] and thus, whenever θ is small, viscous dissipation dominates the total dissipation. In Fig. 16, viscous dissipation is more significant at low speeds (later stages of the spreading) when the contact angle is relatively small. We have used Voinov's equation (6) which is a better approximation than the one used to obtain the hydrodynamic term in equation (14). During the initial stages of spreading, the dynamic contact angle is rather large and, although contact line speed is high, molecular dissipation prevails.

Recent work by Fetzer and Ralston [91] on the receding contact line formed when an emerging bubble encounters a hydrophobic solid surface provides another example following the above scheme. In that case, hydrodynamic dissipation was predominant at high speed, while molecular dissipation came into play when the contact line moved relatively slowly. In qualitative agreement with equation (14), viscous dissipation dominated at high speeds because dynamic contact angles (as measured through the water) were small. As contact line speed decreased, during the dewetting of the hydrophobic substrate, θ steadily increased and molecular dissipation became dominant.

E. Conclusion

Droplets of ionic liquids (Rmim.BF$_4$ and bmim.X) were immersed in an immiscible liquid (n-hexadecane) and electrowet on a flat electrode insulated with a microscopic layer of an amorphous fluoropolymer (Teflon AF1600). The static contact angles at zero voltage are very large ($\sim$150°) and decrease when voltage is applied according to the Young–Lippmann equation. The lowest contact angles achievable through electrowetting were 48° (with DC voltage) and 15° (with AC voltage). The reversibility of the electrowetting curves was excellent and contact angle hysteresis was very small ($\leqslant$2°). AC electrowetting is very effective and robust, and thus offers significant advantages in microfluidic devices. The saturation contact angle could not be predicted with the zero-interfacial tension theory most probably because of the existence of an alkane film trapped between the droplet and the insulator. Experiments with bmim.BF$_4$-water mixtures of various compositions showed that electrowetting with dilute and concentrated electrolytes proceeds macroscopically in a similar fashion.

Electrowetting was very fast (initial contact line speed reached 0.08 m/s) during spreading and retraction. The base area of the droplet (at a given DC voltage) increased exponentially during advancing (exponential saturation) and decreased exponentially during receding (exponential decay, after the voltage was switched off). A characteristic time was related to the viscosity of the ionic liquid and was shorter during spreading (wetting) and longer during retraction (dewetting). The dynamics of wetting was examined by comparing the dependence of the dynamic contact angle on the speed of the contact line with the hydrodynamic model (Voinov's equation) and the molecular-kinetic model (Blake's equation). Viscous dissipation dominated at small dynamic contact angles and molecular dissipation prevailed at large angles in agreement with the scheme outlined by Brochard-Wyart and de Gennes.

F. Acknowledgements

The experimental results discussed here were obtained by Mani Paneru during his PhD candidature. Financial support from the Australian Research Council is gratefully acknowledged (Special Research Centre Scheme and DP110103391). RS acknowledges the support obtained from the Department of Further Education, Employment and Training (Government of South Australia) and the University of South Australia (RLDP 2009-10 and TRGGS 2009-10). This work was also supported by the Department of Innovation, Industry, Science and Research (Australian Government) through the Australia-India Strategic Research Fund (ST020050).

G. References

1. Welters, W. J. J. and Fokkink, L. G. J., Langmuir, 14 (1998) 1535.
2. Quinn, A., Sedev, R. and Ralston, J., J. Phys. Chem. B, 109 (2005) 6268.
3. Mugele, F. and Baret, J.-C., J. Phys.: Condens. Matter, 17 (2005) R705.
4. Tabeling, P., Introduction to Microfluidics. Oxford University Press, Oxford, 2005.
5. Jackel, J. L., Hackwood, S., Veselka, J. J. and Beni, G., Applied Optics, 22 (1983) 1765.
6. Berge, B. and Peseux, J., European Physical Journal E: Soft Matter, 3 (2000) 159.
7. Satoh, W., Yokomaku, H., Hosono, H., Ohnishi, N. and Suzuki, H., J. Appl. Phys., 103 (2008) 034903/1.
8. Khare, K., Herminghaus, S., Baret, J.-C., Law, B. M., Brinkmann, M. and Seemann, R., Langmuir, 23 (2007) 12997.
9. Hayes, R. A. and Feenstra, B. J., Nature, 425 (2003) 383.
10. Hayes, R. A., Feenstra, B. J., Camps, I. G. J., Hage, L. M., Roques-Carmes, T., Schlangen, L. J. M., Franklin, A. R. and Valdes, A. F., Digest of Technical Papers—Society for Information Display International Symposium, 35 (2004) 1412.
11. Sun, B. and Heikenfeld, J., Journal of Micromechanics and Microengineering, 18 (2008) 025027/1.
12. Nisar, A., Afzulpurkar, N., Mahaisavariya, B. and Tuantranont, A., Sens. Actuators B, B130 (2008) 917.
13. Chen, L., Lee, S., Choo, J. and Lee, E. K., Journal of Micromechanics and Microengineering, 18 (2008) 013001/1.
14. Fair, R. B., Microfluid. Nanofluid., 3 (2007) 245.
15. Adamson, A. W. and Gast, A. P., Physical Chemistry of Surfaces, 6th ed. Wiley, New York, 1997.
16. de Gennes, P. G., Brochard-Wyard, F. and Quéré, D., Capillarity and Wetting Phenomena: Drops, Bubbles, Pearls, Waves. Springer, New York, 2004.
17. Sedev, R., Eur. Phys. J. Spec. Top. (2011) in press.
18. Quinn, A., Sedev, R. and Ralston, J., J. Phys. Chem. B, 107 (2003) 1163.
19. Moon, H., Cho, S. K., Garrell, R. L. and Kim, C.-J., J. Appl. Phys., 92 (2002) 4080.
20. Li, Y., Parkes, W., Haworth, L. I., Ross, A. W. S., Stevenson, J., T. M. and Walton, A. J., Journal of Microelectromechanical Systems, 17 (2008) 1481.
21. Lin, Y.-Y., Evans, R. D., Welch, E., Hsu, B.-N., Madison, A. C. and Fair, R. B., Sensors and Actuators, B: Chemical, B150 (2010) 465.
22. Berry, S., Kedzierski, J. and Abedian, B., J. Colloid Interface Sci., 303 (2006) 517.
23. Kedzierski, J. and Berry, S., Langmuir, 22 (2006) 5690.
24. Berry, S., Kedzierski, J. and Abedian, B., Langmuir, 23 (2007) 12429.
25. Verheijen, H. J. J. and Prins, M. W. J., Langmuir, 15 (1999) 6616.
26. Vallet, M., Vallade, M. and Berge, B., European Physical Journal B: Condensed Matter Physics, 11 (1999) 583.
27. Seyrat, E. and Hayes, R. A., J. Appl. Phys., 90 (2001) 1383.
28. Shapiro, B., Moon, H., Garrell, R. L. and Kim, C.-J., J. Appl. Phys., 93 (2003) 5794.
29. Papathanasiou, A. G. and Boudouvis, A. G., Appl. Phys. Lett., 86 (2005) 164102/1.
30. Papathanasiou, A. G., Papaioannou, A. T. and Boudouvis, A. G., J. Appl. Phys., 103 (2008) 034901/1.
31. Janocha, B., Bauser, H., Oehr, C., Brunner, H. and Goepel, W., Langmuir, 16 (2000) 3349.
32. Quilliet, C. and Berge, B., Europhysics Letters, 60 (2002) 99.
33. Kuo, J. S., Spicar-Mihalic, P., Rodriguez, I. and Chiu, D. T., Langmuir, 19 (2003) 250.

34. Srinivasan, V., Pamula, V. K. and Fair, R. B., Lab on a Chip, 4 (2004) 310.
35. Antelmi, D. A., Connor, J. N. and Horn, R. G., J. Phys. Chem. B, 108 (2004) 1030.
36. Staicu, A. and Mugele, F., Phys. Rev. Lett., 97 (2006) 167801/1.
37. Seddon, K. R. and Rogers, R. D., Ionic Liquids: Industrial Applications for Green Chemistry. American Chemical Society, Washington, D.C., 2002.
38. Seddon, K. R. and Rogers, R. D., Ionic Liquids III: Fundamentals, Progress, Challenges, and Opportunities. ACS, Washington, D.C., 2005.
39. Rogers, R. D. and Seddon, K. R., Science, 302 (2003) 792.
40. Rogers, R. D. and Seddon, K. R., Ionic Liquids as Green Solvents. ACS, Washington, D.C., 2003.
41. Lockett, V., Sedev, R., Bassell, C. and Ralston, J., Phys. Chem. Chem. Phys., 10 (2008) 1330.
42. Lockett, V., Sedev, R., Harmer, S., Ralston, J., Horne, M. and Rodopoulos, T., Phys. Chem. Chem. Phys., 12 (2010) 13816.
43. Plechkova, N. V. and Seddon, K. R., Chemical Society Reviews, 37 (2008) 123.
44. Sedev, R., Curr. Opin. Colloid Interface Sci. (2011) doi:10.1016/j.cocis.2011.01.011.
45. Pham, T. P. T., Cho, C.-W. and Yun, Y.-S., Water Research, 44 (2010) 352.
46. MacFarlane, D. R., Forsyth, M., Izgorodina, E. I., Abbott, A. P., Annat, G. and Fraser, K., Phys. Chem. Chem. Phys., 11 (2009) 4962.
47. Ohno, H., Electrochemical Aspects of Ionic Liquids. Wiley-Interscience, Hoboken, N.J., 2005.
48. Millefiorini, S., Tkaczyk, A. H., Sedev, R., Efthimiadis, J. and Ralston, J., J. Am. Chem. Soc., 128 (2006) 3098.
49. Paneru, M., Priest, C., Sedev, R. and Ralston, J., J. Phys. Chem. C, 114 (2010) 8383.
50. Paneru, M., Priest, C., Sedev, R. and Ralston, J., J. Am. Chem. Soc., 132 (2010) 8301.
51. Paneru, M., Priest, C., Ralston, J. and Sedev, R., Journal of Adhesion Science and Technology, (2011) in press.
52. Ralston, J., Popescu, M. and Sedev, R., Ann. Rev. Mater. Res., 38 (2008) 23.
53. Voinov, O. V., Fluid Dyn., 11 (1976) 714.
54. Cox, R. G., Journal of Fluid Mechanics, 168 (1986) 169.
55. de Gennes, P. G., Colloid and Polymer Science, 264 (1986) 463.
56. Blake, T. D. and Haynes, J. M., J. Colloid Interface Sci., 30 (1969) 421.
57. Blake, T. D., Surfactant Science Series, 49 (1993) 251.
58. Blake, T. D., J. Colloid Interface Sci., 299 (2006) 1.
59. Brochard-Wyart, F. and de Gennes, P. G., Adv. Colloid Interface Sci., 39 (1992) 1.
60. Pollack, M. G., Fair, R. B. and Shenderov, A. D., Appl. Phys. Lett., 77 (2000) 1725.
61. Zimmermann, R., Dukhin, S. and Werner, C., J. Phys. Chem. B, 105 (2001) 8544.
62. Creux, P., Lachaise, J., Graciaa, A., Beattie, J. K. and Djerdjev, A. M., J. Phys. Chem. B, 113 (2009) 14146.
63. Sacher, E. and Klemberg-Sapieha, J. E., Journal of Vacuum Science & Technology, A: Vacuum, Surfaces, and Films, 15 (1997) 2143.
64. Zisman, W. A., Adv. Chem. Ser., 43 (1964) 1.
65. Werner, C., Konig, U., Augsburg, A., Arnhold, C., Korber, H., Zimmermann, R. and Jacobasch, H. J., Colloids and Surfaces, A: Physicochemical and Engineering Aspects, 159 (1999) 519.
66. Briggs, D., Surface Analysis of Polymers by XPS and Static SIMS. Cambridge University Press, Cambridge, U.K., New York, 1998.
67. Bain, C. D. and Whitesides, G. M., J. Am. Chem. Soc., 110 (1988) 5897.
68. Lockett, V., Sedev, R., Ralston, J., Horne, M. and Rodopoulos, T., J. Phys. Chem. C, 112 (2008) 7486.

69. Rasband, W., in: http://rsbweb.nih.gov/ij/index.html, NIH, Bethesda, 2009.
70. van Oss, C. J., Interfacial Forces in Aqueous Media, 2nd ed. Taylor & Francis, Boca Raton, FL, 2006.
71. Quéré, D., Annu. Rev. Mater. Res., 38 (2008) 71.
72. Maillard, M., Legrand, J. and Berge, B., Langmuir, 25 (2009) 6162.
73. Van Wagenen, R. A., Coleman, D. L., King, R. N., Triolo, P., Brostrom, L., Smith, L. M., Gregonis, D. E. and Andrade, J. D., J. Colloid Interface Sci., 84 (1981) 155.
74. Fowkes, F. M. and Hielscher, F. H., Organic Coatings and Plastics Chemistry, 42 (1980) 169.
75. Cross, J. A., Electrostatics: Principles, Problems and Applications. Adam Hilger, Bristol, 1987.
76. Kornyshev, A. A., J. Phys. Chem. B, 111 (2007) 5545.
77. Bazant, M. Z., Storey, B. D. and Kornyshev, A. A., Phys. Rev. Lett., 106 (2011) 046102/1.
78. Israelachvili, J. N., Intermolecular and Surface Forces, 2nd ed. Academic Press, London, 1991.
79. Herminghaus, S., Phys. Rev. Lett., 83 (1999) 2359.
80. Atkin, R. and Warr, G. G., J. Phys. Chem. C, 111 (2007) 5162.
81. Hayes, R., El Abedin, S. Z. and Atkin, R., J. Phys. Chem. B, 113 (2009) 7049.
82. Wakeham, D., Hayes, R., Warr, G. G. and Atkin, R., J. Phys. Chem. B, 113 (2009) 5961.
83. Ueno, K., Kasuya, M., Watanabe, M., Mizukami, M. and Kurihara, K., Phys. Chem. Chem. Phys., 12 (2010) 4066.
84. Churaev, N. V. and Sobolev, V. D., Adv. Colloid Interface Sci., 61 (1995) 1.
85. Starov, V. M., Velarde, M. G. and Radke, C. J., Wetting and Spreading Dynamics. CRC, Boca Raton, FL, 2007.
86. Hong, J. S., Ko, S. H., Kang, K. H. and Kang, I. S., Microfluid. Nanofluid., 5 (2008) 263.
87. Lavi, B. and Marmur, A., Colloids and Surfaces, A: Physicochemical and Engineering Aspects, 250 (2004) 409.
88. Dodge, F. T., J. Colloid Interface Sci., 121 (1988) 154.
89. Saiz, E. and Tomsia, A. P., Proc. 3rd Int. Brazing Soldering Conf., 203 (2006).
90. Phan, C. M., Nguyen, A. V. and Evans, G. M., Langmuir, 19 (2003) 6796.
91. Fetzer, R. and Ralston, J., J. Phys. Chem. C, 113 (2009) 8888.
92. Blake, T. D. and De Coninck, J., Adv. Colloid Interface Sci., 96 (2002) 21.

Single Drop Impacts of Complex Fluids: A Review

V. Bertola [a] and M. Marengo [b]

[a] University of Liverpool, School of Engineering, Brownlow Hill, Liverpool L69 3GH, UK
[b] Università degli Studi di Bergamo, Dipartimento di ingegneria Industriale Viale Marconi 5, 24044 Dalmine (BG), Italy

Abstract

The present review deals with the impact of drops of fluids with complex microstructure (non-Newtonian) on solid surfaces, with the aim to highlight a number of relevant differences with respect to the impact of drops of Newtonian liquids. Attention is focused on shear-thinning fluids, viscoplastic (or yield-stress) fluids, and viscoelastic fluids, in particular dilute polymer solutions.

Contents

A. Introduction 267
B. Impact of power-law drops]Power-Law Drops 269
C. Impact of Yield-Stress Drops 272
D. Impact of Viscoelastic Drops 277
    D.1. Impact on Homo-Thermal, Hydrophobic Surfaces 277
    D.2. Impact on heated surfaces]Heated Surfaces 283
Appendix. constitutive models]Constitutive Models for Non-Newtonian Fluids 289
    A1.1. power-law fluids]Power-Law Fluids 289
    A1.2. yield-stress fluids]Yield-Stress Fluids 290
    A1.3. Viscoelastic Fluids 292
E. References 294

A. Introduction

Drop impact phenomena have attracted the interests of scientists and engineers for more than one century [1–3], both because of their complex physical mechanisms and because they have a very large number of applications of practical relevance. Examples are inkjet printing, internal combustion engines, aerosol drug delivery, distribution of agrochemicals, and of course all spray applications, which are ubiquitous in industrial as well as in domestic processes, from painting or cleaning surfaces to spray cooling and firefighting.

Drops and Bubbles in Contact with Solid Surfaces
© Koninklijke Brill NV, Leiden, 2012

The impact morphology of liquid droplets onto solid, dry surfaces, is well known (see e.g., [4, 5]). Upon impact, the liquid spreads on the surface taking the form of a disk; for low impact velocity, the disk thickness is approximately uniform, while for higher impact velocities the disk is composed of a thin central part (often called 'lamella') surrounded by a circular rim. This initial spreading stage is typically very fast (≈5 ms). After the drop has reached maximum spreading, two qualitatively different outcomes are possible. If the initial kinetic energy exceeds a threshold value capillary forces are insufficient to maintain the integrity of the drop, which disintegrates into smaller satellite droplets jetting out of its outermost perimeter (splashing). If splashing does not occur, the drop is allowed to retract under the action of capillary forces, which tend to minimize the contact with the surface; in some cases, retraction is so fast that the liquid rises in the middle forming a Worthington jet, which may subsequently result in the complete rebound of the drop from the surface.

Impacts onto smooth and chemically homogeneous surfaces, for low or moderate impact kinetic energy, are controlled by three key factors: inertia, viscous dissipation and interfacial energy [6, 7]. During the initial stages of impact with the surface, the vertical inertia of the falling drop is converted into the horizontal motion of the fluid, and as the drop spreads kinetic energy is partly stored as surface energy. This balance is characterized by the Weber number, $We = \rho v_i^2 D_0/\sigma$, where ρ and σ are the fluid density and surface tension, respectively, D_0 is the equilibrium drop diameter, and v_i the normal impact velocity. As the fluid spreads across the surface, the kinetic energy of the fluid is partly dissipated by viscous forces in the fluid, which is described using the Reynolds number, $Re = \rho v_i D_0/\mu$, where μ is the fluid viscosity. This is sometimes used in combination with the Weber number to yield the Ohnesorge number, $Oh = We^{0.5}/Re$. Finally, the retraction stage is governed by the balance between interfacial energy and viscous dissipation, expressed by the Capillary number, $Ca = \mu v_r/\sigma$, where v_r is the retraction velocity.

Whilst there exists a significant volume of literature about single drop impacts of simple (Newtonian) fluids, the number of works about fluids with complex microstructure (polymer melts or solutions, gels, pastes, foams and emulsions, etc.) is comparatively very small. However, these fluids are frequently used in common applications, such as painting, food processing, and many others. Moreover, with a better understanding of the microscopic structure of complex liquids, industries have realized that working fluids can be tailored specifically to optimize existing industrial processes, by altering their formulation (e.g., by means of chemical additives) in such a way as to change one or more physical properties. An example of industrial optimization is the use of polymer additives in agrochemical formulations, which improves the application efficiency of agrochemical sprays and reduces the environmental impact from ground contamination [8].

The microscopic structure of fluids is described from the macroscopic point of view of continuum mechanics by constitutive equations, which express the relationship between the stress tensor and the velocity gradient. Simple liquids, such as water, are generally characterized by a Newtonian constitutive equation, where the stress tensor is a linear function of the velocity gradient, whereas in complex (or non-Newtonian) fluids the stress tensor is a generic function of the velocity gradient and of its derivatives [9, 10].

Thus, the study of non-Newtonian drops requires a completely different modeling approach, as well as different sets of experimental data, depending on the form of the constitutive equation of the fluid. In particular, due to the large spatial and temporal gradients observed during drop impacts, small changes in the constitutive equation result into large macroscopic effects (for example, most Newtonian drop impact models become inaccurate for large fluid viscosities [11]).

The present review describes some recent developments about the study of drop impact of the most common types of non-Newtonian fluids: power-law fluids, yield-stress (or viscoplastic) fluids, and viscoelastic fluids. In particular Section B is about power-law fluids, with focus on shear-thinning fluids; Section C presents recent works about viscoplastic drop impacts. Finally, Section D reviews the literature about viscoelastic drop impacts onto both homo-thermal and heated surfaces. An overview of constitutive equations and of the main features of complex liquids is presented in an extended Appendix. We advise readers who are not familiar with complex fluids to read the Appendix before the following chapters.

B. Impact of Power-Law Drops

Existing research into shear-thinning fluid drops focuses mainly on detachment dynamics from capillary nozzles [12] and the spreading behavior of sessile drops on solid surfaces [13–15] rather than on impacting drops. One of the difficulties arising in the study of shear-thinning drop impacts is that during any dynamic process the fluid viscosity will vary both spatially and temporally as a function of the local shear-rate. This additional complexity increases the difficulty in establishing relationships between the macroscopic drop dynamic behavior and the underlying viscometric properties of the fluid. Moreover, common parameters used to characterize drop behavior, such as the Reynolds, the Ohnesorge, and the Capillary numbers, cannot be defined adequately due to the variation of the viscosity term.

Recently, an experimental study about the impact of shear-thinning drops [16, 17] was carried out using a set of model shear-thinning fluids consisting of aqueous solutions of Xanthan gum at mass concentrations of polymer ranging from 0.125% to 0.1%, whose experimental flow curves were fitted using the Ostwald–De Waele equation to determine the values of the consistency coefficient, K, and of the power-law index, n (see Appendix). The effect of surface wettability was taken into account by comparing impacts on a hydrophilic surface (glass) with impacts on a hydrophobic surface (Parafilm, equilibrium contact angle: $\approx 105°$). This study shows that although the impact morphology is qualitatively similar to that of Newtonian drops of comparable average viscosity (Fig. 1), it is not immediately obvious how the consistency coefficient K and the power law index n independently

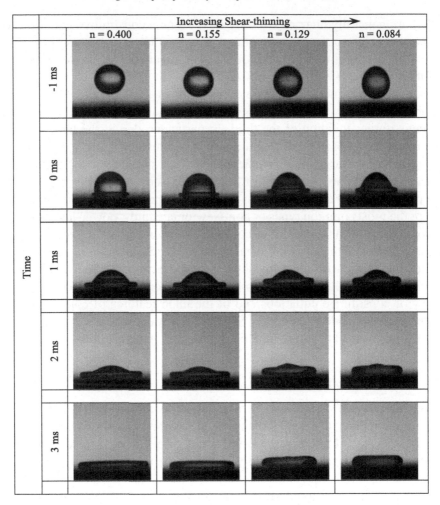

Figure 1. Impact morphology of shear-thinning drops (equilibrium diameter: 3.4 mm), on a parafilm substrate from a fall height of 50 mm [16].

influence the impacting drop dynamics. From Fig. 1 it seems that an increase of the power law index increases the maximum spreading diameter, and this would be against the fact that, on the contrary, for lower power law index, the viscosity will be lower for high velocity gradients (at the beginning of the impact for example) and, since the viscous dissipation is related to maximum spreading, this will cause less dissipation and then a greater spreading diameter. Equivalent impacts of shear-thinning drops on glass and parafilm substrates are very similar during the inertial expansion phase, as shown in Fig. 2, which displays the dimensionless drop diameter, β, with respect to time for different impact Weber numbers.

Significant differences can only be observed after maximum inertial spread: drops impacting on the hydrophobic surface exhibit a clear retraction stage, which

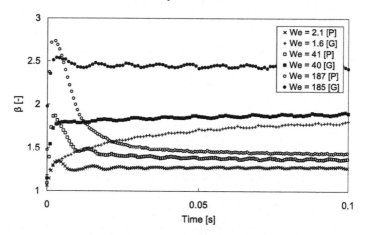

Figure 2. Dimensionless drop diameter, β, plotted against time for shear-thinning drops ($K = 5.064$ Pas^n, $n = 0.084$) impacting on glass and parafilm-M surfaces [16].

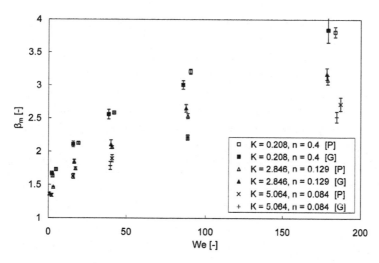

Figure 3. Maximum spreading diameter of shear-thinning drops impacting on substrates of different wettability (glass and parafilm) [16].

is barely noticeable for impacts on the glass surface, where the slow capillary driven spreading continues directly after the fast spreading of the inertial expansion phase.

In particular, the maximum spreading diameter, reported with respect to the Weber number in Fig. 3, can be considered the same for the two surfaces, within experimental error.

Increasing the mass fraction of Xanthan gum increases the consistency coefficient K, however it also decreases the power law index n. In other words, fluids become thicker but at the same more shear-thinning. Again, fluids exhibiting large degrees of shear-thinning (i.e., with a value of the power-law exponent close to

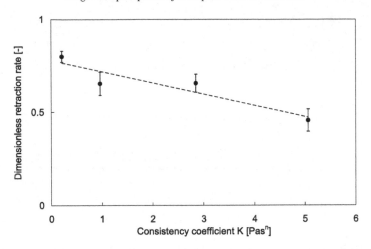

Figure 4. Average dimensionless retraction rate of shear-thinning drops as a function of the consistency coefficient [16].

zero) would be expected to exhibit lower local viscosities during inertial expansion and therefore an increased maximum spreading diameter.

Measurements shown in Fig. 3 however show the opposite trend: the maximum spreading diameter is higher for less shear-thinning fluids. Together with Fig. 1 this suggests that whilst shear-thinning effects affect the impact behaviour, the influence of the consistency coefficient K, which gives an indication of the maximum viscosity of the fluid for $\dot{\gamma} \to 0$ hence determines the average apparent viscosity during the process, appears to dominate.

Drop retraction on the hydrophobic substrate can be studied by measuring the retraction rate, defined as:

$$\dot{\chi} = \frac{v_r}{R_{max}} = -\max\left\{\frac{\dot{\beta}(t)}{\beta_m}\right\}. \tag{1}$$

When this quantity is nondimensionalized with the inertial timescale given by:

$$T_I = \sqrt{\frac{4}{3}\frac{\pi\rho R_0^3}{\sigma}} \tag{2}$$

one finds it decreases linearly with the consistency coefficient, as shown in Fig. 4. This trend is qualitatively similar to the retraction rate dependence on viscosity in Newtonian drops, and further confirms the dominant influence of the consistency coefficient with respect to the power-law exponent.

C. Impact of Yield-Stress Drops

Despite yield-stress fluids have been studied for about one century, and one of the most important spray applications, painting, makes use of a yield stress fluid (paint), the first investigation of yield-stress drops was published only recently [18].

This work studies the impact of a model viscoplastic fluid (vaseline) on a Plexiglas surface, for different impact velocities. The rheological behavior of the fluid was modeled using a Cross model, modified to include a yield-stress component. This was found to provide a good fit with viscometric measurements in the range $10^{-2} \leqslant \dot{\gamma} \leqslant 10^3 \text{ s}^{-1}$.

Upon impact at low velocities ($v_i = 0.67 \text{ ms}^{-1}$), drops decrease sharply in height, however inertial spreading is small with no lamella formation: drops appear to be gently deposited on the surface and behave like a deformable solid. A slow creeping flow is observed thereafter until drops assume a sessile state after approximately 5 minutes. At slightly larger velocities ($v_i = 0.85 \text{ ms}^{-1}$), drops retract after maximum inertial spreading and oscillations of the drop height are observed. After retraction, the diameter continues to spread at a near constant rate. At high impacts ($v_i = 2.3 \text{ ms}^{-1}$), recoil is inhibited with wetted area retractions of no greater than 5% of the maximum spreading. Again, creeping flow is observed, however this is small and does not affect the maximum drop diameter.

The variation in final drop shape with respect to the impact velocity, which becomes noticeable only above the value $v_i = 0.85 \text{ ms}^{-1}$, was characterized with respect to the Bingham number:

$$\text{Bm} = \frac{\tau_c D_0}{\mu v_i}, \tag{3}$$

where μ is defined as the zero shear rate viscosity μ_0; however, such definition of the viscosity term is not well posed because whilst the Bingham number characterizes the ratio of viscous to yield-stress forces, viscous dissipation only occurs during fluid motion therefore the μ_0 term is only valid at zero shear rate.

Because surface forces play an important role in all drop impact phenomena, it is interesting to observe what happens when the yield stress magnitude is comparable with the capillary (Laplace) pressure [16, 17]. This leads to the definition of a capillary regime and a viscoplastic regime, which can be characterized through the Bingham-capillary number [19]:

$$\hat{B} = \frac{\tau_c D_0}{\sigma}. \tag{4}$$

Whilst in the capillary regime the impact morphology is qualitatively similar to that of simple liquids, in the viscoplastic regime one can sometimes observe permanent deformations that do not disappear upon impact or under the action of surface forces.

For example, if drops are produced from a capillary nozzle, the prolate shape that creates during the fluid extrusion [20–22] remains partly visible after impact, as shown in Fig. 5, which displays the impact morphology of hairgel-water drops for different yield stress magnitudes. This phenomenon is also influenced by inertia, and becomes less and less pronounced at higher impact Weber numbers. The droplets symmetry can be improved significantly if the dispensing nozzle has a very

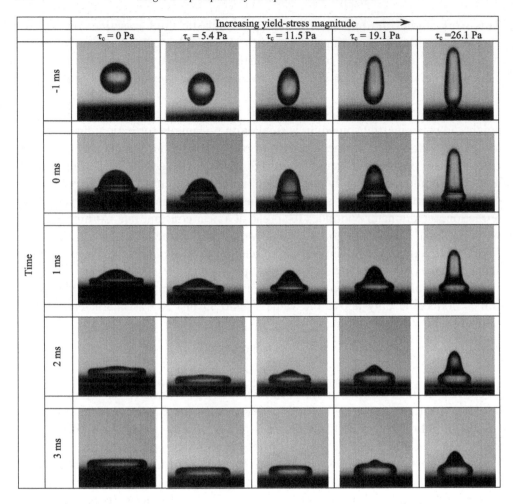

Figure 5. Impact morphology of viscoplastic drops (equilibrium diameter: ∼2.5 mm), on a parafilm substrate from a fall height of 50 mm [16].

small diameter [23], however for high yield stress magnitudes and low impact velocities drops still preserve the original form they have after detachment from the capillary.

Another interesting feature of viscoplastic drop impacts is the linear dependence of the maximum spreading diameter on the yield stress magnitude, displayed in Fig. 6, whereas in Newtonian drops the same quantity depends on viscosity according to a power law [6, 17, 24]. Similar to shear-thinning fluids, the influence of surface wettability on viscoplastic drop impacts is only noticeable after the end of the inertial expansion stage, as shown in Fig. 7, which compares drop diameters of viscoplastic drops impacting respectively on a hydrophobic Parafilm surface (equilibrium contact angle: ≈105°) and a hydrophilic glass surface, for different Weber numbers. In the viscoplastic regime ($\hat{B} > 1$), drops exhibit only small retrac-

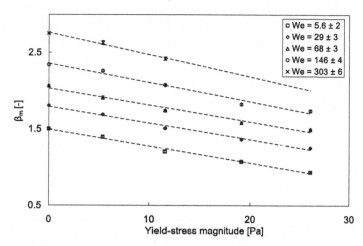

Figure 6. Influence of the yield-stress magnitude on the maximum spreading diameter of viscoplastic drops, for different impact Weber numbers [16].

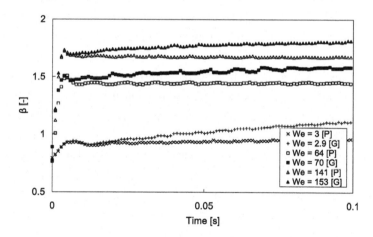

Figure 7. Dimensionless drop diameter, β, plotted against time for drops of a viscoplastic fluid ($K = 7.936$ Pas^n, $n = 0.3727$, $\tau_c = 26.1$ Pa) impacting from fall heights of 10, 50 and 200 mm on glass [G] (solid symbols) and parafilm-M [P] (open symbols) solid substrates [16].

tion phases similar to those observed for high viscosity Newtonian fluids, whereas impacts on glass show no significant retraction phase, and slow capillary-driven spreading follows directly on from the fast spreading of inertial expansion for low impact velocities. At higher Weber numbers, drop diameters remain nearly constant after maximum spreading [16, 17].

These results are substantially confirmed by a more detailed study of the effect of surface wettability and roughness on viscoplastic drop impacts [24], which compares two smooth substrates with distinct surface energies and three substrates with similar surface energy but different roughness. The same work also attempts at a quantification of the effects of apparent wall slip [19, 26] on drop impact, however without being conclusive since it was not possible to disentangle the effects of wall slip and surface wettability during experiments. It is speculated that at low inertia, where a gravitational subsidence is observed, the creeping movement amplitude is governed by interfacial effects rather than wall slip, while at high impact velocities wall slip effects become appreciable only in the last moments of the recoil, when shear rates become very low.

When the drop radius is much larger than the capillary length, $a = \sqrt{\sigma/\rho g}$, surface tension effects can be neglected in comparison with those of gravity; furthermore, large diameters also imply large Weber numbers, so that impacts are dominated by inertia and by the rheological properties of the fluid only. Such experimental conditions are explored in a recent work [27], which describes the impact of relatively large bits (characteristic sizes between 10 and 30 mm) of various viscoplastic fluids, with yield stress magnitudes ranging from 4 to 124 Pa, and capillary lengths of the order of a few millimeters. Although these fluids include many aqueous Carbopol dispersions, it must be observed that their yield stress magnitudes are significantly smaller than the values reported in the open literature for similar fluids [28].

By comparing impacts on a glass surface and on a super-hydrophobic surface (contact angle of nearly 180°), these experiments confirm that the maximum spreading diameter of viscoplastic drops is weakly dependent on the surface wettability, and smaller than the capillary limit as defined in Ref. [24]; unfortunately, the latter result can also be obtained with high-viscosity Newtonian fluids [11], so that it is not possible to establish whether the yield stress has an independent influence.

The most interesting finding of this work is the strong and rapid recoil, which may even be followed by a complete rebound, observed after the spreading phase of Carbopol drops impacting on the super-hydrophobic surface. Since both a recoil driven by surface tension and a purely elastic rebound (the flow threshold corresponds to a shear deformation of about 25%, whereas deformations during impact vary between 100% and 500%) must be ruled out, it is suggested that at such high velocity gradients (We ≈ 1400) Carbopol solutions may exhibit a viscoelastic behaviour: during the rapid spreading phase, flow is faster than the fluid relaxation time, resulting into giant elastic deformations on short time scales. This conjecture is supported by the comparison of experimental results with a minimal model of elasto-viscoplastic inertial spreading, where elasticity is tentatively accounted for by the storage modulus measured below the flow threshold (indeed, a very rough approximation). However, it appears that in order to obtain independent evidence in support of this picture, dynamic rheometric tests with characteristic frequency comparable with the inverse of the impact time scale are necessary.

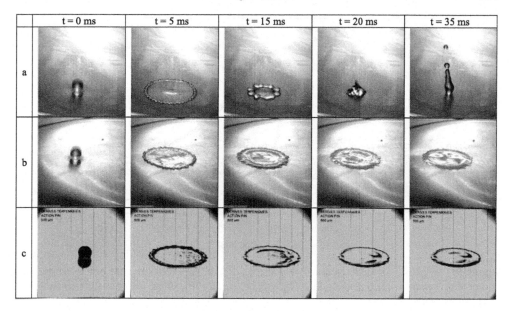

	t = 0 ms	t = 5 ms	t = 15 ms	t = 20 ms	t = 35 ms
a					
b					
c					

Figure 8. Impact of a liquid droplet ($D_0 = 2.5$ mm, We $= 150$) on a hydrophobic Parafilm surface: a drop of pure water (a) bounces off the surface, while both a dilute solution of polyethylene oxide (b) and a solution of a high-mobility surfactant can prevent drop rebound.

D. Impact of Viscoelastic Drops

D.1. Impact on Homo-Thermal, Hydrophobic Surfaces

The study of viscoelastic drops received considerable attention after the accidental discovery that small amounts (of the order of 100 ppm) of Polyethylene Oxide (PEO) can reduce the tendency of drops to rebound after impacting on hydrophobic surfaces, which can be exploited to control many spray applications [8, 29, 30].

A qualitatively similar result can be achieved, as shown in Fig. 8, by using certain surfactants hence changing the dynamic wetting properties of the liquid [31, 32]; however, the surface tension and shear viscosity of aqueous solutions of these flexible polymers in the dilute regime are known to be almost identical to those of the pure solvent. A similar independent study demonstrated that polymer additives also increase the critical Weber number for drop splashing [33]; moreover, drop impact control can be achieved by means self-assembling surfactant systems creating rod- or worm-like micelles [34].

Having ruled out the influence of capillarity and viscosity, it was suggested [31] that these effects are related to the elongational viscosity of the fluid, defined as the ratio of the first normal stress difference to the rate of elongation of the fluid, which for a polymer solution can be two or three orders of magnitude higher than that of water [35]. In particular, it was argued that the rapid deformation of drops upon impact can induce a coil-stretch transition [36] of the molecules in the bulk fluid, resulting into a transient increase of the elongational viscosity; the large energy

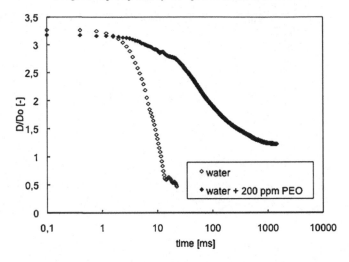

Figure 9. Comparison of the retraction dynamics of a drop of pure water and one of a dilute polymer solution after impact and spreading on a hydrophobic surface: the dilute polymer solution exhibits a much slower retraction rate.

dissipations caused by elongational viscosity leave nothing of the impact kinetic energy available to propel the drop off the surface [30].

This seemed to be confirmed by the retraction velocity of the drop after maximum spreading, which for polymer solutions is one or two order of magnitudes smaller than that measured for the pure solvent [29, 30], as shown in Fig. 9.

A careful experimental study aiming at establishing the independent influence of dynamic surface tension and elongational viscosity on drop impact [37] indeed demonstrated the existence of a correlation between the transient elongational viscosity, measured with an opposed jets rheometer, and the drop rebound. Figure 10 shows in fact that both the retraction velocity after maximum spreading and the maximum height reached by the drop decrease as the ratio between the elongational and the shear viscosity η_E/η (called Trouton ratio) of the fluid grows. For this reason, energy dissipation associated to elongational viscosity has been trusted to be the cause of the anti-rebound effect of polymer additives on drops impacting onto hydrophobic surfaces until recently [38]. However, an empirical correlation between two phenomena does not necessarily imply a cause-effect relationship: in fact, this scenario was not supported by the results of other independent investigations, which demonstrated that in the absence of direct contact between the liquid and the substrate the anti-rebound effect is no longer observed.

A study of drops impacting on small targets [39] suggests that polymer additives do not change the retraction velocity at all, while independent experiments on Leidenfrost drops (where the liquid is separated from the target surface by a thin vapor layer, as discussed more in detail in Section D.2) show that they cause only a slight reduction of this quantity [40]. Because in these experiments wetting effects are absent or negligible, one must conclude that the retraction velocity reduction ob-

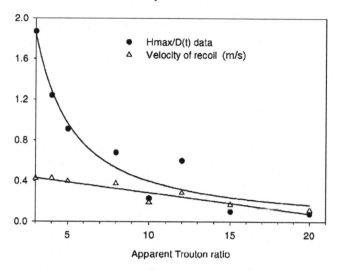

Figure 10. Effect of the fluid elasticity (Trouton ratio) on the retraction velocity and on the bouncing height of a polymer solution droplet [37].

served in drops containing flexible polymers impacting on solid surfaces is due to the drop-surface interaction rather than to an increased energy dissipation connected to the elongational viscosity of the fluid. Furthermore, measurements of the maximum bouncing height of Leidenfrost drops (indicative of the fraction of the initial kinetic energy which is not dissipated during impact) suggest that in some cases polymer additives indeed reduce instead of increasing the overall energy dissipation [40]. This result is consistent with recent direct numerical simulations showing a reduction in small-scale convective motions in dilute polymer solutions hence in bulk viscous dissipation [41]. Therefore, one must conclude that the real cause of this phenomenon is to be sought in the dynamic wetting behavior of dilute polymer solutions, and not in some bulk property of the fluid such as the elongational viscosity.

In order to explain the anti-rebound effect in terms of wetting, Bartolo *et al.* [42] suggest that it can be related to the normal stresses arising near the contact line during drop retraction. Using an equation that generalizes the lubrication theory for thin films, and accounting for capillarity and normal stresses in addition to shear stresses, they find that for large contact angles the retraction velocity, v_r, assumed to be identical to the average velocity on the drop height, is related to the normal stress coefficient, Ψ, and to the contact angle, θ, according to the following equation:

$$\sigma(\cos\theta - \cos\theta_{eq}) = \frac{4\Psi v_r^2}{l_m},\qquad(5)$$

where θ_{eq} is the equilibrium contact angle, and l_m the microscopic cut-off height of the drop in the neighborhood of the contact line. Equation (5) is found to be in good agreement with experimental data obtained from polyethylene oxide and polyacrylamide solutions with concentrations ranging between 200 and 2000 ppm.

Unfortunately, this approach is deeply flawed for several reasons. First of all, whilst the anti-rebound effect is observed for drops of dilute polymer solutions (i.e., when the polymer concentration is less than the overlap concentration), Eq. (5) has been validated with experimental data based on solutions in the semi-dilute regime whose properties, and in particular viscosity, are different from the dilute regime [35]. Thus, the slow retraction observed in these experiments can be simply due to a higher shear viscosity of the solution. Second, Eq. (5) is derived for large contact angles, but this is in open contradiction with the fact that it is used only in the case of small contact angles, which correspond to the experimental situation as also confirmed by more recent and accurate measurements [43].

It is not clear at all how the velocity gradients responsible for normal stresses have been measured or estimated: according to rheometric data, for the most dilute solution normal stresses are significant only for shear rates much greater than 10^3 s^{-1}, corresponding to a timescale <1 ms, whereas drop retraction takes times of the order of one second. The total confusion about basic kinematics is further confirmed when the microscopic lengths extracted from the fit of experimental data to Eq. (5) are compared with the lengths of the fully extended polymer chains: this suggests that polymer chains are thought to stretch in the vertical direction, whereas the main velocity component is parallel to the impact surface. Finally, if it was true that large normal stresses arise in the fluid near the contact line, they should work in the direction of increasing the contact angle with respect to the case with no normal stresses (Newtonian fluid), which is exactly the opposite of what can be observed during experiments [43].

Figure 11 shows the difference between the dynamic contact angles measured during retraction on a hydrophobic substrate for a drop of pure water and one of a 200 ppm PEO solution. The smaller dynamic contact angle observed during retraction of polymer solution drops indicates that the horizontal component of the driving force required in order to move the contact line is much larger than in the case of pure water, which is in contrast with the fact that the contact line velocity is one order of magnitude smaller. However, this cannot be explained by the modest reduction of surface tension ($\sim$0.2 mN/m) and the modest increase of shear viscosity ($\sim$0.23 mPa s). Thus, one must conclude that if polymer solution drops exhibit a smaller dynamic contact angle, there must be an additional dissipative force opposed to the force at the surface/air interface.

Direct experimental evidence to rule out the dissipation mechanism provided by bulk elongational viscosity is provided by recent particle velocimetry measurements inside impacting drops [44–46], showing that the local velocities measured at different times during expansion and retraction are similar for the drops of polymer solution and for those of pure water.

Velocity gradients in the fluid measured with respect to time during expansion (a parameter which determines whether a polymer undergoes a coil-stretch transition) are almost identical for the dilute PEO solutions and for water, suggesting similar amounts of dissipation throughout the spreading of the droplet.

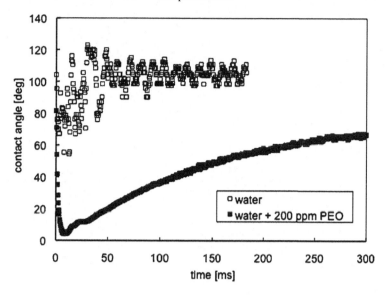

Figure 11. Dynamic contact angles observed during the retraction of drops of pure water and of a dilute polymer solution after impact on a hydrophobic surface.

During retraction, radial velocity gradients in both cases are small ($\ll 1,000 \text{ s}^{-1}$), and the fluid elements are in compression rather than extension, making the stretching of molecules in the drop interior unlikely.

A comparison of the fluid velocity in the bulk of the droplet during retraction with the velocity extracted from macroscopic observations of the contact line shows a dramatic difference between water and PEO drops. Figure 12 shows that the motion of the contact line for droplets of pure water is similar to that of the fluid inside the drop. By contrast, the motion of the contact line for PEO drops is one order of magnitude slower than that of the corresponding bulk velocity measurements, and further confirms that the difference between the behavior of the two fluids lies solely at the droplet edge.

The nature of the dissipative force arising on the contact line was revealed by direct visualization of fluorescent λ-DNA molecules [44, 45]: immediately after the transit of the receding contact line, DNA molecules can be observed on the impact surface in a stretched conformation, oriented in the direction perpendicular to the contact line. Thus, it was suggested that during drop retraction polymer molecules are stretched by a combination of hydrodynamic and surface forces arising when a liquid meniscus moves on a solid surface, which is a phenomenon known with the name of 'molecular combing' and can be exploited, for example, in order to stretch DNA molecules using a flow field [47]. This mechanism is illustrated qualitatively in Fig. 13. The ensemble of polymer molecules stretching as the drop edge sweeps the surface provide the dissipative force necessary to retard the displacement of the contact line. This also explains the reduction of the dynamic contact angle observed in experiments: to overcome the action of polymer molecules on the contact line,

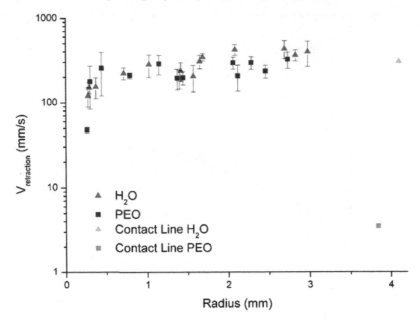

Figure 12. Comparison between the fluid velocity measured inside liquid droplets impacting on hydrophobic surfaces and the contact line velocity during the recoil phase [44].

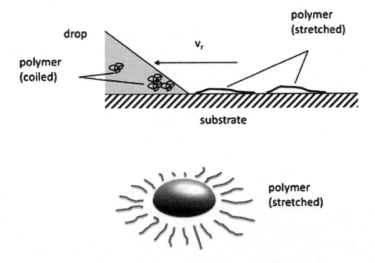

Figure 13. Tentative model for the dissipative force arising at the contact line during drop retraction: as the meniscus recedes, polymer molecules accumulating in the liquid wedge are stretched by molecular combing.

the horizontal component of the surface force driving the droplet retraction must be larger than in a Newtonian fluid, therefore the apparent dynamic contact angle must be smaller.

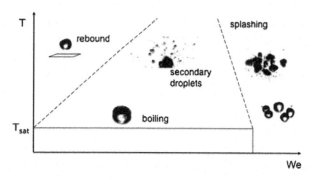

Figure 14. Impact regimes of drops impacting on heated surfaces.

D.2. Impact on Heated Surfaces

The morphology of drop impact is even more complex if the target surface is heated above the saturation temperature of the liquid, both because of convective heat transfer within the liquid, and because of the liquid–vapor phase transition. For example, the process of drop break-up (splashing) can be enhanced by boiling initiated at the contact points between the liquid and the hot wall. Therefore, the critical Weber number at which splashing begins is expected to be lower than in the case of cold wall.

A qualitative map of drop impact regimes as a function of the Weber number and the wall temperature [48] is presented in Fig. 14. When the wall temperature grows above the saturation temperature of the liquid, one can observe nucleate boiling, with vapor bubbles migrating from the wall to the free surface of the liquid. If the Weber number grows beyond a critical threshold, the drop kinetic energy becomes too large and is no longer balanced by the combined effect of capillary forces and viscous dissipation so that splashing occurs; as the target surface temperature grows, the liquid receives more thermal energy, and consequently splashing occurs at a smaller critical value of the Weber number.

For intermediate values of the wall temperature and of the Weber number, secondary droplets bursting from the free surface can be observed even if the Weber number is well below the splashing threshold [49, 50]. Such droplets can be very small, so that they look like mist.

This phenomenon is due to the synergetic action of heating, convective currents, and fast-rising vapor bubbles, resulting into local peaks of the dynamic pressure of the liquid near the free surface. A similar phenomenon can be observed in sessile drops, and is sometimes called drop miniaturization [51].

For high wall temperatures and low Weber numbers, reflection of drops off the wall occurs, due to the formation of a continuous vapor layer between the liquid and the solid surface [49], which is known as Leidenfrost phenomenon [52]. Although this phenomenon looks quite similar to drop bouncing on hydrophobic surfaces, the vapor film elasticity is thought to play a fundamental role in the mechanism of drop reflection from hot walls [48].

Drop rebound can be observed either with or without emission of satellite droplets, the latter case being often referred to as 'dry rebound'. Conventionally, one defines a 'dynamic Leidenfrost temperature' as the smallest temperature for which the drop bounces off the wall without breaking-up or scattering secondary droplets [53]. Its value has been shown to increase with the Weber number [54].

The impact morphology of viscoelastic drops can be significantly different from that of Newtonian drops, even when viscoelasticity is obtained through the addition of small amounts of polymer to a Newtonian solvent [55]. In particular, the fluid viscoelasticity has three major effects, which are illustrated in Fig. 15: (i) inhibition of drop splashing (or, equivalently, translation of the splashing threshold to higher Weber numbers); (ii) suppression of secondary atomization in sessile drops; (iii) suppression of secondary atomization during drop bouncing, which in turn affects the dynamic Leidenfrost temperature [56].

It should be observed that qualitatively similar effects could be achieved by simply increasing the fluid viscosity, which reduces the mechanical energy available for drop break-up or for the ejection of satellite droplets by increasing the overall energy dissipation. However, since the viscosity of dilute polymer solutions is almost identical to that of the pure solvent, these effects can be interpreted solely in terms of the polymer chains elasticity.

In particular, a high elongational viscosity is known to change substantially the breakup dynamics of free-surface flows and their decay into drops [57]: thus, elongational viscosity opposes to the scattering of secondary droplets from the free surface of the impacting drop.

However, one can speculate about at least two other effects of the polymer additive that contribute simultaneously to suppress secondary atomization. First of all, it is likely that the additive improves not only the stability of the surface between the drop and the surrounding atmosphere, but also that of the surface in contact with the vapor cushion that separates the drop from the hot wall: this reduces the chances that the liquid may locally touch the wall and start boiling. Second, even if the liquid makes contact with the wall, the presence of the polymer can significantly affect the process of growth, detachment, and rise of vapor bubbles [58, 59], hence prevent their bursting on the drop free surface.

Elongational viscosity can also be invoked to explain the inhibition of drop splashing observed in dilute polymer solutions, both because the fluid elasticity improves the stability of the liquid rim during drop spreading, counter-acting the growth of perturbations, and because it prevents the break-up of liquid bridges which may form between different parts of the drop in case of large deformations [57].

The transition temperature where drop bouncing occurs without secondary atomization, which conventionally defines the dynamic Leidenfrost temperature, is plotted with respect to the impact Weber number in Fig. 16, for both Newtonian and viscoelastic drops impacting on a polished aluminum surface.

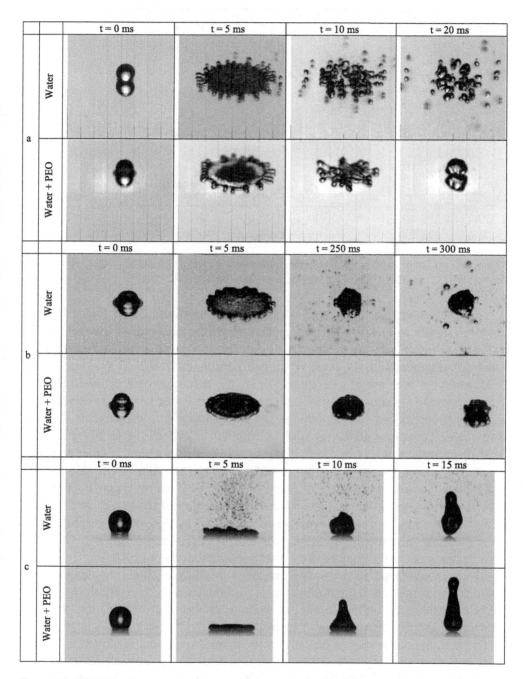

Figure 15. Effect of polymer additives (200 ppm PEO) on drops impacting onto hot surfaces: (a) drop splashing inhibition ($T = 180°C$, We = 220); (b) secondary atomization suppression ($T = 140°C$, We = 120); (c) secondary atomization suppression during drop rebound (dynamic Leidenfrost phenomenon) ($T = 250°C$, We = 40).

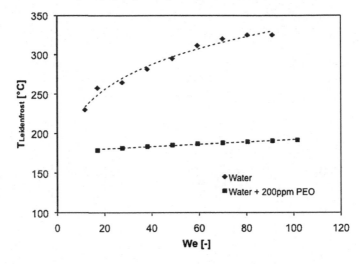

Figure 16. 'Dynamic' Leidenfrost temperature observed for water drops and dilute polymer solution drops ($D_0 = 3.2$ mm) impacting on a polished aluminum surface.

This plot shows that the dynamic Leidenfrost temperature of polymer solution drops is significantly lower than that of water drops, and almost independent of the impact Weber number; moreover, its value is close to that of the conventional Leidenfrost temperature for water on polished aluminum [60].

However, in the case of viscoelastic fluids the definition of a dynamic Leidenfrost temperature is arguable. In fact, for drops of pure water, secondary atomization actually disappears when a continuous and stable vapor cushion prevents the drop from making contact with the hot surface, which is indeed analogous to the Leidenfrost phenomenon in sessile drops. This is no longer true when polymer additives are dissolved into the impacting drop: in fact, even if the film is unstable and the liquid locally touches the hot wall, there are other physical mechanisms that prevent scattering of satellite droplets from the free surface of the liquid, as discussed above. In this case, using the expression 'dynamic Leidenfrost temperature' may be misleading, because it would suggest that the impacting drop never wets the surface, whereas wetting might occur without secondary atomization.

Above the dynamic Leidenfrost point, the vapor film between the drop and the hot surface is stable, and the liquid is never in contact with the wall: thus, one can study drop bouncing in the absence of wetting effects and wall friction, for both Newtonian and viscoelastic drops [40, 61].

The maximum diameter of drops after the inertial spreading is plotted in Fig. 17 with respect to the impact Weber number. Theories based on the conservation of energy [6, 24] suggest that this quantity should scale as $We^{1/2}$. However, it was proposed that for $We \gg 1$ a momentum conservation approach is more accurate, because it is difficult if not impossible to quantify energy dissipation during impact [62, 63]. Whilst the latter approach, which leads to a scaling of the maximum spreading diameter as $We^{1/4}$, seems to be confirmed by the trends obtained for

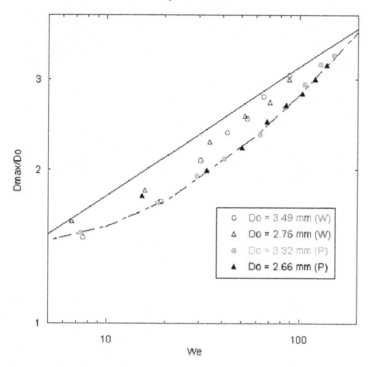

Figure 17. Maximum spreading diameter of Leidenfrost drops of different diameter: comparison between water (W) and a 200 ppm PEO solution (P).

water drops, in particular for $We > 30$, the maximum spreading diameter of viscoelastic drops follows the trend predicted by the energy conservation approach.

In any case it must be observed that for $We > 20$ the maximum spreading diameter of viscoelastic drops is systematically smaller than that of water drops having the same impact Weber number. This means that the fraction of impact kinetic energy (which is proportional to the Weber number) converted into surface energy (which is proportional to the area of the drop surface at maximum spreading) is smaller.

In principle, such reduction of the surface energy at maximum spreading could be interpreted as the consequence of increased dissipation during the expansion stage due to the higher elongational viscosity of the polymer solution, providing an independent confirmation of the theory originally proposed to explain the anti-rebound effect of polymer additives [29, 30, 33, 37, 38]. However, this does not exclude that the missing surface energy may be stored elsewhere, for example as elastic energy.

In order to get a deeper insight of the energy re-distribution upon impact, one can observe what happens during the rebound stage, when surface energy is converted back to kinetic energy and propels the drop off the surface. In particular, the maximum height of the centre of mass of the drop allows one to calculate exactly the fraction of surface energy recovered as mechanical energy during this stage, $E_{rec} = mgH_{max}$. Figure 18 shows the maximum height of the drop centre of

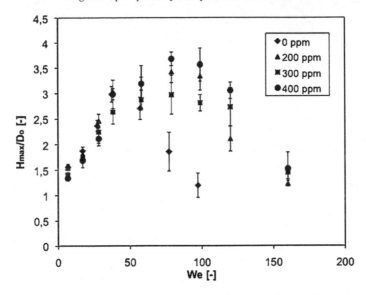

Figure 18. Maximum bouncing height of Leidenfrost drops ($D_0 = 2.7$ mm) impacting on a polished aluminum surface heated at 400°C: comparison between water and PEO solutions at various concentrations.

mass during rebound. For We $\leqslant$ 50, no significant differences can be observed between drops of pure water and viscoelastic drops. On the contrary, for We > 50 the maximum height reached by viscoelastic drops is significantly larger than that of Newtonian drops, irrespective of the drop diameter. These results clearly show that viscoelastic drops can recover a higher fraction of the initial impact kinetic energy even if they store less in the form of surface energy.

Neglecting the momentum transferred to the drop by evaporation and the elasticity of the vapor cushion [48] because their effects are similar for both the Newtonian and the viscoelastic drops at the same impact Weber number, one can write a comparative energy balance for two drops (one Newtonian and one viscoelastic) at maximum spreading after impacting on a hot surface with the same velocity, i.e., with the same initial kinetic energy [61]:

$$E_S^{(W)} + E_{D,e}^{(W)} = E_S^{(P)} + E_{D,e}^{(P)} + E_E^{(P)}, \qquad (6)$$

where E_S, $E_{D,e}$, and E_E denote respectively the surface energy, the energy dissipated during expansion, and the elastic energy stored in the polymer solution, while the superscripts indicate the water (W) and the polymer solution (P) drops. Since Fig. 17 suggests that the surface energy of viscoelastic drops is smaller than that of Newtonian drops, one can write:

$$E_S^{(W)} = E_S^{(P)} + \delta_S, \qquad (7)$$

where δ_S is the difference between the surface energies of the two drops, which can be estimated from the maximum spreading diameter data.

The mechanical energy available for drop rebound and entirely converted into potential energy at the maximum bouncing height, is given by the contributions of elastic energy (only for the polymer solution drop) and surface energy, minus the dissipation during the retraction phase, $E_{D,r}$. However, Fig. 18 shows that viscoelastic drops exhibit higher rebounds than Newtonian drops, so that one can write the following inequality:

$$E_S^{(P)} + E_E^{(P)} - E_{D,r}^{(P)} > E_S^{(W)} + E_{D,r}^{(W)}.$$
(8)

Introducing Eqs (21) and (22) into this inequality yields:

$$E_{D,e}^{(W)} + F_{D,r}^{(W)} > E_{D,e}^{(P)} + E_{D,r}^{(P)}.$$
(9)

Thus, one can conclude that the total energy dissipation during expansion and retraction in the viscoelastic drop is smaller than the total dissipation in the Newtonian drop.

Appendix. Constitutive Models for Non-Newtonian Fluids

A1.1. Power-Law Fluids

The simplest type of non-Newtonian fluid behavior occurs when the viscosity coefficient is not a constant, but is a monomial function of the shear velocity gradient (power-law, or Ostwald–De Waele fluid):

$$\mu = K\dot{\gamma}^{n-1},$$
(A1)

where the consistency coefficient K and the power-law index n are empirical constants. The power-law index is indicative of the shear-thinning ($n < 1$) or shear-thickening ($n > 1$) behavior of the fluid, whereas for $n = 0$ the Newtonian behavior is retrieved. The consistency coefficient describes the fluid viscosity at low shear rates, and coincides with the Newtonian viscosity for $n = 0$. Power-law fluids are time-independent, i.e., the shear stress does not depend on the previous deformation history. Physically, shear-thinning is usually explained by the breakdown of structure formed by interacting particles within the fluid, while shear-thickening is often due to flow-induced jamming [64].

The Ostwald–De Waele equation implies that viscosity will change indefinitely for any values of the shear rate; to account for a more realistic behavior, where viscosity varies between minimum and a maximum value respectively at very low and very high shear rates, a number of constitutive equations have been proposed, such as the Cross model [65]:

$$\frac{\mu - \mu_\infty}{\mu_0 - \mu_\infty} = \frac{1}{1 + (C\dot{\gamma})^{1-n}}$$
(A2)

and the Carreau model [66]:

$$\frac{\mu - \mu_\infty}{\mu_0 - \mu_\infty} = \frac{1}{[1 + (C\dot{\gamma})^a]^{(1-n)/a}},$$
(A3)

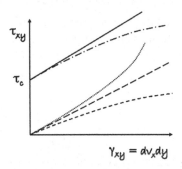

Figure 19. Qualitative flow curves for Newtonian fluids (long dashed), shear-thinning (short dashed) and shear-thickening (dotted) fluids, Bingham fluids (continuous), and Herschel–Bulkley fluid (dash–dot).

where μ_0 and μ_∞ are the viscosities at zero shear and infinite shear, respectively, and C is the Cross time constant. This parameter is the reciprocal of the strain rate at which the zero-strain rate component and the power-law component of the flow curve intersect.

Finally, the Ellis model is obtained by setting $\mu_\infty = 0$ in the Cross model:

$$\frac{\mu_0}{\mu} = 1 + \left(\frac{\tau}{\tau_{1/2}}\right)^{n-1}, \tag{A4}$$

where τ is the applied shear stress, and $\tau_{1/2}$ is the shear stress at which μ is exactly half of the zero-shear viscosity value.

Qualitative flow curves displaying the shear stress with respect to the shear rate for shear-thinning and shear-thickening fluids, as compared with Newtonian fluids, are shown in Fig. 19.

A1.2. Yield-Stress Fluids

An important type of non-Newtonian fluid is the viscoplastic or yield-stress fluid, which responds like elastic solids for applied stresses lower a certain threshold value, called the yield stress, and flows only when the yield stress is overcome. Practically, such flow behavior occurs in many situations, including slurries and suspensions, certain polymer solutions, crystallizing lavas, muds and clays, heavy oils, avalanches, cosmetic creams, hair gel, liquid chocolate, pasty materials, foams and emulsions.

The simplest constitutive model describing viscoplastic fluids was introduced by Bingham to characterize the behavior of paints [67], and represents the shear stress component as a linear function of the velocity gradient, with the intercept τ_c corresponding to the threshold yield point:

$$\begin{cases} \dot{\gamma} = 0, & \tau \leqslant \tau_c, \\ \tau = \tau_c + \mu\dot{\gamma}, & \tau > \tau_c. \end{cases} \tag{A5}$$

A more refined model is the Herschel–Bulkley equation [68], given by:

$$\begin{cases} \tau = G\gamma, & \tau \leqslant \tau_c, \\ \tau = \tau_c + K\dot{\gamma}^n, & \tau > \tau_c, \end{cases} \tag{A6}$$

where G is the shear modulus, γ is the shear deformation and τ_c is the yield-stress magnitude. Whilst this model is well established and amongst the most widely used when analyzing yield-stress behavior, another popular model was proposed by Casson [69]:

$$\sqrt{\tau} = \sqrt{\tau_c} + \sqrt{\mu\dot{\gamma}}, \quad \tau > \tau_c. \tag{A7}$$

Figure 1 shows the qualitative flow curves for Bingham fluids and Herschel–Bulkley fluids, as compared with power-law fluids.

According to Eqs (A5)–(A7), the transition from the elastic regime to the fluid regime is abrupt, which means that the shear stress derivative with respect to the shear rate exhibits a first-order discontinuity. This represents a major technical issue when the yield stress fluid constitutive equation is implemented to find analytical or (especially) numerical solutions of fluid flow problems. To remove this discontinuity, Papanastasiou proposed a constitutive equation featuring a smooth transition between the two regimes [70], which provides a better description of real materials:

$$\tau = \tau_c[1 - \exp(-m\dot{\gamma})] + C\dot{\gamma}^n, \tag{A8}$$

where m is a material-dependent constant with values of the order of 10^2.

Research into viscoplastic fluids, their measurement and characterization is extensive and has been summarized in numerous reviews [71–74]. One matter still subject of debate is the definition of yield-stress fluid itself, that is, whether fluids can actually exhibit such a physical property as the yield-stress. The review by Barnes [71] examines the evidence for and against its existence, and argues that whereas the concept of a definable yield-stress has proven and continues to prove useful in a whole range of applications, if viscosity is plotted as a function of the shear stress, one can clearly identify a Newtonian plateau when the velocity gradient tends to zero (typically less than 10^{-5} s^{-1}), which implies that the material continues to creep although this can be observed only on very long timescales. However, in many practical situations time frame of observation is much shorter than the time necessary for viscoplastic fluids to exhibit measurable flow characteristics.

Whilst several fluids exhibit an apparent yield stress, Carbopol dispersions are probably the most thoroughly studied model viscoplastic fluid system. Carbopol consists of highly cross-linked polymer particles, with dangling free ends of polymer gel strands that strongly interact with adjacent microgel particles, resulting into to a very high viscosity at low shear stress [28, 73]. Carbopol dispersions and gels are found in dozens of everyday products, ranging from toothpastes, through hair and shower gels, to artificial tears.

A1.3. Viscoelastic Fluids

In viscoelastic fluids, such as polymer melts or solutions, a part of the deformation energy is stored as elastic energy, and released with a certain delay depending on the relaxation time of the fluid. The basic feature that essentially all viscoelastic fluids share is the occurrence of elastic stress effects: when the shear rate is sufficiently strong, the forces along the normals of a little cubical fluid element are different in different directions, unlike what happens for a Newtonian fluid where the pressure is isotropic. From the microscopic point of view, this behavior is usually related to conformational rearrangements of the macromolecules which compose the fluid under the action of hydrodynamic forces. Viscoelasticity manifests itself in a variety of phenomena, including creep (the time-dependent strain resulting from a constant applied stress), stress relaxation resulting from a steady deformation, the Weissenberg rod-climbing effect due to nonzero normal stress difference, and many others [74–76].

The earliest constitutive model to describe linear viscoelastic fluids was introduced by Maxwell:

$$\frac{\mathrm{d}\tau}{\mathrm{d}t} = E\dot{\varepsilon} - \frac{\tau}{t_0}, \tag{A9}$$

where τ is the one-dimensional stress, ε the one-dimensional strain, t_0 is a time constant, and E the elastic modulus. This equation can be derived using a lumped parameter model where the elastic and the viscous element are in series, i.e., they are subject to the same force but experience different elongations.

In a similar fashion, one can derive a constitutive model where the elastic and the viscous element are connected in parallel, with the same elongation rate but resisting forces of different magnitude:

$$\tau = G\gamma + \eta\dot{\gamma}, \tag{A10}$$

where G is the elastic modulus, γ the shear strain, and η the fluid viscosity.

These concepts were generalized by Boltzmann's approach, where the stress does not depend only on the current deformation, but also on the deformation history:

$$\tau = \int_{-\infty}^{t} G(t - \xi)\dot{\gamma}(\xi)\,\mathrm{d}\xi, \tag{A11}$$

where $G(t)$ accounts for the fact that the past deformations contribute less to build up the current stress than the more recent deformations.

The most popular constitutive equation for viscoelastic fluids is the Oldroyd-B model, which captures the main features of viscoelastic flows but at the same time is simple enough to allow finding the analytical solution for the flow field in many circumstances. In this model, the total stress tensor, $\mathbf{T}$, is decomposed into the Newtonian solvent component, $2\eta_S\mathbf{D}$ (where η_S is the solvent viscosity, and $\mathbf{D}$ is the velocity gradient tensor), and the viscoelastic polymeric component, for which one can write the relation with the velocity gradient as:

$$(\mathbf{T} - 2\eta_S\mathbf{D}) + \lambda_1\frac{\delta}{\delta t}(\mathbf{T} - 2\eta_S\mathbf{D}) = 2\eta_P\mathbf{D}, \tag{A12}$$

where λ_1 is the relaxation time, and η_P the polymer viscosity. Re-arranging Eq. (A12) into a more compact form gives:

$$\mathbf{T} + \lambda_1\frac{\delta\mathbf{T}}{\delta t} = 2(\eta_S + \eta_P)\left(\mathbf{D} + \lambda_2\frac{\delta\mathbf{D}}{\delta t}\right), \tag{A13}$$

where $\lambda_2 = \lambda_1\eta_S/(\eta_S + \eta_P)$ is the retardation time. The symbol $\delta/\delta t$ denotes one of the derivatives introduced by Oldroyd to ensure that the stress tensor remains unchanged under any change of the reference coordinate system [74, 76].

The non-isotropic principal components of the stress tensor allow one to characterize the elasticity of polymer solutions and melts through normal stress differences: in particular, the first normal stress difference, $N_1 = \tau_{xx} - \tau_{yy}$, is the first-order elastic effect, and is usually interpreted as being due to the stretching and/or orientation of the polymer chains by the flow. The entropic tendency of polymers that are stretched by the flow to recover their equilibrium chain conformation generates an elastic stress, the macroscopic manifestation of which is a difference in stress between the flow direction and the direction normal to it [75]. For the Oldroyd-B fluid in steady-state shear flow, the first normal stress difference is a quadratic function of the shear rate, $\dot{\gamma}$:

$$N_1 = \Psi_1\dot{\gamma}^2. \tag{A14}$$

The dissipation of energy associated to the process of stretching and relaxation of macromolecules is described by introducing the concept of elongational (or extensional) viscosity, the ratio of the first normal stress difference to the rate of elongation of the fluid:

$$\eta_E = \frac{\tau_{xx} - \tau_{yy}}{\varepsilon_{xx}}. \tag{A15}$$

For a Newtonian incompressible fluid, one can easily verify that the elongational viscosity is three times the shear viscosity. For a polymer solution the ratio η_E/η, also known as the Trouton ratio [77], can be of the order of 10^3–10^4.

Quantitative measurements of elongational viscosity are not easy [72], especially for dilute polymer solutions in low-viscosity solvents, because they require the creation of a steady-state elongational flow. This is difficult to achieve in practice because it is not possible for a volume of fluid to stretch to infinity, since it will get thinner and thinner and eventually break-up. Furthermore, the stiffness of polymer chains is not constant, but grows as they approach the maximum elongation: thus, the instantaneous values of elongational viscosity are not constant during measurements.

Measurements that reasonably approach steady state have been obtained for certain polymer solutions by means of the filament-stretching technique [78, 79].

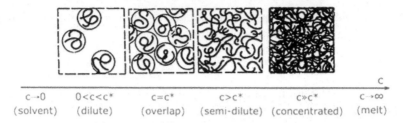

c→0	0<c<c*	c=c*	c>c*	c»c*	c→∞
(solvent)	(dilute)	(overlap)	(semi-dilute)	(concentrated)	(melt)

Figure 20. Solution regimes of flexible polymers in a Newtonian solvent.

Unfortunately, this technique is only applicable to relatively viscous liquids, as the filament breaks up too rapidly for low-viscosity samples. The alternative stagnation point devices, such as the opposed nozzle rheometer [80], do offer a stationary flow, but the residence time of a polymer chain in the elongational flow field is typically quite short, has large statistical fluctuations and depends on the rate of elongation. Therefore, a steady-state value for the elongational viscosity is very hard to obtain.

At a molecular level, the energy dissipation mechanism for elongational flows can be explained in terms of the interaction between the additive and the surrounding fluid, which is essentially due to hydrogen bonds between water molecules and monomers. Thus, when the polymer is coiled, the only monomers affected by the interaction are those located in the external shell, and the polymer molecule behaves like a spherical particle advected by the flow. As the velocity gradient increases the polymer stretches [81], and therefore more of its monomers become affected by the interaction with the fluid, increasing molecular friction and hence viscous dissipation.

Common viscoelastic fluids are solutions of long-chain polymers in a Newtonian solvent (e.g., water). The macroscopic physical properties of these solutions are strongly dependent on the polymer concentration, as illustrated in Fig. 20. For low concentrations, the average distance among polymer molecules is larger than their size, so that they do not interact with one another: polymers exhibit a random coil conformation and can be described as spherical particles suspended in the fluid (dilute regime). As the polymer concentration grows, the average distance between polymer molecules reduces until the monomers placed on the external shell of one coil can interact with monomers of other coils. This happens for a critical value of concentration (the overlap concentration), and marks the beginning of the semi-dilute regime, where polymer chains are randomly entangled and can be no longer described as coils. Dilute solutions are of particular importance in many engineering applications, because due to the low polymer concentration their viscosity is very similar to that of the pure solvent, although the solution exhibits viscoelasticity.

E. References

1. Reynolds, O., On the action of the rain to calm the sea. Papers on Mechanical and Physical Subjects, vol. 1, pp. 86–88. Cambridge University Press, 1900.

2. Worthington, A. M., On the forms assumed by drops of liquids falling vertically on a horizontal plate. Proc. Roy. Soc. Lond., 25 (1876) 261–272.

3. Edgerton, H. E. and Killian, J. R., Flash!: Seeing the Unseen By Ultra High-Speed Photography. Hale-Cushman & Flint, Boston, 1939.

4. Rioboo, R., Tropea, C. and Marengo, M., Outcomes from a drop impact on solid surfaces. Atomization & Sprays, 11 (2001) 155–166.

5. Rioboo, R., Marengo, M. and Tropea, C., Time evolution of liquid drop impact onto solid, dry surfaces. Exp. Fluids, 33 (2002) 112–124.

6. Rein, M., Phenomena of liquid-drop impact on solid and liquid surfaces. Fluid Dynamics Research, 12(2) (1993) 61–93.

7. Yarin, A. L., Drop impact dynamics: Splashing, spreading, receding, bouncing. Annu. Rev. Fluid Mech., 38 (2006) 159–192.

8. Vovelle, L., Bergeron, V. and Martin, J. Y., Use of polymers as sticking agents. World Patent WO 0008926, 2000.

9. Larson, R. G., The Structure and Rheology of Complex Fluids. Oxford University Press, New York, 1999.

10. Astarita, G. and Marrucci, G., Principles of Non-Newtonian Fluid Mechanics. McGraw-Hill, New York, 1974.

11. German, G. and Bertola, V., Review of drop impact models and validation with high-viscosity Newtonian fluids. Atomization and Sprays, 19 (2009) 787–807.

12. Davidson, M. R. and Cooper-White, J. J., Pendant drop formation of shear-thinning and yield stress fluids. Appl. Math. Model., 30 (2006) 1392–1405.

13. Weidner, D. E. and Schwartz, L. W., Contact-line motion of shear-thinning liquids. Phys. Fluids, 6 (1994) 3535–3538.

14. Carré, A. and Eustache, F., Spreading kinetics of shear-thinning fluids in wetting and dewetting modes. Langmuir, 16 (2000) 2936–2941.

15. Starov, V. M., Tyatyushkin, A. N., Velarde, M. G. and Zhdanov, S. A., Spreading of non-Newtonian liquids over solid substrates. J. Colloid Interface Sci., 257 (2003) 284–290.

16. German, G., Yield-stress drops, Ph.D. Thesis, University of Edinburgh, Edinburgh, 2009.

17. German, G. and Bertola, V., Impact of shear-thinning and yield-stress drops on solid substrates. Journal of Physics: Condensed Matter, 21 (2009) 375111.

18. Nigen, S., Experimental investigation of the impact of an (apparent) yield-stress material. Atom. Sprays, 15 (2005) 103–117.

19. Bertola, V., Wicking with a yield-stress fluid. J. Phys. Condens. Matter, 21 (2009) 035107.

20. Coussot, P. and Gaulard, F., Gravity flow instability of viscoplastic materials: The ketchup drip. Phys. Rev. E, 72 (2005) 031409

21. German, G. and Bertola, V., Formation of viscoplastic drops by capillary breakup. Physics of Fluids, 22 (2010) 033101.

22. German, G. and Bertola, V., The free fall of viscoplastic drops. Journal of Non-Newtonian Fluid Mechanics, 165 (2010) 825–828.

23. Saidi, A., Martin, C. and Magnin, A., Influence of yield stress on the fluid droplet impact control. J. Non-Newtonian Fluid Mech., 165 (2010) 596–606.

24. Chandra, S. and Avedisian, C. T., On the collision of a droplet with a solid surface. Proc. R. Soc. Lond. A, 432 (1991) 13–41.

25. Saidi, A., Martin, C. and Magnin, A., Effects of surface properties on the impact process of a yield stress fluid drop. Exp. Fluids, 51 (2011) 211–224.

26. Barnes, H. A., A review of the slip (wall depletion) of polymer solutions, emulsions and particle suspensions in viscometers: its cause, character, and cure. J. Non-Newton. Fluid Mech., 56 (1995) 221–251.

27. Luu, L. and Forterre, Y., Drop impact of yield-stress fluids. J. Fluid Mech., 632 (2009) 301–327.

28. Roberts, G. P. and Barnes, H. A., New measurements of the flow-curves for Carbopol dispersions without slip artefacts. Rheol Acta, 40 (2001) 499–503.

29. Bergeron, V., Bonn, D., Martin, J. Y. and Vovelle, L., Controlling droplet deposition with polymer additives. Nature, 405 (2000) 772–775.

30. Bergeron, V., Designing intelligent fluids for controlling spray applications. C. R. Phys., 4 (2003) 211–219.

31. Zhang, X. G. and Basaran, O. A., Dynamic surface tension effects in impact of a drop with a solid surface. J. Colloid Interface Sci., 187 (1997) 166–178.

32. Mourougou-Candoni, N., Prunet-Foch, B., Legay, F., Vignes-Adler, M. and Wong, K., Influence of dynamic surface tension on the spreading of surfactant solution droplets impacting onto a low-surface-energy solid substrate. J. Colloid Interface Sci., 192 (1997) 129–141.

33. Crooks, R. and Boger, D. V., Influence of fluid elasticity on drops impacting on dry surfaces. J. Rheol., 44 (2000) 973–996.

34. Cooper-White, J., Crooks, R. and Boger, D. V., A drop impact study of worm-like viscoelastic surfactant solutions. Colloids and Surfaces A: Physicochemical and Engineering Aspects, 210 (2002) 105–123.

35. Bird, R. B., Armstrong, R. C. and Hassager, O., Dynamics of Polymeric Liquids. Wiley, New York, 1987.

36. De Gennes, P.-G., Coil-stretch transition of dilute flexible polymers under ultrahigh velocity gradients. J. Chem. Phys., 60 (1974) 5030–5042.

37. Crooks, R., Cooper-White, J. and Boger, D. V., The role of dynamic surface tension and elasticity on the dynamics of drop impact. Chem. Eng. Sci., 56 (2001) 5575–5592.

38. Williams, P. A., English, R. J., Blanchard, R. L., Rose, S. A., Lyons, L. and Whitehead, M., The influence of the extensional viscosity of very low concentrations of high molecular mass water-soluble polymers on atomization and droplet impact. Pest Management Sci., 64 (2008) 497–504.

39. Rozhkov, A., Prunet-Foch, B. and Vignes-Adler, M., Impact of water on small targets. Phys. Fluids, 14 (2002) 3485–3501.

40. Bertola, V., An experimental study of bouncing Leidenfrost drops: comparison between Newtonian and viscoelastic liquids. International Journal of Heat and Mass Transfer, 52 (2009) 1786–1793.

41. Boffetta, G., Mazzino, A., Musacchio, S. and Vozella, L., Polymer heat transport enhancement in thermal convection: the case of Rayleigh–Taylor turbulence. Phys. Rev. Lett., 104 (2010) 184501.

42. Bartolo, D., Boudaoud, A., Narcy, G. and Bonn, D., Dynamics of non-Newtonian droplets. Phys. Rev. Lett., 99 (2007) 174502.

43. Bertola, V., The effect of polymer additives on the apparent dynamic contact angle of impacting drops. Colloids and Surfaces A: Physicochemical and Engineering Aspects, 363 (2010) 135–140.

44. Smith, M. I. and Bertola, V., Effect of polymer additives on the wetting of impacting droplets. Physical Review Letters, 104 (2010) 154502.

45. Smith, M. I. and Bertola, V., The anti-rebound effect of flexible polymers on impacting drops, in: Proc. 23rd European Conference on Liquid Atomization and Spray Systems, Brno, Czech Republic, 6–8 September, 2010.

46. Smith, M. I. and Bertola, V., Particle velocimetry inside Newtonian and non-Newtonian droplets impacting a hydrophobic surface. Experiments in Fluids, 50 (2011) 1385–1391.

47. Kim, J. H., Shi, W. and Larson, R. G., Methods of stretching DNA molecules using flow fields. Langmuir, 23 (2007) 755–764.
48. Rein, M., Interactions between drops and hot surfaces, in: M. Rein (Ed.), Drop-Surface Interactions, CISM Courses and Lectures, vol. 456, Chap. 6. Springer, Wien, New York (2003).
49. Wachters, L. H. J. and Westerling, N. A. J., The heat transfer from a hot wall to impinging water drops in the spheroidal state. Chem. Eng. Sci., 21 (1966) 1047–1056.
50. Naber, J. D. and Farrell, P. V., Hydrodynamics of droplet impingement on a heated surface. SAE, 930919 (1993) 1–16.
51. Inada, S. and Yang, W.-J., Mechanism of miniaturization of sessile drops on heated surfaces. Int. J. Heat Mass Transfer, 36 (1993) 1505–1515.
52. Leidenfrost, J. G., De aquae communis nonullis qualitatibus tractatus—On the fixation of water in divers fire. A tract about some qualities of common water. Int. J. Heat Mass Trans., 9 (1966) 1153–1166.
53. Wang, A.-B., Lin, C.-H. and Chen, C.-C., The critical temperature of dry impact for tiny droplet impinging on a heated surface. Phys of Fluids, 12 (2000) 1622–1625.
54. Yao, S. C. and Cai, K. Y., The dynamics and leidenfrost temperature of drops impacting on a hot surface at small angles. Experimental Thermal and Fluid Sci., 1 (1988) 363–371.
55. Bertola, V., Drop impact on a hot surface: effect of a polymer additive. Experiments in Fluids, 37 (2004) 653–664.
56. Bertola, V. and Sefiane, K., Controlling secondary atomization during drop impact on hot surfaces by polymer additives. Physics of Fluids, 17 (2005) 108104.
57. Eggers, J., Nonlinear dynamics and breakup of free-surface flows. Rev. Mod. Phys., 69 (1997) 865–929.
58. Hartnett, J. P. and Hu, R. Y. Z., Role of rheology in boiling studies of viscoelastic liquids. Int. Commun. Heat Mass Transfer, 13 (1986) 627–637.
59. Ki, Y. M., Kang, S. L. and Kwak, H. Y., Bubble nucleation and growth in polymer solutions. Polym. Eng. Sci., 44 (2004) 1890–1899.
60. Bernardin, J. D. and Mudawar, I., The Leidenfrost point: experimental study and assessment of existing models. ASME Journal of Heat Transfer, 121 (1999) 894–903.
61. Bertola, V., Viscoelastic Leidenfrost drops, in: Proc. 22nd European Conference on Liquid Atomization and Spray Systems, Como, Italy, 8–10 September, 2008.
62. Clanet, C., Béguin, C., Richard, D. and Quéré, D., Maximal deformation of and impacting drop. J. Fluid Mech., 517 (2004) 199–2008.
63. Biance, A.-L., Chevy, F., Clanet, C., Lagubeau, G. and Quéré, D., On the elasticity of an inertial liquid shock. J. Fluid Mech., 554 (2006) 47–66.
64. Frith, W. J., d'Haene, P. and Buscall, R., Shear thickening in model suspensions of sterically stabilized particles. J. Rheol., 40 (1996) 531–548.
65. Cross, M. M., Rheology of non-Newtonian fluids: a new flow equation for pseudoplastic systems. J. Colloid Sci., 20 (1965) 417–437.
66. Carreau, P. J., Ph.D. Thesis, University of Wisconsin, Madison, 1968.
67. Bingham, E. C., An investigation of the laws of plastic flow. U.S. Bur. of Standards Bull., 13 (1916) 309–353.
68. Herschel, W. H. and Bulkley, R., Konsistenzmessungen von Gummi-Benzol-Lösungen. Kolloid Z., 39 (1926) 291–300.
69. Casson, N., in: C. C. Mill (Ed.), Rheology of Disperse Systems, Chap. 5. Pergamon, London (1959).
70. Papanastasiou, T. C., Flows of materials with yield. Journal of Rheology, 31 (1987) 385–404.

71. Barnes, H. A., The yield stress—a review or 'παητα ρει' — everything flows? J. Non-Newton. Fluid Mech., 81 (1999) 133–178.
72. Coussot, P., Rheophysics of pastes: a review of microscopic modeling approaches. Soft Matter, 3 (2007) 528–540.
73. Nguyen, Q. D. and Boger, D. V., Measuring the flow properties of yield stress fluids. Annu. Rev. Fluid Mech., 24 (1992) 47–88.
74. Macosko, C. W., Rheology Principles, Measurements and Applications. Wiley-VCH, 1994.
75. Carnali, J. O. and Naser, M. S., The use of dilute solution viscometry to characterize the network properties of carbopol microgels. Colloid Polymer Sci, 270 (1992) 183–193.
76. Joseph, D. D., Fluid Dynamics of Viscoelastic Liquids. Springer-Verlag, New York, Berlin, Heidelberg, 1990.
77. Phan-Thien, N., Understanding viscoelasticity. Springer-Verlag, New York, Berlin, Heidelberg, 2002.
78. Trouton, F. T., On the coefficient of viscous traction and its relation to that of viscosity. Proc. R. Soc. Lond. A, 77 (1906) 426–440.
79. Sridhar, T., Tirtaatmadja, V., Nguyen, D. A. and Gupta, R. K., Measurement of extensional viscosity of polymer solutions. J. Non-Newtonian Fluid Mech., 40 (1991) 271–280.
80. Bazilevskii, A. V., Entov, V. M. and Rozhkov, A. N., Breakup of an Oldroyd liquid bridge as a method for testing the rheological properties of polymer solutions. Polym. Sci. Ser. A+, 43 (2001) 716–726.
81. Fuller, G. G., Cathey, C. A., Hubbard, B. and Zebrowski, B. E., Extensional viscosity measurements for low-viscosity fluids. J. Rheol., 31 (1987) 235–249.

Wetting at High Temperature

A. Passerone, F. Valenza and M. L. Muolo

Inst. for Energetics and Interphases—IENI CNR, Via de marini, 6, 16149 Genoa, Italy

Abstract

Processes and equipments that must operate at high temperatures require materials able to face high thermal fluxes, severe stresses, high chemical reactivity. Ultra High Temperature Ceramics (UHTC's) represent a class of materials most promising for such applications. In particular it is often necessary, to obtain the best performances, to join these ceramic parts one to the other or to special metallic alloys by means of brazing processes, where liquid alloys realize the bonding between the two phases. Thus, also in relation to metallurgical, crystal growth and composite production processes, the knowledge of wettability, interfacial tensions and interfacial reactions is mandatory to understand what happens when a liquid metal comes into contact with a ceramic surface. In this paper, recent systematic studies, addressing both basic (wettability, interfacial tension, phase equilibria determination) and application (joining) aspects, are critically reviewed with a particular reference to transition metal diborides and silicon carbide. A special attention is paid to studies aimed at elucidating the role that dissolution, chemical reactions, additions of active metal elements to the molten matrix have in the wetting process and on the solid–liquid adhesion, and, eventually, on the mechanical characteristics of the brazed joints, as the main final 'sensitive' parameter for application purposes.

Contents

A. Introduction 300
    1. Characteristics of high temperature wetting]High Temperature Wetting 300
    2. Exchange Phenomena at Solid–Liquid–Vapour Interfaces 300
B. Wetting Typologies 303
    1. Non-reactive (or Adsorptive) Wetting 303
    2. dissolutive wetting]Dissolutive Wetting 304
    3. reactive wetting]Reactive Wetting 305
C. Role of the Solid Surface in Modifying the Wetting Behaviour 306
    1. roughness]Roughness 306
    2. anisotropy]Anisotropy 307
    3. Stoichiometry 308
D. Examples of Interacting Systems at High Temperature 310
    1. transition metals diborides]Transition Metals Diborides 310
    2. Silicon Carbide 318
    3. Remarks 322

Drops and Bubbles in Contact with Solid Surfaces
© Koninklijke Brill NV, Leiden, 2012

E. Experimental Techniques for High Temperature Wettability Measurements 322

F. References 328

A. Introduction

1. Characteristics of High Temperature Wetting

A large number of industrial processes, extremely relevant for the world economy, are governed, to a great extent, by interfacial interactions. Among them, those involving the interaction of solid bodies with molten phases at high temperatures pertain to metallurgy, glass making, electronics, aerospace, joining processes and so on. In all these categories, the control of the interactions of molten metals with higher melting point metals and ceramics is the critical step to assure good quality of products and reliability in their performances. These characteristic interactions can be classified, from a general point of view, under the term 'wetting phenomena'.

Wetting phenomena at high temperatures are characterized by a number of processes which are specific of this peculiar, aggressive, environment [1]. At variance with 'room temperature' processes, where water or organic liquids are mainly involved, at high temperatures the atomic mobility is high; diffusion processes are very active, and reactivity between the different phases in presence can put into play a large amount of chemical energy. Thus, even if capillarity still plays an important role, its impact on the final equilibrium configuration of the system under study cannot be foreseen only on the basis of the Young and Young–Dupré equations (Eqs (1) and (2)).

$$\cos\theta = (\sigma_{SV} - \sigma_{SL})/\sigma_{LV}, \tag{1}$$

$$W_a = \sigma_{LV}(1 + \cos\theta), \tag{2}$$

where σ_{SV}, σ_{SL}, σ_{LV} and θ represent, as usual, the solid–vapour, solid–liquid, liquid–vapour surface tensions and θ the contact angle at the triple line (TL); W_a is the thermodynamic work of adhesion. It should be noted that Eq. (1) is strictly valid only if the solid surface can be considered 'infinitely' rigid, or if the atomic mobility is not sufficient to rearrange the surface structure.

The main processes which can take place at the various interfaces can be described following the schematic representation shown in Fig. 1.

2. Exchange Phenomena at Solid–Liquid–Vapour Interfaces

As, usually, the solid–liquid interactions are studied by the sessile drop method, reference is made here to this special configuration. The first point to be underlined is that the solid and the liquid phases are 'immersed' in a specific environment, namely 'high vacuum' or an atmosphere whose chemical composition and flow conditions can affect to a high extent the liquid bulk composition and all the three solid–vapour, solid–liquid and liquid–vapour interfacial tensions.

The role of the solid–vapour interface is twofold: (a) it can interact with the vapour phase (containing both gaseous elements and metal vapours) giving rise to

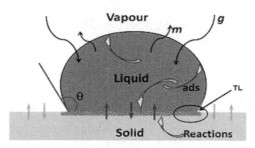

Figure 1. Transport processes at solid–liquid–vapour interfaces.

the release of volatile compounds (oxides, sub-oxides, etc.); (b) adsorbed layers or even new compounds can form at the solid surface, originated from metal vapours. In the first case, the released components can enter the liquid phase, modifying the liquid surface and interfacial tensions, while in the second case these new surface layers can modify the contact angle, as a result of a new balance among the three interfacial tensions, as shown by the Young equation (1).

When working with molten metallic phases, one of the most relevant effects is caused by oxygen: this element, ubiquitous in high temperature processes, has a tensioactive effect on the liquid–vapour as well as on the solid–liquid surface tensions, through adsorption processes and reactions [2–6]. In particular, studies taking into account both thermodynamics and transport phenomena have shown that a molten metal surface can be kept oxygen-free despite the thermodynamic driving forces foresee surface oxidation conditions [7–11]. This important effect, which is substantiated by the definition of an 'effective oxygen pressure' [8], is mainly due to oxygen and suboxides flux in the opposite direction to the oxygen adsorption/reaction. On the other hand, once oxygen reaches the liquid surface, it diffuses through the liquid phase, eventually reaching the solid–liquid interface modifying its energetics by adsorption and/or formation of new phases [12].

Indeed, oxygen can also play a positive effect on the wetting of solid oxides, and this effect is currently exploited in the 'air brazing technique' which makes use of the strong effect of oxygen to increase the wettability of oxides (mainly those to be used in force of their electrical properties which depend on oxygen stoichiometry) by means of processes running in air or in atmospheres with high oxygen content [13–15].

Under equilibrium conditions, the variation of surface and interfacial tensions as a function of oxygen concentration in the liquid phase (M) can be interpreted in terms of the Gibbs adsorption equation

$$d\sigma = -\Gamma_O \, d\mu_O - \Gamma_M \, d\mu_M \qquad (3)$$

referred to the liquid/vapour or to the solid/liquid interfaces. Following the development used by Chatain *et al.* [16] and provided that the dissolution of the metal cation from the oxide substrate into the liquid matrix can be considered negligible, the extent of oxygen adsorption Γ_O with respect to M is given by: at the liquid/vapour interface:

$$\Gamma_O^{LV} = -\frac{1}{RT}\frac{d\sigma_{LV}}{d\ln X_O} \tag{4}$$

and, at the solid/liquid interface:

$$\Gamma_O^{SL} = -\frac{1}{RT}\frac{d\sigma_{SL}}{d\ln X_O}. \tag{5}$$

Here X_O represents the oxygen mole fraction in the liquid phase, and is calculated after the Sievert's law (Eq. (6))

$$X_O = K(PO_2)^{1/2}, \tag{6}$$

where K is the equilibrium reaction constant of the dissolution reaction of oxygen into the liquid metal.

In addition to oxygen, other elements can play a fundamental role in modifying the wetting characteristics of an otherwise non-wetting metal. Thermodynamics provides criteria for selection of the most effective solute for promoting wetting: experiments and theoretical models are today available which demonstrate this effect, also trying to understand, from more basic points of view, the underlying principles [17–22].

Thus, a complex situation occurs at the solid–liquid interface.

In Fig. 1, a schematic picture is shown, where the liquid phase rests, at equilibrium, on a modified solid–liquid interface. In order to understand what may happen at this specific interface, the following processes must be taken into consideration: (a) dissolution of the solid into the liquid, (b) penetration/diffusion of the liquid components into the solid, (c) adsorption of components of the liquid phase at the solid–liquid interface, (d) reaction of some component of the liquid with the solid and formation of new phases, (e) dynamic restructuring of the solid surface.

As all these processes depend strongly on atoms movements, each one of them has its characteristic time (depending on diffusion coefficients and/or reaction rates and, thus, on temperature) which in turn affects the equilibration kinetics, i.e., the overall wetting process.

From the large amount of experimental results and from their thermodynamic interpretation, it appears clearly that the adhesion between liquid metals and ceramic materials is mainly due to two kinds of chemical bonds: those at the metal–ceramic interface and the metal–metal ones. An efficient way to study these interactions and to quantify their effects in terms of adhesion energy is offered by modelling [23–25], through molecular dynamics approaches or by applying the Density Functional Theory (DFT) [26–32]. Up to now, most efforts have been made to model metal/oxides interfaces, but a few calculations exist on metal/carbides [33, 34], metal/nitrides [35] and metal/borides [36] systems, arriving at a correct estimation of the metal–ceramic bonding mechanisms. These efforts have indeed opened new important insights into the basic solid–liquid interactions at high temperatures.

B. Wetting Typologies

Three main wetting categories can be recognised, in general: (a) Non-reactive (or adsorptive) wetting, (b) Dissolutive wetting, (c) Reactive wetting.

1. Non-reactive (or Adsorptive) Wetting

Non-reactive wetting is a modality that only seldom can be invoked to describe completely high temperature processes, and is verified if certain conditions are met. Indeed, if the interplay of the three interfacial tensions allows a low contact angle to be established (as computed by Eq. (1)), and if, at the same time, no dissolution nor reactions occur between the solid and the liquid phases, the liquid spreads over the solid which retains its original plane geometry. This happens, nearly always, with low temperature systems, but also at high temperatures in the following cases: (a) the solid and the liquid phases are mutually immiscible, (b) the liquid phase is in chemical equilibrium with the solid (i.e., it is saturated of the solid elements), (c) the temperature is not high enough to allow for a sufficiently fast diffusion of atoms from the solid into the liquid phase.

It may also happen that the spreading process, i.e., the displacement as a function of time of the triple line, is fast if compared to dissolution processes, so that the liquid front moves fast on a flat, unmodified surface [37]. In this case the friction at the triple line is the main factor governing the kinetics of the process. The study of the kinetics of spreading is a very complex subject, because the laws governing this phenomenon depend closely on the real physical phenomena occurring at the solid–liquid–vapour interfaces, so that different laws can be applied to the different possible cases which may happen in a single experiment. A systematic overview of the kinetics of spreading is outside the scope of this paper; however, the interested reader is referred to recent specialized papers [38–43].

Another process, which can be classified in the non-reactive category, is the wetting 'assisted' by the adsorption, at the solid–liquid interface, of active metal elements. This possible process is especially relevant when dealing with liquid–metal/oxides systems, as the one found in metal/ceramic and ceramic/ceramic brazing processes. It is usual, in these cases, to add some transition metal elements (such as Ti, Zr, but also oxygen can play the same role) which migrate to the solid–liquid interface adsorbing on it and, eventually, reacting with it. There is a great deal of discussion on whether the wetting process is governed only by the adsorption process or whether the reaction(s) involving the 'active' element supply the full driving force for spreading [19, 44]. The study of the kinetics of spreading from the very first instant of solid–liquid contact can offer a means to understand which phenomenon is prevalent in the specific process. However, especially at high temperatures, the experimental part is very difficult, even if high-speed recording is today at hand easily. What is still lacking, is the possibility to have access to an '*in-situ*' analysis of the interface, which is usually studied after the drop solidification, thus when the wetting process is finished and a large number of chemical reactions and/or precipitation phenomena have occurred.

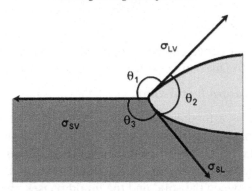

Figure 2. Equilibrium configuration in the presence of dissolution of the solid.

2. Dissolutive Wetting

As already mentioned, at high temperature, the atom mobility is also high so that at least some dissolution of the solid phase into the molten matrix can take place until the chemical potential of the diffusing species is the same in the solid and liquid phases [37, 43, 45–51]. Even if no reactions occur, the composition of the liquid phase changes with time; as a consequence, also its surface tension changes and a dynamic condition is set up, where both contact angle and drop dimensions (base diameter, height and also volume) are a function of time. The solid–liquid interface does not lie anymore on a plane, but a groove under the drop forms with the 'classical' shape of a (nearly) spherical cup (Fig. 2). With the immediate consequence that the Young law (Eq. (1)) is no longer valid, as the triple line equilibrium must take into consideration also the vertical components of the surface tension vectors. Indeed, at the new triple line, the interfacial tension vectors reach an equilibrium configuration expressed by the so-called Neumann rule (Fig. 2):

$$\frac{\sigma_{SL}}{\sin \theta_1} = \frac{\sigma_{SV}}{\sin \theta_2} = \frac{\sigma_{LV}}{\sin \theta_3}. \tag{7}$$

Thus, in this case, what is measured in a sessile drop test is the 'apparent' contact angle, that is the one 'above' the plane defined by the solid surface ($[\pi - \theta_1]$ in Fig. 2). This dissolutive stage of the wetting process is usually very fast; indeed, as noted for ex. in [45], dissolution is often increased by solutocapillary Marangoni movements which arise near the triple line, due to differences in surface tension between the drop apex (richer in atoms of the metal matrix) and the triple line, where the concentration of atoms dissolved out of the solid support is high. These processes give rise to 'Grooves' and/or 'Ridges' of the micrometer size (Figs 3 and 4). They have been shown to occur, at the microscopic scale, in many systems at high temperature, even with high melting point ceramic materials and with systems which, at the macroscopic scale, are considered non-reactive such as Ni/sapphire [44, 52–57].

As the dissolution process changes the liquid–vapour and the solid–liquid surface tensions, a new contact angle should result. If the new angle is lower than the

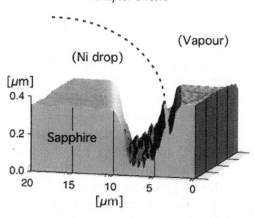

Figure 3. AFM scan of the reaction ring produced by molten Ni on a sapphire substrate, after complete removal of the Ni drop. The Ni drop free surface is marked schematically by a dashed curve. The scan reveals the presence of a relative large *ridge* located at the triple line junction, and of a deep *groove* adjacent to it, located beneath the drop, due to dissolution into liquid Ni of sapphire (reprinted from [57] with permission).

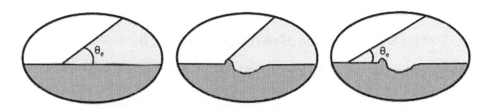

Figure 4. Ridge and groove formation and pinning effects.

Figure 5. Evolution of a drop in the presence of dissolution.

initial one, as soon as it reaches a critical value which allows for 'depinning' to take place (see below) (Fig. 4), the (saturated) liquid drop spreads over the solid surface producing a final profile similar to that shown in Fig. 5, where the resulting solid–liquid interface can assume a characteristic sigmoidal profile [58].

3. Reactive Wetting

Reactive wetting, proper, means that chemical interactions occur at the solid–liquid interface, with the production of new species which in turn can modify profoundly the wetting behaviour of the system, through the formation of interfacial new phases that have different wetting characteristics. Much debate can be found in the literature, especially as concerns the role that the free energy of reaction can have on the final contact angle. Good reviews/viewpoints on this topic can be found in Refs [12, 19, 21, 41, 44, 59–61]. It is clear that wetting, especially in reactive systems, is a very complex phenomenon. What can safely be said, is that the final equilibrium configuration, in the presence of reactions, should be determined by the formation of continuous layers of specific reaction products, such as the M_6X metal-like compounds in metal-oxide systems [60, 62, 63]. Which compound can form depends on the chemistry of the system: the presence in the liquid phase of minor elements, such as the active elements used in brazing processes (e.g., Ti, Zr, Cr, V, Nb, Al), can give rise to interfacial reactions whose type and extent depend on the activity of these elements in the experimental conditions. But the kinetics of the spreading process is a much more complex process. From the first contact of the liquid phase with the solid support, diffusion processes take place, so that diffusion-controlled reactive wetting is one of the basic mechanisms to be considered [42]. The triple line velocity in reactive-limited wetting has been shown to depend on the instantaneous contact angle, whose variation with time follows the relation:

$$\cos\theta_F - \cos\theta = (\cos\theta_F - \cos\theta_0)\exp(-kt), \tag{8}$$

where θ_F and θ_0 are the final and 'initial' contact angles [39].

C. Role of the Solid Surface in Modifying the Wetting Behaviour

Wetting of solid surfaces is affected, in addition to the mechanisms which are behind the three wetting typologies already discussed, by 'physical' and 'chemical' factors intrinsic of the solid surface that are briefly discussed below.

1. Roughness

The solid surfaces used in wetting experiments are not ideal. Especially when working at high temperature with polycrystalline materials, surfaces are not perfectly smooth. Ceramic materials are usually produced by sintering; this means that a residual porosity is always present and that, even after a very accurate polishing procedure, a residual surface roughness is still present when the liquid phase is put into contact with the solid. In addition, with temperature rising, micro-faceting and thermal grooving at grain boundaries takes place, so that it may be even impossible to evaluate the correct roughness value of the solid surface at the test temperature. It is well known that roughness affects the contact angle value by various mechanisms: (a) by increasing the solid–liquid contact area (increase of interfacial energy), (b) by providing 'pockets' of entrapped vapours, so that forming a composite interface, (c) by offering specific nucleation points for reaction and growth when reactions can occur, (d) by offering specific 'pinning' points to the movement of the triple line.

Wenzel [64] proposed, for rough surfaces, the following equation linking the 'macroscopic' contact angle θ_m to the 'intrinsic' one:

$$\cos\theta_m = \mathbf{r}\cos\theta. \tag{9}$$

Here **r** is the average roughness ratio (actual area/projected area). As **r** is always ≥ 1, this means that if $\theta > 90°$, θ_m increases with increasing roughness; the contrary when $\theta_m < 90°$.

The triple line can also be 'pinned' by the presence of surface heterogeneities. In this case, the local situation must be analysed. Grooves, ridges are all imperfections which must be overridden by the moving triple line. They alter the local geometrical profile of the surface, with an orientation of the contact plane different from the original reference one. On the basis of energetic arguments [65], it is possible to establish the maximum value for advancing contact angles and the minimum value for receding contact angles (the difference between the two is called hysteresis).

In the case of plane, heterogeneous surfaces, i.e., surfaces formed by two different materials or different phases, the macroscopic contact angle can be calculated using the Cassie equation [66]:

$$\cos \theta_c = f_1 \cos \theta_1 + f_2 \cos \theta_2, \tag{10}$$

where θ_1 and θ_2 represent the contact angles on fractions f_1 and f_2 of the total contact area, respectively.

2. Anisotropy

The Young (Eq. (1)) or Neumann (Eq. (7)) equations, dictate that the equilibrium solid–liquid configuration at the triple line depends on the values assumed by the solid–fluid interfacial energies. It is also well known that these energies are a function of crystallographic orientation at the surface, so that any anisotropy of the solid–fluid interfacial energies should result in anisotropy of wetting. A thorough discussion of this phenomenon can be found in recent papers by D. Chatain [67], with specific reference to surfaces and by Rohrer [68] with respect to grain boundaries.

Surface energy anisotropy affects wetting in different ways. Following the variation of interfacial tensions with orientation, also the Young contact angle should vary with the orientation along the surface, with the macroscopic consequence that a sessile drop is no longer axisymmetric.

But, more subtly, the change in surface energy has a certain influence on the adsorption and segregation processes which enter into play in the solid–liquid interactions [69]. Thus, also the kinetics of spreading and the final contact angle can be affected. Another point is related to the equilibrium 'shape' of the solid surface. At high temperature, due to the increased atom mobility, the surface, if not part of a monocrystal, tends to re-orient itself locally to expose the most stable crystallographic planes. This means that local faceting can occur which increases the surface roughness with the consequences on spreading and equilibrium contact angles already discussed. As pointed out in Ref. [67], the more the crystal orientations emerging at the surface are stable, the more the solid surface remains unchanged. The 'resistance' of a certain crystallographic plane to rotate can be measured by the

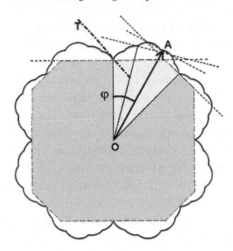

Figure 6. The Wulff construction (redrawn from [70]).

'torque terms' appearing in the Herring relationship [70] describing, in vectorial form, the equilibrium at the triple line:

$$\sum_{ij} \sigma_{ij} + \sum_{ij} \frac{d\sigma_{ij}}{d\varphi} = 0, \tag{11}$$

where i and j stay for solid, liquid or vapour and φ for the orientation of the surface energy vector σ. The torque term $(d\sigma/d\varphi)$ is a measure of the steepness of the cusp defining the stability of a certain orientation in the Wulff diagram [71]. This polar diagram, Fig. 6, is constructed by drawing, from the center **O**, vectors with radius **OA** proportional to σ and polar angle φ. The envelope of the perpendiculars to these vectors in the points **A** (dashed lines) define the limits of the most stable planes and thus the form of the equilibrium crystal. The line **T** shows the slope of the Wulff line in a cuspidal point: the more this line is steep, the more the cusp is deep and the relative (h, k, l) plane is stable.

3. Stoichiometry

Another important parameter which can play a leading role in determining the wetting behaviour of a particular solid–liquid system is the solid surface stoichiometry. A typical example is offered by TiN, TiC and TiB$_2$ when in contact with liquid metals. These systems have been studied many years ago by Samsonov [72] and Naidich [73], and more recently some contributions have appeared [74–78], in which contact angles and interfacial reactions have been reported for (Cu, Ag)/TiN$_x$, Cu–Ti/TiN$_x$ and (Cu, Ag, Au, Sn)/TiC$_x$ as a function of the nitride or carbide stoichiometry. These studies, notwithstanding a certain scatter of the experimental data, mainly due to the difficulty to characterise the solid surface properties and the real experimental conditions, seem to converge to the conclusion that the higher the ipo-stoichiometry in the solid surface the better its wettability by liquid metals, due to an enhanced metal-like electronic structure of the substrate. In particular, it has been shown [79] that two non-stoichiometric titanium diboride substrates ($TiB_{1.9}$ and $TiB_{1.95}$) show surprisingly good wetting in contact with liquid Cu and Au. This feature was attributed to a departure of titanium diborides composition from its stoichiometry. A thermodynamic analysis of the Me–Ti–B system has shown that an extremely limited boride dissolution takes place at the TiB_x/Me interface while the ratio of B/Ti in the interacting TiB_x layers could change substantially, improving wetting. For these reasons, it was suggested that the scatter of the experimental results of the wetting experiments using titanium diboride substrates reported in the literature could be attributed to the uncertainty regarding the substrate composition. Other interesting systems are represented by metal/oxide samples. In a dedicated study [80] the influence of stoichiometry on the wetting of zirconia by metals was investigated. The wetting behaviour of stoichiometric (white) and of ipo-stoichiometric (black) zirconia by different metal melts was studied. The results show that black zirconia is wetted by inert (Cu, Sn, Ag, Au, Pd, Pt, Cu–Ga, Pd–Rh) and by low active (Al, Cu–5Zr, (Cu–17.5Ga)–10Ti) melts better than the white one, while reactive melts have contact angles on black and on white substrates close to each other. Moreover, the ipo-stoichiometrization of the substrate can occur not only by annealing under either a vacuum or reducing atmospheres but also in presence of active metals in the melts.

The same problem has been also investigated by means of theoretical models. In one case [81], the wetting of sapphire by aluminium was simulated using a potential applying a combination of the Embedded Atom Method (EAM) and Coulomb potentials on a system of atoms with defined electronegativity. In the simulations of high-temperature wetting, the formation of an oxygen-deficient reaction layer between the liquid and the substrate was observed. The driving force for the creation of this layer has been attributed to the partial oxidation of the metallic aluminium, which results in partial reduction of the aluminium ions in the substrate and diffusion of oxygen from the substrate to the reaction layer. This change in stoichiometry at the metal–ceramic interface may influence diffusivities in the surface of the ceramic and may therefore have an important effect on the kinetics of the evolution of surface features observed experimentally during reactive wetting.

In a recent study [32], ab-initio Density Functional Theory (DFT) calculations have been applied to a Ag(111)/sapphire interface with a surface oxygen concentration which is intermediate between the Al-terminated stoichiometric interface and the O-rich, O-terminated interface. It has been shown that, in a large range of oxygen partial pressures in the gas phase, the Al-terminated/O-rich-Ag(111) interface is more stable with respect to the other two interfaces. On the basis of the electronic structure of the two most important configurations, a mechanism has been proposed to explain the gradual increase in adhesion observed experimentally as due to an increasing amount of oxygen at the interface. Such oxygen does not represent a particular termination of the alumina surface, instead probably diffuses through the molten Ag drop from the surrounding atmosphere.

D. Examples of Interacting Systems at High Temperature

Ceramic/metal joining is of great technological significance because, through this process, the individual characteristics of the two types of materials can be used to produce new components with improved performances. Liquid phase bonding processes, including brazing and Transient Liquid Phase bonding (TLPB) technique [82–88], are the most promising techniques for joining ceramics to metals. Recently, a revival of metal-ceramic bonding *via* glass- and glass-ceramic phases appeared in the literature, due to their thermal expansion characteristics, oxidation resistance, easy of processing [89, 90]. Therefore, investigations on wetting, spreading and interfacial behaviour in metal/ceramic [1, 19–21, 36, 91–96] or metal/metal systems [97–101], joining processes and joint performances [102–113] are essential.

In the following, the attention will be focused on two classes of materials, borides and carbides, less extensively studied with respect to oxides: they are classified within the class of Ultra High Temperature Ceramics (UHTC's), of particular interest for advanced applications, which are extensively studied also in our Laboratory.

Transition metals ceramic diborides, such as hafnium, titanium and zirconium diborides, as well as silicon carbide, are members of a family of materials with extremely high melting temperatures, high thermal conductivity, good electrical properties, excellent thermal shock resistance, high hardness and chemical inertness. These materials are promising candidates for use in high performance applications, where high temperatures, high thermal fluxes and severe surface stresses are involved [114, 115]. The unique combination of their properties makes them suitable for the extreme chemical and thermal environments associated with hypersonic flight, atmospheric re-entry, and rocket propulsion [116, 117]. Applications include nuclear plants [118, 119], refractory linings [120], electrodes [121, 122], microelectronics [123] and cutting tools [124].

In the following, a few examples are shown, reporting an overview of old and more recent results on the wettability and joining of these materials.

1. Transition Metals Diborides

Due to the high melting point of Ti-, Zr- and Hf-diborides ($3216°C < T_m < 3380°C$), the main technological barrier to the production of these materials is the sintering stage. As, generally, using pure borides the final microstructures are coarse due to the residual porosity, higher densities are achieved by pressure-assisted sintering procedures at temperatures approaching 2100–2300°C with the addition of appropriate sintering aids [125–129]. On the other hand, sintering aids modify not only the ceramic structure, but can also form new phases, grain-boundary precipitates and new solid solutions: in all cases they change the nature of the ceramic body, so that the contact angle measurements should be referred to the new system, which is often hard both to characterize or to reproduce [130]. Indeed, especially in the case of diborides, residual porosity together with oxide grains are commonly found, resulting in physical and chemical heterogeneity of substrate surface, both contributing to the scatter in contact angle measurements, as discussed in [131], and in the following Section E.

Other interactions can occur at the solid–vapour and solid–liquid interfaces, mostly due to the reduction or evaporation of surface oxides, or due to the dissolution of the boride into the liquid metal, as suggested by SIMS measurements [132].

Already some 40 years ago, a large amount of work was done on the high-temperature wettability and reactivity (in particular oxidation) of transition metal borides. In an extensive study [133] it was found that non-transition metals do not wet Group IV–VI metal diborides. The same authors [134] found that the diborides of group IV metals are more 'inert' in contact with liquid Fe and Ni than those of V and VI groups, which show very low contact angles often arriving to complete spreading. However, a better wetting behaviour was reported for Ni on TiB_2 ($\theta = 46°$) and ZrB_2 ($\theta = 54°$) by Tumanov *et al.* [135] under vacuum conditions (10^{-3} Pa).

Dissolution and chemical reactions have been found to occur in these systems, where infiltration of Ni in the solid ZrB_2 has been observed together with the dissolution of the diboride into the molten Ni drop, leaving, upon solidification, well recrystallized diboride crystals; the presence of a Ni_xB phase has also been consistently detected in the liquid drop [136].

In another study [137], Ag has been found to wet TiB_2 and ZrB_2 only at high temperatures, the contact angle decreasing from 125° for TiB_2 and 115° for ZrB_2 at 1100°C to 92° and 74°, respectively, at 1600°C, authorising as an extrapolation to foresee a good wetting behaviour also in the case of HfB_2.

More recently, Voytovich *et al.* [138] addressed the study of the role of Ni in modifying the wettability of Au on ZrB_2. It was shown that the Au–40at%Ni alloy had different interfacial reactivity with the substrate, leading to an 'equilibrium' contact angle $\theta = 25°$ and $\theta < 10°$ at 1170°C and 980°C respectively. This 'abnormal' behaviour (i.e., increasing wetting with decreasing T) was attributed to dissolutive wetting at 1170°C and compound formation (Ni_2B) at the interface at 980°C.

In the recent years, our Team has undertaken a systematic study, by the sessile-drop technique, of wettability, reactivity and interfacial properties of Group IV diborides, both from the basic (wettability, interfacial tension) and the application (joining) points of view [36, 94, 132, 139–141] in two main temperature ranges, i.e., at around 1100°C and 1500°C.

Wetting of ZrB_2 by Cu, Ag and Au. A series of wetting experiments for Zr diboride in contact with liquid Ag, Cu and Au have been performed under high vacuum (10^{-3} Pa) in the presence of a Zr getter [141] to keep low the oxygen partial pressure (10^{-26} Pa at 1100°C).

These tests have shown that, at $T = 1.05T_m$, at the end of the spreading process, the contact angles of Ag, Cu and Au decrease in substantial way ($\theta = 153°$; 80° and 34° respectively). For Cu and Au a kinetics exists, leading to a stationary situation

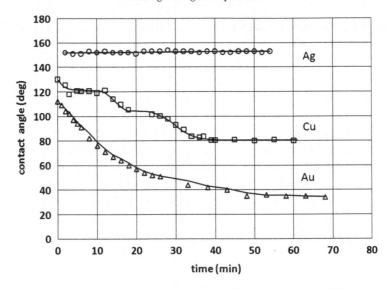

Figure 7. Kinetics of wetting of Ag, Au and Cu on ZrB$_2$ at $T = 1.05T_m$ [K].

only after a long time (35–60 minutes) whereas Ag maintains a very high non-wetting angle for all the experimental time (Fig. 7).

In the case of Cu and Au, interactions should occur at the interface, most probably due to the reduction or evaporation of surface oxides, and to a limited dissolution of the boride into the liquid metal, as suggested by SIMS measurements and thermodynamic calculations. In particular, boron can segregate to the liquid metal surface, suggesting a surfactant action. Recent surface tension measurements on the Cu–B system have shown that this effect is of the order of 10% [142].

In particular, gold not only wets well the ZrB$_2$ but also can penetrate along the solid grains, as shown in Fig. 8.

Another series of experiments were aimed at elucidating the role of the addition of active metal elements on the wetting kinetics of ZrB$_2$ by Ag.

As shown in Fig. 9, while pure Ag does not show any evolution of θ with time, a sharp decrease in contact angle is obtained by adding an 'active' element, Ti, Zr or Hf, to the molten Ag matrix.

However, the extent of this effect is different for the various metals. Zr is the most effective in promoting wettability of the boride. The Work of Adhesion, computed using the Young–Dupré eq. (2) (Table 1, where the alloys surface tensions have been computed by the Quasi-Chemical Solution Model [143, 144]), reaches quite high values in these systems, despite the fact that the added elements raise the liquid alloy surface tension with respect to the matrix, confirming that the relevant phenomenon which promotes wetting is the segregation of the active element to the solid–liquid interface. The term ($\sigma_{lv} \cos \theta$) represents the amount ('adhesion tension') by which the solid surface tension is decreased by the solid–liquid interactions to equal the interfacial tension value ($\sigma_{sv} - \sigma_{lv} \cos \theta = \sigma_{sl}$).

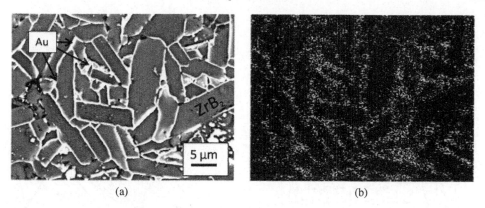

(a) (b)

Figure 8. SEM image (Back-scattering Electron—BE mode) of a cross sectioned Au/ZrB$_2$ sample: (a) Au penetration along ZrB$_2$ grains at the surface; (b) Au map.

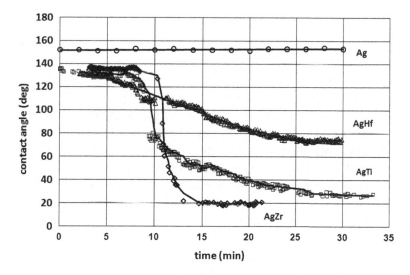

Figure 9. Kinetics of wetting of different Ag–X alloys on ZrB$_2$ at $T = 1050°$C.

Table 1.
Surface tension and contact angles for Ag–X alloys on ZrB$_2$ ($T = 1050°$C)

Metal/alloy (at%)	σ_{lv} (mN/m)	θ	W_a (mN/m)	$\sigma_{lv} \cos\theta$ (mN/m)
Ag	913	153°	995	−813
Ag–Ti 2.4	931	27°	1761	830
Ag–Zr 2.4	929	21°	1796	867
Ag–Hf 2.4	933	75°	1175	242

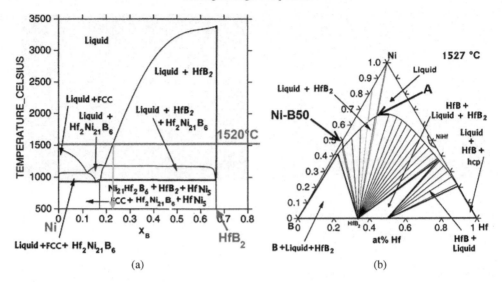

Figure 10. Calculated Ni–B–Hf phase diagram: (a) Ni–HfB$_2$ isopleth; (b) isothermal section at 1527°C [58].

These values (Table 1) clearly show that Zr is very efficient in adsorbing/reacting at the interface and promoting wetting.

Ni–HfB$_2$ System—Evaluation of the Phase Diagram. As already said, transition metal diborides, are ideal materials for applications where extreme conditions are encountered. Thus, brazing processes making use of filler alloys able to withstand temperatures higher that those characteristic of Ag- or Cu-based alloys should be studied. We present here the particular case of the Ni–HfB$_2$ system.

In order to foresee and understand the solid–liquid interactions in this system, a detailed estimation of the various phase equilibria existing at different temperatures is strongly needed. Due to the lack of literature data, [145, 146], we have recently computed, by CALPHAD methods, the complete Ni–Hf–B ternary diagram [58, 147]. The resulting ternary sections at temperatures between 800°C and the melting point of HfB$_2$ at 3380°C, display the binary phases, the ternary liquid as well as the Hf$_2$Ni$_{21}$B$_6$ 'tau' solid phase.

When molten Ni is brought into contact with HfB$_2$, an equilibrium condition is established between solid HfB$_2$ and Ni-based phases which, when temperature is high enough, are in the liquid state. On the basis of the computed phase diagrams, pure Ni, put in contact with HfB$_2$ melts at a temperature (about 1050°C) well below its melting point (1453°C), giving rise to a ternary liquid phase due to an important dissolution of HfB$_2$ (Fig. 10a); as an example, at 1527°C, the composition (in at%) of this liquid solution results Ni 66%–B 22%–Hf 12% (point A in Fig. 10b). But the same diagram shows also that, at this same temperature, the dissolution of the boride is greatly reduced and even suppressed if, instead of pure Ni, a Ni–B alloy with $X_B > 50$ at% is put in contact with HfB$_2$.

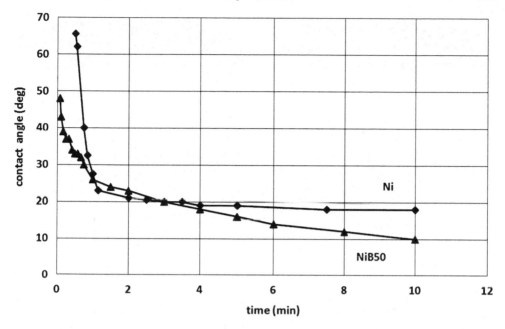

Figure 11. Kinetics of wetting of Ni (♦) and NiB50 (▲) on HfB_2 at 1520°C.

Wetting Results. On the basis of these thermodynamic considerations, sessile drop experiments have been made, at 1520°C, using different Ni–B alloys composition. In particular, the results regarding pure Ni and the Ni–50 at%B intermetallic compound (NiB50) which undergoes a peritectic reaction at 1035°C [93] are briefly discussed here.

The behaviour of contact angles *vs.* time for the Ni alloys in contact with the HfB_2 substrate at 1520°C is shown in Fig. 11.

Two results must be underlined: (1) the largest part of the solid–liquid interactions occur during the first minute of contact, where, in the case of pure Ni, strong dissolution of the substrate takes place, and, in the case of the NiB50 alloy, where no dissolution at all has been found; (2) in the case of dissolution, the contact angles remain nearly constant after about two minutes of contact, while in the case of wetting without dissolution the contact angle keeps decreasing for a long time and seems not having reached an equilibrium value even after 10 minutes, with a final contact angle < 10°.

As shown in Fig. 10b, the isothermal section at $T = 1527$°C shows that the NiB50–HfB_2 line crosses the liquidus line at $X_{Hf} \cong 0.03$. This means that a very little amount of HfB_2 dissolves into the liquid phase, which however reduces to nearly zero already at 1130°C as shown by other diagrams not reported here [58].

The drop–ceramic cross sections are presented in Fig. 12: a striking difference is seen in the interface structure for the two cases. With pure Ni, the formation of a sigmoidal interface profile, is the result of the competition between the rapid dissolution of HfB_2 in the liquid phase, the pinning of the liquid drop at the triple line

(a)   (b)

Figure 12. Optical micrographs of sessile drop cross sections: (a) Ni/HfB$_2$ couple showing the particular sigmoidal interface; (b) NiB50/HfB$_2$ with a flat solid–liquid interface.

(a)   (b)

Figure 13. SEM images (BE mode) of Ni–HfB$_2$ system. Details of the internal structure of the solidified drop: (a) pure Ni: solid–liquid interface, with big HfB$_2$ crystals grown inside the liquid phase. (b) HfB$_2$ crystals, from primary solidification at the surface of the drop.

and the subsequent de-pinning followed by fast drop spreading along the substrate surface. No new products are seen at the interface, except for a layer of HfB$_2$ phase newly formed by the dissolution-precipitation process (dissolutive wetting).

Compared to pure Ni, the NiB50 alloy shows no dissolution of HfB$_2$ with the consequent formation of a flat, 'undisturbed' interface as predicted by the calculated phase diagram.

The SEM image (Fig. 13a) of the pure Ni/HfB$_2$ couple shows relevant recrystallisation of HfB$_2$ crystals at the solid–liquid interface, about ten times larger than the original sintered ones and (Fig. 13b) at the liquid surface in the form of thin, hexagonal platelets. The solidified Ni-based alloy has an eutectic structure with a large part composed of the intermetallic compound Hf$_2$Ni$_{21}$B$_6$ (theoretical composition B 21%, Ni 72%, Hf 7%) which is foreseen to appear below 1220°C and an intermetallic phase whose composition is close to Ni$_3$B.

The NiB50 alloy (Fig. 14) reveals, after solidification, no boride recrystallization both at the interface and in the bulk alloy, and an eutectic structure with a NiB light grey phase and a Ni$_4$B$_3$ dark grey one. No Hf has been detected in the liquid phase.

Joining. Joining of transition metal diborides is a topic which is increasingly attracting research in force of its intrinsic interest for advanced applications, mainly in transport, energy and aerospace industry. In recent years a number of papers has appeared, addressing different aspects of bonding technology of such materials [148–156].

Also in our laboratory, on the basis of the experimental and theoretical findings reported in the previous paragraphs, ZrB$_2$ and HfB$_2$ joining tests have been made,

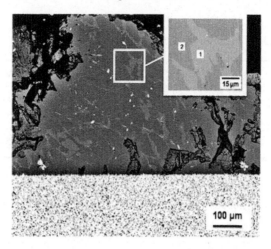

Figure 14. NiB50 alloy: solid–liquid interface showing the eutectic structure and a negligible recrystallization of HfB$_2$ at the interface.

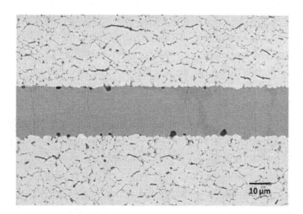

Figure 15. Cross section SEM micrographs (BE mode) of a HfB$_2$//NiB50//HfB$_2$ joint: $T = 1300°$C, Ar/H$_2$ 10^{+5} Pa, 10 min [157].

operating at 1300°C under the same conditions of the wetting tests using the NiB50 alloy as filler metal [139, 140, 157, 158].

Figure 15 shows the metallographic section of a HfB$_2$/Ni50B/HfB$_2$ joint, where a very good penetration of the filler alloy can be observed, without any voids and any dissolution of the solid ceramic phase. Some exploratory shear tests have been made, using a specific attachment device to an INSTRON 9501 mechanical testing machine, to evaluate the interfacial shear strength. Very good results have been obtained, ranging from 70 to 170 MPa [157].

For all specimens a typical brittle fracture happens in the ceramic body and not along the joint with cracks propagating in the ceramic and sometimes across the interfacial metallic layer.

Table 2.
Reactivity of pure metals in contact with SiC [167–169]

Category	Reactivity type	Metal
1	No reaction	Au, Ag, Sn, Pb, Ge
2	Me + SiC → silicide + C	Ni, Fe, Cu, Co
3	Me + SiC → Si + carbide	V, Al, Nb
4	Me + SiC → silicide + carbide	Zr, Ti, Hf, Cr, Ta, W, Mo

2. Silicon Carbide

Silicon carbide is of great technological significance due to its good overall chemical, thermal and mechanical properties, which make it a suitable candidate for aggressive environments. It is used both as a structural and as a functional material; indeed, it is currently used in high-temperature applications, such as metal-matrix composites, heat exchangers, aerospace and electronic applications, etc. [159–161]. A case in point is nuclear applications, with demands in thermal shock resistance, radiation stability and low neutron-induced activation [162–166]. Therefore, joining (brazing) SiC to SiC or to SiC_f, to refractory metals or to superalloys, can lead to optimized highly performing applications.

The reactivity of the pure metals with SiC directly affects the interfacial reactions, wetting and bonding. Based on the research results of Kurokawa and Nagasaki [167] and the reviews of Rabin [168] and Li and Zhang [169], the pure metals may be classified into four categories according to their reactivity with SiC, as shown in Table 2. The metals of the first category show almost no chemical reactions with SiC under the experimental conditions. In the second category, only stable silicides can form at the interface; the silicon and carbon originated from the SiC matrix tend to diffuse into the metal component; the carbon can exist as a solid solution provided the solubility of carbon in the metal is high enough, otherwise it will segregate as graphite in the liquid phase or even at the solid–liquid interface after saturation. In the third one, the carbides are the main products, as most silicon dissolves into the metal and forms carbides, although sometimes a small quantity of silicide can form [170]. In the fourth category, a mixed layer containing silicides and carbides often forms at the interface by reaction and mutual diffusion between the metal and the ceramic.

A recent review on SiC wetting by liquid metals and alloys [171] has shown that the wettability of SiC depends on three main factors:

(1) the condition of the SiC surface,

(2) the presence of adsorption phenomena at the solid–liquid interface,

(3) the presence of reaction/dissolution phenomena between the liquid phase and the solid substrate.

The presence of a SiO_2 layer at the SiC external surface inhibits wetting by liquid metals and adhesion. Only the removal of this oxide layer allows wetting to take place. As a consequence, the spreading kinetics is controlled by the kinetics of removing of these wetting barriers on SiC surface, which can be due to the presence of a specific reducing agent such as Si, or to a sufficiently low oxygen partial pressure. The related chemical reactions may be expressed as follows:

$$SiO_2 \text{ (s)} + Si \text{ (l)} \rightarrow 2SiO \text{ (g)} \uparrow \tag{12}$$

$$SiC \text{ (s)} + 2SiO_2 \text{ (s)} \Leftrightarrow 3SiO \text{ (g)} \uparrow + CO \text{ (g)} \uparrow \tag{13}$$

$$SiO_2 \text{ (s)} \Leftrightarrow SiO \text{ (g)} + 1/2O_2 \text{ (g)} \tag{14}$$

Reaction (14) shows that low oxygen partial pressures shift the equilibrium towards silicon monoxide and molecular oxygen formation.

Furthermore, the spreading arrives at equilibrium more slowly in the gas mixture than under a vacuum, and sometimes it can even not be attained in the test time. This can be explained by the fact that the reaction (12) and the forward reaction (13) are slower in the gas mixture than under vacuum due to the presence of gaseous products.

Thus, experimental or processing conditions should be chosen where this surface layer can be eliminated. This can be obtained by fixing an oxygen partial pressure PO_2 for which the transition between passive (formation of SiO_2) to active (formation of SiO) oxidation takes place (in the case of an Ar atmosphere, with $PO_2 \approx 1$ Pa, $T \approx 1100°C$). But, on the other hand, the SiO_2 surface layer can also be eliminated through reaction (12), by adding Si to the molten alloy. In both cases, wetting is made to occur on a clean SiC surface.

At present, the techniques for joining SiC to itself or to metals for high temperature applications include mainly brazing (including TLPB) [102–105, 172–174], diffusion bonding [175–179], self-propagating high temperature synthesis (SHS) welding [111, 180] and glass- or glass-ceramic sealing [89, 90].

In the case of (non-reactive pure metal + Si)/SiC systems, the good wetting and adhesion on clean SiC surfaces is attributed to silicon chemisorption at the metal/SiC interface with formation of strong covalent-like bonds between Si and SiC, to the electronic properties of these alloys and to their interaction with SiC, as well as the high adhesion energy of the pure metal itself on SiC. Therefore, many researchers endeavoured to seek for some non-reactive joining processes of SiC using binary or ternary compounds with high silicon content [181–187]. Actually, many investigations showed that a series of binary silicon alloys, such as Ni–Si [188–191], Co–Si [205], Fe–Si [192], Au–Si [193, 194], Ag–Si [195, 196] and Cu–Si [197–199], have this good non-reactive wettability in contact with SiC.

In the (reactive pure metal + Si)/SiC systems [200–204], it has been shown that the Si concentration in the liquid, reacting metal is a critical parameter in determining the solid–liquid behaviour. Indeed, pure metals like Ni, Co, Fe, etc. when put in contact with SiC react with it and dissolve a large amount of Si and C into the liquid phase. However, it has been found in the Ni–40Si/SiC [190], Co–72.5Si/SiC

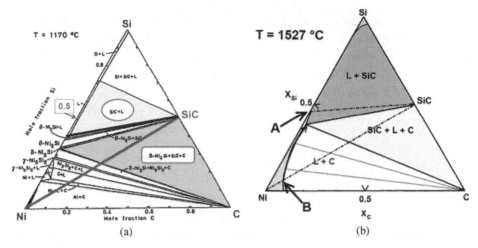

Figure 16. Ternary phase diagrams computed at 1170°C and 1527°C (adapted from [206]).

[205] and Ag–5Si/SiC [195] systems that, when the original content of Si in Ni is above ∼40%, the equilibrium is established between the Ni–Si alloy and SiC. After a small dissolution of SiC the saturated liquid alloy does not react anymore with the solid substrate, leaving an undisturbed interface.

In close similarity with what already discussed in the case of diborides systems, this behaviour can be interpreted by means of the ternary phase diagrams X–Si–C. In the specific case of Ni–Si–C diagram [206], it is seen (Fig. 16) that the tie-line between the pure Ni and the SiC compound crosses first of all, in point B the liquidus line: here, the liquid phase has already dissolved a small quantity of SiC. From this point onwards, the composition of the liquid follows the arrow becoming enriched in Si and segregating C until a three-phase region is reached with the simultaneous presence of SiC + C + Liquid. It is then clear that when pure Ni is put in contact with SiC both dissolution of the substrate and reprecipitation of C takes place. However, the same diagram shows that additions of Si from $X_{Si} \approx 0.4$ upwards can limit or even suppress the substrate dissolution leaving only two phases at equilibrium, i.e., the nearly pure Ni–Si alloy and the solid SiC. It is thus clear that ternary Me–Si–C phase diagrams should be established for a number of systems as large as possible in order to allow a wider access to their isothermal sections related to the experimental temperatures.

These results have been recently confirmed in our Laboratory, in a study of the wetting of SiC by Si-rich Ni alloys [204]. The spreading kinetics of Ni–56Si on SiC at three temperatures (1100°C, 1200°C and 1350°C) is shown in Fig. 17. The spreading rate in the first, fast stage obviously increases with temperatures increasing (Fig. 17a). According to the changes of drop dimensions (base diameter and drop height) and contact angle, the spreading does not reach the contact equilibrium completely at 1200°C after a holding time of 1 hour, and the contact angle curve can be separated into three parts, I, II and III (Fig. 17b). The spreading rates in the three

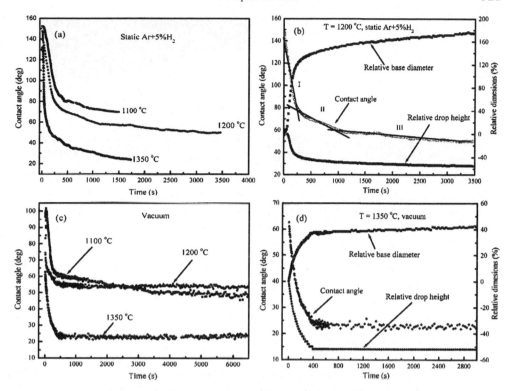

Figure 17. Contact angle (and relative drop dimensions) as a function of time. Spreading kinetics of Ni–56Si on SiC: (a) at three temperatures in static Ar + 5%H$_2$; (b) at 1200°C in static Ar + 5%H$_2$; (c) at three temperatures in a vacuum and (d) at 1350°C in a vacuum, showing the effects of temperature and atmosphere on the wettability and spreading process (reprinted from [191] with permission).

stages fall in three ranges, namely, $\sim 2.9 \times 10^{-1}$, 2.7×10^{-2} and 3.7×10^{-3} (deg/s), respectively. As shown in Fig. 17a and c, the spreading processes show some differences between the tests made in a gas mixture (Ar + 5%H$_2$) and under a vacuum. In vacuum the spreading arrives at the equilibrium in 500 seconds at 1200°C and 1350°C (Fig. 17c and d), however, at the lower temperature (1100°C) the spreading does not arrive at the equilibrium after the fast spreading and keeps going at a low rate for ~ 50 min (Fig. 17c). These facts can be seen as a confirmation of the mechanisms of 'surface cleaning' explained by reactions (12)–(14).

Figure 18 shows the SEM micrographs of cross-sectioned Ni–Si/SiC couples at two temperatures. It can be seen that the solid–liquid interface in the Ni–56Si/SiC system is smooth and clean without any precipitates also in the region close to the triple line. Thus, from the applications point of view, it should be noted that the Ni–Si alloys possess the best wettability ($\theta \approx 20$–$30°$) and the greater potential for joining SiC ceramics. However, as shown also in Fig. 18b, some cracks are present at the interfaces and in the bulk phases due to the presence of the brittle silicides

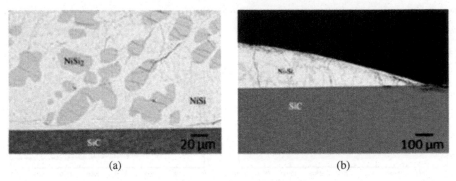

Figure 18. Cross-section SEM micrographs (BE mode) of the Ni–Si56/SiC couples at (a) 1100°C and (b) 1350°C, showing the non-reactive wetting characteristics and some cracks formed during cooling. The white and gray phases in the drop are NiSi and NiSi$_2$, respectively.

(NiSi and NiSi$_2$) and to the thermal expansion coefficient difference between the Ni–Si alloy and SiC.

3. Remarks

The high temperature studies of the interactions between molten alloys with ceramic materials, have shown that the dissolution phenomena, occurring between the molten pure metal and the ceramic, can be nearly completely suppressed by ad-hoc additions of a second element which saturates the molten phase. This result and the interpretation of the complex phenomena occurring at high temperatures between the molten phase and the ceramic substrate (dissolution, reaction as well as the formation of new phases during the solidification process), are made possible by using reliable phase diagrams of the ternary and higher-order systems, which can be computed on the basis of known thermodynamic quantities using advanced databases and dedicated softwares.

The resulting isothermal sections and the isopleths connecting the ceramic composition to that of the filler alloy permit a deeper understanding of the experimental phenomena observed both during the wetting tests and in the cooling process.

E. Experimental Techniques for High Temperature Wettability Measurements

Surface tension and wettability measurements at high temperature are increasingly requested to characterize the behavior of systems of industrial interest, both to define precisely their thermophysical properties to be used in the modelling of production processes and to determine their behavior in highly aggressive environments, be them of chemical or mechanical origin.

Good suggestions and rules to observe when setting up a measurement apparatus (furnaces, optical lines, data acquisition, etc.) or when designing a series of specific tests can be found in specialized books and papers [1, 54, 91, 131, 207].

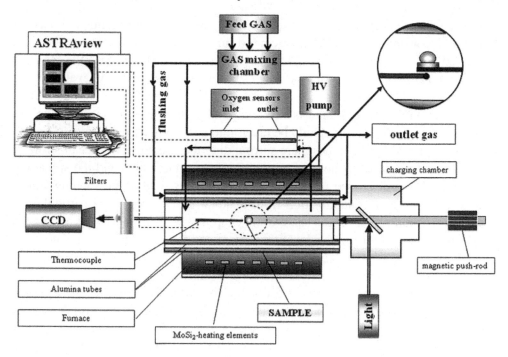

Figure 19. Schematic view of our sessile drop apparatus in the horizontal tubular configuration. Block diagram without flanges and fittings.

For a fixed metal/ceramic couple, the wetting behaviour can be influenced by many factors: to obtain reliable Young contact angle measurements, first of all some essential information about the substrate surface must be given in detail. The substrate quality should include surface chemistry state, such as surface oxidation, metallization and, as already discussed, non-stoichiometry [74, 76], surface roughness or surface patterning [208], surface polarity [190] or structure and substrate phase composition (such as, presence of sintering aids [94] and of different phases [41]), but also surface anisotropy [67]. In addition, as some products often form at the interface due to interfacial reactions during the wetting process, the presence of this new interface has to be demonstrated by suitable analytical methods.

Experiments are conducted using in most of cases the sessile-drop technique in conjunction with ad-hoc designed image analysis softwares, which allow surface tension and contact angle data to be obtained during each experimental run (e.g., [209, 210]). Especially designed furnaces are used which can reach 1600°C or higher temperatures (Fig. 19).

In Fig. 19a schematic view of our experimental apparatus is shown, where: (1) the specimen can be introduced into the test zone only when temperature and gas composition are constant at the preset values, (2) the oxygen content of the gas mixture is measured at the entrance and at the exit of the furnace, (3) a gas (Ar) is made to flux between the inner and the outer alumina tube in order to keep at a minimum oxygen diffusion in the experimental 'cell', (4) the specimen is back-lit to obtain a constant contrast in the recorded image, whereas the self-emitted light is filtered out, (5) vacuum turbomolecular pumps are used to produce the required high-vacuum conditions, (6) the specimen support can be accurately leveled at any moment.

To perform reliable wetting tests a certain number of requirements should be followed strictly. First of all, it should be taken into account that an extremely important role is played by the atmosphere surrounding the specimens as it can strongly affect the wetting results (the final or equilibrium contact angle and the spreading kinetics [211]) as discussed in the previous paragraphs. Ultra high vacuum conditions can assure that contamination from residual gases (oxygen in particular) is kept to a minimum, but, at the same time, evaporation from the liquid and/or the solid surfaces can alter the residual atmosphere and thus the interfacial chemistry. For this reason, a getter (e.g., Zr), is often placed near the specimen to assure the right oxygen pressure to obtain stable oxide-free surfaces.

In particular, under low pressures, also the materials forming the experimental cell can enter into play, for example giving rise to volatile oxides which impose a specific oxygen partial pressure, as in the presence of silica-based materials. If gas mixtures are used, their oxygen content should be monitored; it is suggested to measure both the inlet and the outlet oxygen partial pressures, when working under flux, if the 'local' values are not accessible (solid-state oxygen gauges do no work above about 900°C). It should be reminded that, when the PO_2 is dictated by an equilibrium reaction in gas phase (buffer systems) the value measured by the pressure gauges must be re-calculated at the experimental temperature.

Another critical point is linked to the sequences leading to the melting of the metallic phases and its coming into contact with the solid surface under study. Indeed, if the metal and the solid support are heated up together, many interaction phenomena can occur during this, usually long, phase. In particular, surface oxidation can easily occur, even if vacuum or protecting gases are used, because the oxygen equilibrium pressure for oxide formation is very low at low temperatures, at levels (often below 10^{-15}–10^{-20} Pa) which cannot be 'corrected' by protective gases. At the same time, chemical interactions can occur between the two solid phases, leading to mutual diffusion phenomena which can arrive at causing local contact melting processes. To avoid this, two possibilities exist: the first one is to introduce the metal/ceramic couple, charged in a pre-evacuated chamber, into the preheated furnace (e.g., by a magnetically operated push-rod) only when all parameters (T, PO_2) have reached equilibrium; the time necessary to complete the melting process is usually of the order of tens of seconds, depending on the set temperature, the atmosphere, the mass of the test pieces and their specific heat. At the end of the pre-set holding period, the couple is moved into the colder part of the experimental chamber without opening it, where it is cooled down to room temperature (this choice is shown in Fig. 19).

Another, more advanced possibility, is to work with a furnace where both the solid ceramic and metal samples are introduced, separately, into the pre-heated chamber and put into contact only when the metal is fully melted. This can be done using either the dispensed drop or the transferred drop methods. In the first one, a ceramic syringe is used from which a molten drop can be squeezed out by a piston or by applying to the molten reservoir a small gas overpressure. This optimized method has the additional, important, advantage of 'cleaning' the liquid metal surface of the oxides which could have formed prior to melting and to allow multiple operations to be made '*in situ*' at the experimental temperature [212, 213], as clearly shown in Fig. 20.

In the transferred drop method, the substrate to be tested is lowered until it just touches a liquid drop, lying on substrate which must be as much as possible inert.

Using these methods, the spreading can be analysed from the very beginning of the process without any prior interaction between the liquid and the substrate during the heating up as in the conventional sessile-drop method.

A further remark: as, quite usually, wetting experiments are performed on poly-crystalline surfaces, and this fact, as shown in Section C, can lead to hysteresis/pinning effects, it is sometimes possible to utilize micrometer-size drops formed on top of single grains to obtain reliable measurements of contact angles (microwetting), assuming the solidification does not alter sensibly the contact angle. In Fig. 21, a tiny Au drop, formed on top a single boride grain, allowed an evaluation of the 'true' contact angle on a monocrystalline ZrB_2 surface, its value of about $\theta = 35°$ is very close to the 'macroscopic' angle measured at the experimental temperature ($\theta = 34° \pm 2°$).

After solidification, the weight of the specimen must be measured to check for the presence of evaporation phenomena, and the top surface of solidified sessile drop/ceramic couples examined by optical and scanning electron microscopy (SEM) coupled with EDS analysis. Next, the specimens are cross-sectioned and polished for structural characterization by optical microscopy (OM) and Scanning Electron Microscopy (SEM), by Energy Dispersion Analysis (EDS) or by Wave-length Dispersion Analysis (WDS) technique, in the presence of light elements (B, C in particular). More sophisticated techniques such as Focused Ion Beam (FIB), Transmission Electron Microscopy (TEM), nano-indentation techniques should be used if the properties of the real metal–ceramic interface has to be studied in a fine way.

To summarize, and in close agreement with the recommendations given in Ref. [131], the following steps should be observed, for reliable wetting and contact angle measurements:

1. The chemical composition of the solid and liquid components should be explicitly given.

2. The wetting experiments must be conducted in controlled furnace conditions obtained either by high vacuum or, if reactive gases (oxygen, sulfur, nitrogen,

Methods / Procedures

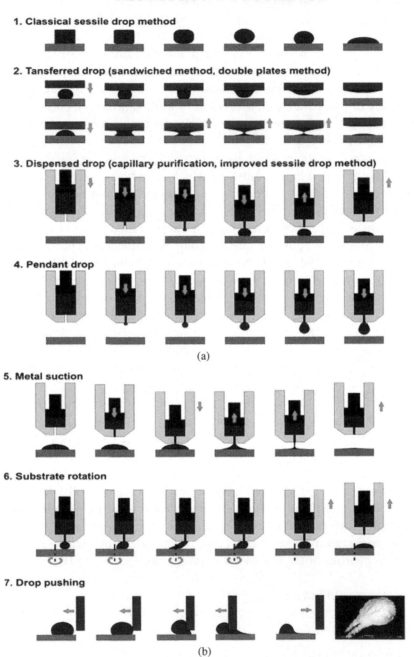

1. Classical sessile drop method

2. Tansferred drop (sandwiched method, double plates method)

3. Dispensed drop (capillary purification, improved sessile drop method)

4. Pendant drop

(a)

5. Metal suction

6. Substrate rotation

7. Drop pushing

(b)

Figure 20. Schematic overview of a number of possible operations by using classical (1), transferred (2), capillary dispenser (3–6) and drop-removal methods. (By courtesy of Prof. N. Sobczak, Cracow [213].)

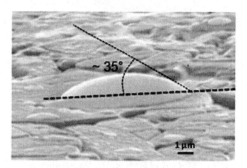

Figure 21. Microscopic contact angle of Au on a ZrB$_2$ single grain.

etc.) are present, by measuring their partial pressure or concentration, and flux rate.

3. The surface roughness has to be quantified: substrates with a roughness parameter $R_a < 100$ nm must be used in order to achieve an acceptable accuracy on θ (a few degrees). In particular, the area on which this measurement has been done should be clearly stated, because measurements along a short, unique, line are not reliable enough.

4. When the substrates are ionocovalent oxides, a heat treatment in high vacuum at temperatures higher than 1000°C prior to wetting allows any adsorbed OH group to be removed from the surface.

5. After the tests a check must be carried out on whether the solid surface remained plane or not. If dissolution has occurred, the measured angle cannot be assigned as the true contact angle of that particular system, but labeled as 'apparent contact angle'.

6. With non-oxide but oxidizable substrates the characterization of their surface chemistry before and after the wetting experiment is often necessary in order to identify clearly the meaning of the measured contact angle.

7. Although, as a general rule, advancing contact angles are involved in wetting experiments, only the simultaneous inspection of $\theta(t)$, $R(t)$ and $h(t)$ curves allows to confirm this feature. By doing this, evaporation and infiltration phenomena leading to variations in the drop volume can be easily detected.

8. When using automatic techniques based on image-analysis softwares, the contact angle should be derived from the solution of the Laplace equation or, if a portion of the drop close to the triple line is used, taking care that the number of coordinate points used is large enough to warrant a sufficient statistical accuracy.

9. Evidence should be offered, at least with a check on the solidified drop, that an axisymmetric profile has been retained during the test: asymmetric spreading

on the solid substrate makes the contact angle to vary from point to point along the triple line.

10. When calculating the work of adhesion W_a (Eq. (2)), the contact angles and surface tension data should come from measurements made under the same conditions (and with the same substrate).

F. References

1. Eustathopoulos, N., Nicholas, M. G. and Drevet, B., Wettability at High Temperatures. Pergamon Materials Series, Oxford, 1999.
2. Chatain, D., Ghetta, V. and Fouletier, J., in: Ceramic Microstructure, Control at the Atomic Level, A. P. Tomsia (Ed.). New York, Plenum Press, 1998, p. 349.
3. Backhaus-Ricoult, M., Acta Mater., 48 (2000) 4365.
4. Laurent, V., Chatain, D., Chatillon, C. and Eustathopoulos, N., Acta Met., 36 (1988) 1797.
5. Arato, E., Ricci, E. and Costa, P., Mater. Chem. Phys., 114 (2009) 809.
6. Ricci, E., Giuranno, D., Arato, E. and Costa, P., Materials Science and Engineering A, 495 (2008) 27.
7. Arato, E., Ricci, E. and Costa, P., Surf. Science, 602 (2008) 349–357.
8. Ricci, E., Arato, E., Passerone, A. and Costa, P., Adv. Colloid Interface Sci., 117 (2005) 15.
9. Novakovic, R., Ricci, E., Gnecco, F., Giuranno, D. and Borzone, G., Surf. Sci., 599 (2005) 15.
10. Ratto, M., Ricci, E., Arato, E. and Costa, P., Met. Trans. B, 32B (2001) 903.
11. Ratto, M., Ricci, E. and Arato, E., J. Crystal Growth, 217 (2000) 233.
12. Saiz, E., Cannon, R. M. and Tomsia, A. P., Annu. Rev. Mater. Res., 38 (2008) 197–226.
13. Wittmer, M., Mat. Res. Soc. Symp. Proc., 40 (1985) 393–398.
14. Kim, J. Y., Hardy, J. S. and Weil, K. S., J. Am. Ceram. Soc., 88 (2005) 2521–2527.
15. Diemer, M., Neubrand, A., Trumble, K. P. and Rodel, J., J. Am. Ceram. Soc., 82 (1999) 2825–2832.
16. Chatain, D., Chabert, F., Ghetta, F. and Fouletier, J., J. Am. Ceram. Soc., 77 (1994) 197–201.
17. Chatain, D., Rivollet, I. and Eustathopoulos, N., J. Chim. Phys., 84 (1987) 201–203.
18. Sangiorgi, R., Muolo, M. L., Chatain, D. and Eustathopoulos, N., J. Am. Ceramic Soc., 71 (1988) 742–748.
19. Eustathopoulos, N., Curr. Opin. Solid State Mater. Sci., 9 (2005) 152–160.
20. Naidich, Y. V., Curr. Opin. Solid State Mater. Sci., 9 (2005) 161–166.
21. Saiz, E. and Tomsia, A. P., Curr. Opin. Solid State Mater. Sci., 9 (2005) 167–173.
22. Wynblatt, P., Acta Mater., 48 (2000) 4439–4447.
23. Benhassine, M., Saiz, E., Tomsia, A. P. and De Coninck, J., Acta Mater., 59 (2011) 1087–1094.
24. Webb III, E. B., Grest, G. S., Heine, D. R. and Hoyt, J. J., Acta Mater., 53 (2005) 3163–3177.
25. Bertrand, E., Blake, T. D. and De Coninck, J., J. Phys.-Cond. Matter, 21 (2009) 464124.
26. Mattsson, A. E., Schultz, P. A., Desjarlais, M. P., Mattsson, T. R. and Leung, K., Modell. Simul. Mater. Sci. Eng., 13 (2005) R1–R31.
27. Wang, X. G. and Smith, J. R., J. Am. Ceram. Soc., 86 (2003) 696–700.
28. Zhang, W. and Smith, J. R., Phys. Rev. Lett., 85 (2000) 3225–3228.
29. Finnis, M. W., J. Phys., Condens. Matter, 8 (1996) 5811–5836.
30. Muolo, M. L., Valenza, F., Passerone, A. and Passerone, D., J. Mater. Sci. Eng. A, 495 (2008) 153–158.
31. Hashibon, A., Elsässer, C. and Rüle, M., Acta Mater., 55 (2007) 1657–1665.

32. Passerone, D., Pignedoli, C. A., Valenza, F., Muolo, M. L. and Passerone, A., J. Mater. Sci., 45 (2010) 4265–4270.

33. Mehandru, S. P. and Anderson, A. B., Surf. Sci., 245 (1991) 333–344.

34. Dudiy, S. V. and Lundqvist, B. I., Phys. Rev. B, 69 (2004) 125421-125421-16.

35. Liu, M. L., Wang, S. Q. and Ye, H. Q., Surf. Interf. Analysis, 35 (2003) 835–841.

36. Passerone, A., Muolo, M. L. and Passerone, D., J. Mater. Sci., 41 (2006) 5088.

37. Saiz, E., Benhassine, M., De Coninck, J. and Tomsia, A. P., Scripta Mater., 62 (2010) 934–938.

38. Voué, V. and De Coninck, J., Acta Mater., 48 (2000) 4405–4417.

39. Dezellus, O., Hodaj, F. and Eustathopoulos, N., Acta Mater., 50 (2002) 4741–4753.

40. Webb III, E. B., Hoyt, J. J. and Grest, G. S., Curr. Opin. Solid State Mater. Sci., 9 (2005) 174–180.

41. Kumar, G. and Prabhu, K. N., Adv. Colloid Interf. Sci., 133 (2007) 61–89.

42. Hodaj, F., Dezellus, O., Barbier, J. N., Mortensen, A. and Eustathopoulos, N., J. Mater. Sci., 42 (2007) 8071–8082.

43. Webb III, E. B. and Hoyt, J. J., Acta Mater., 56 (2008) 1802–1812.

44. Saiz, E., Cannon, R. M. and Tomsia, A. P., Acta Mater., 48 (2000) 4449.

45. Protsenko, P., Garandet, J. P., Voytovych, R. and Eustathopoulos, N., Acta Mater., 58 (2010) 6565–6574.

46. Kozlova, O., Voytovych, R., Protsenko, P. and Eustathopoulos, N., J. Mater. Sci., 45 (2010) 2099–2105.

47. Villanueva, W., Boettinger, W. J., Warren, J. A. and Amberg, G., Acta Mater., 57 (2009) 6022–6036.

48. Su, S., Yin, L., Sun, Y., Murray, B. T. and Singler, T. J., Acta Mater., 57 (2009) 3110–3122.

49. Yin, L., Chauhan, A. and Singler, T. J., Mater. Sci. Eng. A, 495 (2008) 80–89.

50. Protsenko, P., Kozova, O., Voytovych, R. and Eustathopoulos, N., J. Mater. Sci., 43 (2008) 5669–5671.

51. Yin, L., Murray, B. T. and Singler, T. J., Acta Mater., 54 (2006) 3561–3574.

52. Saiz, E., Tomsia, A. P. and Cannon, R. M., Acta Mater., 46 (1998) 2349–2361.

53. Saiz, E., Cannon, R. M. and Tomsia, A. P., Acta Mater., 47 (1999) 4209–4220.

54. Levi, G. and Kaplan, W. G., Acta Mater., 51 (2003) 2793–2802.

55. Saiz, E., Tomsia, A. P. and Cannon, R. M., Scripta Mater., 44 (2001) 159–164.

56. Champion, J. A., Keene, B. J. and Sillwood, J. M., J. Mater. Sci., 4 (1969) 39–49.

57. Levi, G., Scheu, C. and Kaplan, W. D., Interf. Sci., 9 (2001) 213–220.

58. Passerone, A., Muolo, M. L., Valenza, F. and Kaufman, L., CALPHAD, 34 (2010) 6–14.

59. Iwamoto, C. and Tanaka, S., Acta Mater., 50 (2002) 749–755.

60. Voytovych, R., Robaut, F. and Eustathopoulos, N., Acta Mater., 54 (2006) 2205–2214.

61. Eustathopoulos, N., Israel, R., Drevet, B. and Camel, D., Scripta Mater., 62 (2010) 966–971.

62. Carim, A., Scripta Metall. Mater., 25 (1991) 51–54.

63. Carim, A. H. and Mohr, C. H., Mater. Letters, 33 (1997) 195–199.

64. Wenzel, R. N., Ind. Eng. Chem., 28 (1936) 988.

65. Eustathopoulos, N., Nicholas, M. G. and Drevet, B., Wettability at High Temperatures. Pergamon Materials Series, Oxford, 1999, pp. 24–42.

66. Cassie, A. B. D., Discuss. Faraday Soc., 3 (1948) 11.

67. Chatain, D., Annu. Re. Mater. Res., 38 (2008) 8.1–8.26.

68. Rohrer, G. S., J. Mater. Sci., 46 (2011) 5881–5895.

69. Wynblatt, P. and Chatain, D., Metall. Mater. Trans. A, 37 (2006) 2595–2620.

70. Herring, C., Phys. Rev., 82 (1951) 87–93.

71. Wulff, G. Z., Krist., 34 (1901) 449.

Wetting at High Temperature

72. Samsonov, G. V. and Vinitskii, I. M., Handbook of Refractory Compounds, IFI/Plenum, 1980.
73. Naidich, Y. V., The wettability of solids by liquid metals, in: Progress in Surface and Membrane Science, Cadenhead, D. A. et al. (Eds). Academic Press, New York, 1981, Vol. 14, pp. 353–492.
74. Xiao. P. and Derby, B., Acta Mater., 44 (1996) 307–314.
75. Lequeux, S., Le Guyadec, F., Berardo, M., Coudurier, L. and Eustathopoulos, N., Wetting and interfacial reactions of Cu and Cu alloys on TiN, in: Eustathopoulos, N. and Sobczak, N. (Eds), Proc. 2nd Int. Conf. High Temp. Capillarity. FRI Cracow, Poland, 1997, pp. 112–117.
76. Muolo, M. L., Bassoli, M., Wollein, B., Lengauer, W. and Passerone, A., Trans. JWRI, 30 (2001) 49–54.
77. Kalogeropoulos, S., Van Deelen, J., Eustathopoulos, N., Le Guyadec, F. and Berardo, M., Trans. JWRI, 30 (2001) 107–112.
78. Frage, N., Froumin, N. and Dariel, M. P., Acta Mater., 50 (2002) 237–245.
79. Aizenshtein, M., Froumin, N., Barth, P. et al., J. Alloys Comp., 442 (2007) 375–378.
80. Durov, A. V., Naidich, Y. V. and Kostyuk, B. D., J. Mater. Sci., 40 (2005) 2173–2178.
81. Swiler, T. P. and, Loehman, R. E., Acta Mater., 48 (2000) 4419–4424.
82. Cook, G. O. III and Sorensen, C. D., J. Mater. Sci., 46 (2011) 5305–5323.
83. MacDonald, W. D. and Eagar, T. W., Annu. Rev. Mater. Sci., 22 (1992) 23–46.
84. Hong, S. M., Bartlow, C., Reynolds, T. B., McKeown, J. T. and Glaeser, A. M., Adv. Mater., 20 (2008) 4799–4803.
85. Shalz, M. L., Dalgleish, B. J., Tomsia, A. P. and Glaeser, A. M., J. Mater. Sci., 28 (1993) 1673–1684.
86. Hong, S. M., Reynolds, T. B., Bartlow, C. C. and Glaeser, A. M., Int. J. Mater. Res., 101 (2010) 133–142.
87. Arafin, M. A., Medraj, M., Turner, D. P. and Bocher, P., Mater. Sci. Eng. A, 447 (2007) 125–133.
88. Pouranvari, M., Ekrami, A. and Kokabi, A. H., J. Alloys Comp., 469 (2009) 270–275.
89. Donald, I. W., Mallinson, P. M., Metcalfe, B. L., Gerrard, L. A. and Fernie, J. A., J. Mater. Sci., 46 (2011) 1975–2000.
90. Smeacetto, F., Salvo, M., D'Herin Bytner, F. D., Leone, P. and Ferraris, M., J. Eur. Ceram. Soc., 30 (2010) 933–940.
91. Sobczak, N., Singh, M. and Asthana, R., Curr. Opin. Solid State Mater. Sci., 9 (2005) 241.
92. Saiz, E. and Tomsia, A. P., Nature Mater., 3 (2004) 903.
93. Passerone, A., Muolo, M. L., Valenza, F., Monteverde, F. and Sobczak, N., Acta Mater., 75 (2009) 356.
94. Muolo, M. L., Ferrera, E., Novakovic, R. and Passerone, A., Scripta Mater., 48 (2003) 191–196.
95. Naidich, Y. V., Zhuravlev, V. S., Gab, I. I., Kostyuk, B. D., Krasovskyy, V. P., Adamovskyy, A. A. and Taranets, N. Y., J. Eur. Ceram. Soc., 28 (2008) 717.
96. Bougiouri, V., Voytovych, R., Dezellus, O. and Eustathopoulos, N., J. Mater. Sci., 42 (2007) 2016–2023.
97. Aluru, R., Gale, W. F., Chitti, S. V., Sofyan, N., Love, R. D. and Fergus, J. W., Mater. Sci. Technol., 24 (2008) 517.
98. Ma, G. F., Zhang, H. L., Zhang, H. F., Li, H. and Hu, Z. Q., J. Alloys Compd., 464 (2008) 248.
99. Xu, J., Liu, X., Bright, M. A., Hemrick, J. G., Sikka, V. and Barbero, E., Metall Mater. Trans. A, 39A (2008) 1382.
100. Brochu, M., Pugh, M. and Drew, R. A. L., Intermetallics, 12 (2004) 289–294.
101. Gauffier, A., Saiz, E., Tomsia, A. P. and Hou, P. Y., J. Mater. Sci., 46 (2007) 9524–9528.
102. Xiong, H. P., Mao, W., Xie, Y. H., Chen, B., Guo, W. L., Li, X. H. and Cheng, Y. Y., J. Mater. Res., 22 (2007) 2727.

103. Xiong, H. P., Mao, W., Xie, Y. H., Chen, B., Guo, W. L., Li, X. H. and Cheng, Y. Y., Mater. Lett., 61 (2007) 4662.

104. Prakash, P., Mohandas, T. and Raju, P. D., Scripta Mater., 52 (2005) 1169.

105. Chen, B., Xiong, H. P., Mao, W., Guo, W. L., Cheng, Y. Y. and Li, X. H., Acta Metall Sini., 43 (2007) 1181.

106. Liu, G. W., Qiao, G. J., Wang, H. J., Yang, J. F. and Lu, T. J., J. Eur. Ceram. Soc., 28 (2008) 2701.

107. Liu, G. W., Li, W., Qiao, G. J., Wang, H. J., Yang, J. F. and Lu, T. J., J. Alloys Compd., 470 (2009) 163.

108. Zhang, C. G., Qiao, G. J. and Jin, Z. H., J. Eur. Ceram. Soc., 22 (2002) 2181.

109. Qiao, G. J., Zhang, C. G. and Jin, Z. H., Ceram. Int., 29 (2003) 7.

110. Hattali, M. L., Valette, S., Ropital, F., Mesrati, N. and Treheux, D., J. Mater. Sci., 44 (2009) 3198.

111. Li, S. J., Zhou, Y., Duan, H. P., Qiu, J. H. and Zhang, Y., J. Mater. Sci., 38 (2003) 4065.

112. Buhl, S., Leinenbach, C., Wegener, K. and Spolenak, R., J. Mater. Sci., 45 (2010) 4358–4368.

113. Schindler, H. J. and Leinenback, C., J. ASTM Int., 7 (2010) 102515.

114. Munro, R. G., J. Res. NIST, 105 (2000) 709–720.

115. Fahrenholtz, W. G., Hilmas, G. E., Talmy, I. G. and Zaykoski, J. A., J. Am. Ceram. Soc., 90 (2007) 1347–1364.

116. Opeka, M. M., Talmy, I. G. and Zaykoski, J. A., J. Mater. Sci., 39 (2004) 5887.

117. Van Wie, D. M., Drewry, D. G. Jr., King, D. E. and Hudson, C. M., J. Mater. Sci., 39 (2004) 5915.

118. Bonal, J. P., Kohyama, A., Van der Laan, J. and Snead, L. L., MRS Bull., 34 (2009) 28–34.

119. Giancarli, L., Bonal, J. P., Caso, A., Le Marois, G., Morly, N. B. and Salavy, J. F., Fusion Eng. Design, 41 (1998) 165–171.

120. Kuwahara, K., Sakamoto, S., Kida, O., Ishino, T., Kodama, T., Nakajima, H., Ito, T. and Hirakawa, Y., Corrosion resistance and electrical resistivity of ZrB2 monolithic refractories, in: The Proceedings of UNITECR 2003. The 8th Biennial Worldwide Conference on Refractories, Osaka, Japan, K. Asano (Ed.). Technical Association of Refractories, Japan, 2003, pp. 302–305.

121. Weimer, A. M., Carbide, Nitride and Boride Materials, Synthesis and Processing. Chapman & Hall, NY, 1997.

122. Jin, Z. J., Zhang, M., Guo, D. M. and Kang, R. K., Key Eng. Mater., 291–292 (2005) 537.

123. Sung, J., Goedde, D. M., Girolami, G. S. and Abelson, J. R., J. Appl. Phys., 91 (2002) 3904.

124. Murata, Y., U.S. Patent, 3 (1970) 487–594.

125. Sciti, D., Melandri, C. and Bellosi, A., Adv. Eng. Mat., 6 (2004) 775–781.

126. Monteverde, F. and Bellosi, A., J. Eur. Cer. Soc., 25 (2005) 1025–1031.

127. Medri, V., Balbo, A., Monteverde, F. and Bellosi, A., Adv. Eng. Mat., 7 (2005) 159–163.

128. Monteverde, F. and Bellosi, A., J. Mater. Res., 19 (2004) 3576–3585.

129. Monteverde, F., J. Mater. Sci., 43 (2008) 1002–1007.

130. Saiz, E., Tomsia, A. P. and Cannon, R. M., Scripta Mater., 44 (2001) 159–164.

131. Eustathopoulos, N., Sobczak, N., Passerone, A. and Nogi, K., J. Mater. Sci., 40 (2005) 2271–2280.

132. Daolio, S., Fabrizio, M., Piccirillo, C., Muolo, M. L., Passerone, A. and Bellosi A., Rapid Comm. Mass. Spectr., 15 (2001) 1–12.

133. Samsonov, G. V., Panasyuk, A. D. and Borovikova, M. S., Poroshkovaya Metallurgiya, 5 (1973) 61.

134. Samsonov, G. V., Panasyuk, A. D. and Borovikova, M. S., Poroshkovaya Metallurgiya, 6 (1973) 51.

135. Tumanov, V. I., Gorbunov, A. E. and Kondratenko, G. M., Russian J. Phys. Chem., 44 (1970) 304.

136. Samsonov, G. V., Sharkin, O. P., Panasyuk, A. D. and D'yakonova, L. V., Poroshkovaya Metallurgiya, 7 (1974) 63.

137. Ukov, V., Esin, O. A., Vatolin, H. A. and Dubinin, E. L., Investigation of wetting of non metallic solids by Pd-based liquid alloys, in: Eremenko, V. N., (Ed.). Physical Chemistry of Surface Phenomena at High Temperature. Naukova Dumka, Kiev, 1971, p. 139.

138. Voytovych, R., Koltsov, A., Hodaj, F. and Eustathopoulos, N., Acta Mater., 55 (2007) 6316–6321.

139. Muolo, M. L., Ferrera, E., Morbelli, L., Zanotti, C. and Passerone, A., ESA SP, 540 (2003) 467.

140. Passerone, A., Muolo, M. L., Morbelli, L., Ferrera, E., Bassoli, M. and Bottino, C., ESA SP, 521 (2003) 295.

141. Muolo, M. L., Ferrera, E. and Passerone, A., Wetting and spreading of liquid metals on ZrB2-based ceramics. J. Mater. Sci., 40 (2005) 2295.

142. Passerone, A., Muolo, M. L., Valenza, F. and Novakovic, R., Surf. Sci., 603 (2009) 2725–2733.

143. Novakovic, R., Ricci, E., Muolo, M. L., Giuranno, D. and Passerone, A., Intermetallics, 11 (2003) 1301.

144. Novakovic, R., Muolo, M. L. and Passerone, A., Surf. Sci., 549 (2004) 281.

145. Stadelmaier, H. H. and Shoemaker, C. A., Metall (Berlin), 20 (1966) 1056.

146. Lugscheider, E., Reiman, H. and Pankert, R., Metall (Berlin), 36 (1982) 247.

147. Kaufman, L., Cacciamani, G., Muolo, M. L., Valenza, F. and Passerone, A., CALPHAD, 34 (2010) 2–5.

148. Yuan, B. and Zhang, G. J., Scripta Mater., 64 (2011) 505–514.

149. Tian, W., Kita, H., Hyuga, H., Kondo, N. and Nagaoka, T., J. Eur. Ceram. Soc., 30 (2010) 3203–3208.

150. Singh, M. and Asthana, R., J. Mater. Sci., 45 (2010) 4308–4320.

151. Ding, W. F., Xu, J. H., Chen, Z. Z., Yang, C. Y. and Fu, Y. C., Mater. Sci. Techn., 25 (2009) 1448–1452.

152. Asthana, R. and Singh, W., Scripta Mater., 61 (2009) 257–260.

153. Singh, M. and Asthana, R., Int. J. Appl. Ceram. Techn., 6 (2009) 113–133.

154. Singh, M. and Asthana, R., Mater. Sci. Eng. A, 460 (2007) 153–162.

155. Berezhinsky, L. I., Berezhinsky, I. L., Grigorev, O. N., Serdega, B. K. and Ukhimchuk, V. A., J. Eur. Ceram. Soc., 27 (2007) 2513–2519.

156. Esposito, L. and Bellosi, A., J. Mater. Sci., 40 (2005) 4445–4453.

157. Muolo, M. L., Valenza, F., Sobczak, N. and Passerone, A., Adv. Sci. Techn., 6 (2010) 98–107.

158. Sobczak, N., Nowak, R., Passerone, A., Valenza, F., Muolo, M. L., Jaworska, L., Barberis, F. and Capurro, M., Trans. Foundry Res. Inst., 2 (2010) 5–14.

159. Ortona, A., Pusterla, S. and Gianella, S., J. Eur. Ceram. Soc., 31 (2011) 1821–1826.

160. Steen, M. and Ranzani, L., Ceram Int., 26 (2000) 849–854.

161. Cabrero, J., Audubert, F. and Pailler, R., J. Eur. Ceram. Soc., 31 (2011) 313–320.

162. Ferraris, M., Salvo, M., Casalegno, V. et al., Ceramic Transactions, 220 (2010) 173–186.

163. Katoh, Y., Snead, L. L., Henager, C. H., Hasegawa, A., Kohyama, A., Riccardi, B. and Hegeman, H., J. Nucl. Mater., 367 (2007) 659.

164. Jones, R. H., Giancarli, L., Hasegawa, A., Katoh, Y., Kohyama, A., Riccardi, B., Snead, L. L. and Weber, W. J., J. Nucl. Mater., 307 (2002) 1057.

165. Riccardi, B., Giancarli, L., Hasegawa, A., Katoh, Y., Kohyama, A., Jones, R. H. and Snead, L. L., J. Nucl. Mater., 329 (2004) 56.
166. Ferraris, M., Salvo, M., Casalegno, V., Ciampichetti, A., Smeacetto, F. and Zucchetti, M., J. Nucl. Mater., 375 (2008) 410–415.
167. Kurokawa, K. and Nagasaki, R., Reactivity of sintered SiC with metals, in: S. Somiya, M. Shimada, M. Yoshimura, R. Watanabe (Eds), Sintering'87. Elsevier, New York, 1988, pp. 1397–1402.
168. Rabin, B. H., EG & G Idaho, Inc., Idaho Falls, Washington, 1991, pp. 1–14.
169. Li, S. J. and Zhang, L., Powder Metall. Technol., 22 (2004) 91–97.
170. Fukai, T., Naka, M. and Schuster, J. C., Trans. JWRI, 26 (1997) 93–98.
171. Liu, G. W., Muolo, M. L., Valenza, F. and Passerone, A., Ceramics Int., 36 (2010) 1177–1188.
172. Mao, W., Li, S. J. and Han, W. B., Rare Metal. Mater. Eng., 35 (2006) 312.
173. McDermid, J. R. and Drew, R. A. L., J. Mater. Sci., 25 (1990) 4804–4809.
174. Iino, Y. J., Mater. Sci. Lett., 10 (1991) 104.
175. Mao, Y. W., Li, S. J. and Yan, L. S., Mater. Sci. Eng. A, 491 (2008) 304–308.
176. Cockeram, B. V., J. Am. Ceram. Soc., 88 (2005) 1892–1899.
177. Liu, H. J., Feng, J. C., Fujii, H. and Nogi, K., Mater. Sci. Technol., 20 (2004) 1069.
178. Larker, R., Nissen, A., Pejryd, L. and Loberg, B., Acta Metall. Mater., 40 (1992) 3129.
179. Yamada, Y., Satoh, M., Kohno, A. and Yokoi, K., J. Mater. Sci., 26 (1991) 2887–2892.
180. Li, S. J., Duan, H. P., Liu, S., Zhang, Y. G., Dang, Z. J., Zhang, Y. and Wu, C. G., Int. J. Refract. Met. Hard Mater., 18 (2000) 33.
181. McDermid, J. R. and Drew, R. A. L., J. Am. Ceram. Soc., 74 (1991) 1855–1860.
182. Koltsov, A., Hodaj, F. and Eustathopoulos, N., Mater. Sci. Eng. A, 495 (2008) 259–264.
183. Riccardi, B., Nannetti, C. A., Woltersdorf, J., Pippel, E. and Petrisor, T., Int. J. Mater. Product. Technol., 20 (2004) 440–451.
184. Riccardi, B., Nannetti, C. A., Woltersdorf, J., Pippel, E. and Petrisor, T., J. Mater. Sci., 37 (2002) 5029.
185. Riccardi, B., Nannetti, C. A., Petrisor, T., Woltersdorf, J., Pippel, E., Libera, S. and Pillonni, L., J. Nucl. Mater., 329 (2004) 562–566.
186. Riccardi, B., Nannetti, C. A., Petrisor, T. and Sacchetti, M., J. Nucl. Mater., 307 (2002) 1237–1241.
187. Li, J. K., Liu, L., Wu, Y. T., Zhang, W. L. and Hu, W. B., Mater. Lett., 62 (2008) 3135.
188. Tsoga, A., Ladas, S. and Nikolopoulos, P., Acta Mater., 45 (1997) 3515–3525.
189. Rado, C., Kalogeropoulou, S. and Eustathopoulos, N., Acta Mater., 47 (1999) 461–463.
190. Rado, C., Kalogeropoulou, S. and Eustathopoulos, N., Scripta Mater., 42 (2000) 203–208.
191. Liu, G. W., Valenza, F., Muolo, M. L. and Passerone, A., J. Mater. Sci., 45 (2010) 4299–4307.
192. Kalogeropoulou, S., Baud, L. and Eustathopoulos, N., Acta Metall. Mater., 43 (1995) 907–912.
193. Drevet, B., Kalogeropoulou, S. and Eustathopoulos, N., Acta Mater., 41 (1993) 3119–3126.
194. Naidich, Y. V., Zhuravlev, V. and Krasovskaya, N., Mater. Sci. Eng. A, 245 (1998) 293–299.
195. Rado, C., Kalogeropoulou, S. and Eustathopoulos, N., J. Mater. Sci. Eng. A, 276 (2000) 195.
196. Li, J. G., Wettability of SiC by liquid Ag and binary Ag–Si alloy. Mater. Lett., 18 (1994) 291–298.
197. Gasse, A., Chaumat, G., Rado, C. and Eustathopoulos, N., J. Mater. Sci. Lett., 15 (1996) 1630–1632.
198. Landry, K., Rado, C. and Eustathopoulos, N., Metall. Mater. Trans. A, 27 (1996) 3181–3186.
199. Rado, C. and Eustathopoulos, N., Interface. Sci., 12 (2004) 85–92.

200. Lemoine, P., Ferraris, M., Salvo, M., Appendino, P. and Montorsi, M., J. Eur. Ceram. Soc., 16 (1999) 1231–1236.
201. Tanaka, S. I. and Iwamoto, C., Mater. Sci. Eng. A, 495 (2008) 168–173.
202. Laurent, V., Rado, C. and Eustathopoulos, N., Mater. Sci. Eng. A, 205 (1996) 1–8.
203. Xiong, H. P., Li, X. H., Mao, W. and Cheng, Y. Y., Mater. Lett., 57 (2003) 3417–3421.
204. Liu, G. W., Valenza, F., Muolo, M. L., Qiao, G. J. and Passerone, A., J. Mater. Sci., 44 (2009) 5590.
205. Maillart, O., Hodaj, F., Chamat, V. and Eustathopoulos, N., Mater. Sci. Eng. A, 495 (2008) 174–180.
206. Du, Y. and Schuster, J. C., Metall Mater. Trans., 30 (1999) 2409–2418.
207. Passerone, A. and Ricci, E., High temperature tensiometry, in: Drops and Bubbles in Interfacial Research, D. Möbius and R. Miller (Eds), Elsevier, 1998, pp. 475–525.
208. Chatain, D., Leseur, C. and Baland, J. P., Langmuir, 22 (2006) 4230–4236.
209. Liggieri, L. and Passerone, A., High Temp. Techn., 7 (1989) 82.
210. Kalantarian, A., David, R., Chen, J. and Neumann, A. W., Langmuir, 27 (2011) 3485–3495.
211. Chatain, D. and Carter, W. C., Nature Mater., 3 (2004) 843–845.
212. Sobczak, N., Nowak, R., Radziwill, W., Budziok, J. and Glenz, A., Mater. Sci. Eng. A, 495 (2008) 43–49.
213. Sobczak, N., Sobczak, J., Asthana, R. and Purgert, R., China Foundry, 7 (2010) 425–437.

Subject Index

3D Rotnc-Prager, 95
3D computations, 153

Acid–Base, 7
adhesion tension, 312
ADSA, 7
adsorption, 16, 42, 301, 302
advancing, 6, 54
advancing and receding contact angle,
    178
advancing contact angle, 195, 327
AFM, 10
Ag, 311
agrochemical formulations, 268
air brazing, 301
air water film, 227
AKD surfaces, 198
alkyl ketene dimmer, 180
aluminum surfaces, 196
amphoteric, 82
anionic, 82
anisotropy, 307
anti-rebound effect, 287
apparent contact, 15
apparent contact angle, 304, 327
Atomic Force Microscope, 161
Au, 311
autophobic, 24
autophobic effect, 186
axi-symmetry, 143
Axisymmetric Drop Shape Analysis,
    7, 191

Baxter, 11
BET Adsorption Isotherms, 16
binary silicon alloys, 319
Bingham model, 290
Bingham-capillary number, 273
bmim.BF$_4$, 250, 252, 257
boron, 312
bovine serum albumin, 122
Brownian forces, 94
bubble oscillation and rebound, 218
bubble rise, 216
bubble rise velocity, 214
bubble tensiometry, 189
bubble trajectory, 219
bubble–surface interaction, 218
bulk flow of solvent, 118

CAC, 59
CALPHAD, 314
capillarity, 12
capillary forces, 40
capillary nozzle, 269, 273
capillary number, 268
capillary pressure, 109
Carbopol, 276
Carreau model, 289
Cassie, 11, 307
Cassie equation, 176
Cassie mode wetting, 187
cationic, 82
ceramic/metal joining, 310
CFD Simulations, 135
characteristic, 260

CMC, 42
cmc, 86
coated aluminum, 203
colloidal particles, 94
complex liquids, 268
conductive droplet, 241
conductive liquids, 245
consistency coefficient, 269
constitutive models, 289
contact angle, 1, 300, 313
contact angle hysteresis, 40, 178, 243
contact angle saturation, 244
contact line, 231, 281
contact mode imaging, 162
contact radius, 109
convective cell, 149
convex microlenses, 114
cooling effect, 149
Cox, 232, 246
crater-like structures, 113
critical aggregation concentration, 59
critical *micelle* concentration, 42
critical surface tension, 5
critical wetting concentration, 59
Cross model, 289
cryofixation, 163
Cu, 311
curvature of the liquid–vapor
    interface, 177
CWC, 59

de Gennes, 234
deformable bubbles, 219
deformation of bubble, 217
dendrites, 49
Density Functional Theory (DFT),
    302, 309
Derjaguin–Frumkin, 39
diffusion, 95, 302
diffusion-controlled, 82
disjoining pressure, 111
disjoining pressures, 222
dispensed drop, 325

dissolution, 302
dissolutive wetting, 304
DLVO theory, 161
DNA, 281
double-layer disjoining pressure, 223
drainage and rupture of wetting films,
    225
drop impact, 267
droplet evaporation dynamics, 135
droplet hydrodynamics, 151
dry rebound, 284
DTAB, 84
DTAS, 84
dynamic contact angle, 261
dynamic elasticity, 229
dynamic surface tension, 85
dynamic wetting, 84
dynamics of electrowetting, 260

effective oxygen pressure, 301
'electrical' work of adhesion, 243
electrowetting, 241
electrowetting curve, 250
electrowetting saturation, 258
ellipsometric studies, 169
Ellis model, 290
elongational viscosity, 277, 284
energy metastable state, 178
enhanced wetting, 205
Equation of State, 5
equations of Wenzel, 176
equilibrium contact angles, 195
equilibrium spreading pressure, 16
ethoxylated alcohol surfactants, 89
evaporation, 105, 129
evaporation kinetics, 108, 148
evaporation of heated droplets, 130
evaporative cooling, 144

far-field, 97, 102
film drainage, 229
film thickness, 226
fingers, 48

Finite Element Method, 135
fluid, 94
fluorinated chemistry, 189
fluoropolymer, 86, 244
Fowkes, 8
freeze fracture techniques, 163
friction, 97
Frumkin–Levich, 215

Gibbs adsorption, 301
Gibbs elasticity, 230
Gibbs' adsorption equation, 181
Good–van Oss–Chaudury, 8
Green operator, 97

Hamaker, 222
Hamaker–Lifshitz function, 222
heat and mass transfers, 132
heated droplet, 148
heated surfaces, 283
Herring, 308
Herschel–Bulkley equation, 291
heterogeneous rough surface, 176
heterogeneous smooth surfaces, 176
hexadecane, 258
hexadecyltrimethylammonium
    bromide, 180
hexamethyldisilazane, 162
HfB_2, 314
high temperature, 322
high temperature wetting, 300
high temperatures, 300
Highly Ordered Pyrolytic Graphite,
    161
homeopathic action, 171
hybrid approaches, 155
hydration forces, 224
hydrodynamic phenomena, 136
hydrodynamic receding contact angle,
    56
hydrodynamics interactions (HI), 94
hydrophilic, 41
hydrophilic surface, 220

hydrophobic, 3, 39, 41
hydrophobic force, 224
hydrophobic surface, 220
hydrophobic surfaces, 159
hysteresis, 123

ink jetting, 114
inkjet etching, 119
interfacial forces, 213
interfacial interaction forces, 222
intrinsic contact angle, 200, 205
ionic liquids, 245, 247, 249
ionic strength, 243
isopropyl acetate, 118

joining, 316
joining SiC, 319

Kelvin Equation, 110
kinematic viscosity, 141
kinetics of spreading, 245, 303

L-77, 51
Langmuir, 16
Langmuir isotherm, 139
Laplace equation, 327
Leidenfrost phenomenon, 283
Leidenfrost temperature, 284
Lifschitz–van der Waals, 8
Lifshitz, 222
liquid films, 225
liquid phase bonding processes, 310
lubrication, 41, 98
(lubrication) theory, 225

magnetoelastic gravimetric sensors,
    171
Marangoni circulation, 152
Marangoni driven convection, 133
Marangoni effect, 229
Marangoni flow, 116
Marangoni number, 231
Marangoni stresses, 138

Marangoni velocity, 150
mass transfer resistance, 121
maximum bubble pressure method, 85
maximum surface charge, 257
Maxwell model, 292
metal/borides, 302
metal/carbides, 302
metal/nitrides, 302
mica surfaces, 161
micro-electronics, 129
microfluidic, 242, 260
microlens arrays, 115
microlenses, 242
microlithography, 116, 119
micropancakes, 167
micropumps, 242
microscopic roughness, 220
microstructure-etching, 118
mixtures of solvents, 115
molecular combing, 281
molecular dynamics, 136
molecular-kinetic model, 246
molecular-kinetic theories, 233
monolayers, 95
Monte Carlo simulations, 122
Morton member, 217

n-alkyl polyoxyethylene surfactants, 61
n-hexadecane, 247
n-decanoyl-n-methylglucamine, 189
nano-droplets, 164
nanobubbles, 159, 168, 225, 229
Navier–Stokes and heat transfer equations, 136
Neumann rule, 304
neutron reflectivity, 169
Ni, 311
Ni–Si–C diagram, 320
non-DLVO forces, 224
non-ionic, 82
non-Newtonian drops, 269
non-Newtonian fluid, 289

non-reactive wetting, 303
Noncontact Probe, 24

Octadecyltrichlorosilane, 189
Ohnesorge number, 268
Oseen kernel, 97
Ostwald–De Waele fluid, 289
oxygen, 301
oxygen content, 243

Parafilm, 89
Particle Image Velocimetry, 154
pendant drop analysis, 189
PEO, 285
PET, 87
phase diagram, 314
Pickering emulsions, 94
PMMA coating, 188
point-particle model, 101
Poisson–Boltzmann equation, 223
poly(tetrafluoroethylene), 189
polydimethylsiloxane, 115, 123
polydispersity of the polymer, 122
polyethylene, 257
Polyethylene Oxide, 277
polyethylmethacrylate, 112
polymer, 85
polymer–solvent systems, 122
polystyrene, 112, 257
polyvinylphenol, 116
porous, 52
power-law drops, 269
power-law fluids, 289
PTFE, 193
PVF, 87

Q2D mobility tensor, 100
quartz crystal microbalance, 171
Quasi-Chemical Solution Model, 312
quasi-steady approximations, 146
quasi-two-dimensional, 94